USS *Los Angeles*

Also by William F. Althoff

Sky Ships: A History of the Airship in the United States Navy
Arctic Mission: By Airship & Submarine to the Far North

USS *Los Angeles*

The Navy's Venerable Airship and Aviation Technology

WILLIAM F. ALTHOFF

Potomac Books, Inc.
Washington, D.C.

Copyright © 2004 by William F. Althoff

Published in the United States by Potomac Books, Inc. (formerly Brassey's, Inc.). All rights reserved. No part of this book may be reproduced in any manner whatsoever without written permission from the publisher, except in the case of brief quotations embodied in critical articles and reviews.

Library of Congress Cataloging-in-Publication Data

Althoff, William F.
  USS Los Angeles : the Navy's venerable airship and aviation technology / William F. Althoff.—1st ed.
       p.   cm.
    Includes bibliographical references and index.
      ISBN 1-57488-620-7 (hardcover : alk. paper)—ISBN 1-57488-621-5 (pbk. : alk. paper)
    1. Los Angeles (Airship)   2. United States Navy—Aviation—History—20th century.   I. Title.
VG93.A86823   2003
359.9'4834—dc21
                                                                    2002156277

Hardcover ISBN 1-57488-620-7

Printed in the United States of America on acid-free paper that meets the American National Standards Institute Z39-48 Standard.

Potomac Books, Inc.
22841 Quicksilver Drive
Dulles, Virginia 20166

Book design and composition by Susan Mark
Coghill Composition Company
Richmond, Virginia

First Edition

10 9 8 7 6 5 4 3 2 1

For
the artisans who created her
the officers and enlisted men who flew her
the civilians who attended her.

I've been privileged to know some of you.

Contents

	Acknowledgments	ix
	Preface	xi
	Significant Chronology	xiii
	Introduction	xv
1	LZ-126: Birthplace Friedrichshafen	1
2	ZR-3: Homeport Lakehurst	39
3	Balloons and Billets: Lighter-than-Air Training	69
4	Rosendahl's Reign	91
5	Testbed for the New Ships	124
6	Grounded	155
7	Denouement	193
	Epilogue	217
	Appendix 1. Performance Data: USS *Los Angeles*	221
	Appendix 2. Graduates, Lighter-than-Air Pilot Training, NAS Lakehurst, 1923–1940	225
	Appendix 3. Flight Log: USS *Los Angeles*, 1924–1932	229
	Appendix 4. Ship's Crew: USS *Los Angeles*, 1931–1932	239
	Appendix 5. Navigational Equipment: USS *Los Angeles*	241
	Notes	243
	Selected Bibliography	277
	Index	283
	About the Author	289

Acknowledgments

This project has incurred numerous obligations, especially to Mr. Charles L. Keller, who made available his research at the National Archives and Records Administration as well as an extended, wide-ranging correspondence with Cdr. Garland Fulton, USN (CC) (Ret.).

Audio-recorded, one-on-one interviews were held with Capt. Ralph S. Barnaby, USN (Ret.), Mr. Kurt Bauch, Lt. Cdr. William A. Baker, USN (Ret.), Capt. Michael M. Bradley, USN (Ret.), Mr. Melvin J. Cranmer, Capt. Marion H. Eppes, USN (Ret.), ADC John A. Iannaccone, USN (Ret.), Mr. William P. Kramer, Mr. John A. Lust (former Seaman 2c), Mr. Norman J. Mayer, Cdr. Joseph P. Norfleet, USN (Ret.) [with Michael C. Miller], Lt. David F. Patzig, USN (Ret.), Rear Adm. Scott E. Peck, USN (Ret.) [by Michael C. Miller], Rear Adm. George E. Pierce, USN (Ret.), Cdr. Francis W. Reichelderfer, USN (Ret.), Rear Adm. Charles E. Rosendahl, USN (Ret.) [notes], Lt. Herbert R. Rowe, USN (Ret.), Lt. Cdr. Leonard E. Schellberg, USN (Ret.), Rear Adm. Leroy C. Simpler, USN (Ret.), Vice Adm. T. G. W. Settle, USN (Ret.) [notes], Capt. Earl K. Van Swearingen, USN (Ret.), Mr. F. W. von Meister, and Cdr. R. H. Ward, USN (Ret.).

Others who provided recorded reminiscences were Rear Adm. Richard S. Andrews, USN (Ret.), Rear Adm. Calvin M. Bolster, USN (Ret.), and Capt. George F. Watson, USN (Ret.). Each prepared taped replies to queries.

The oral-history research conducted by the writer was supported in part by generous grants (1975 and 1996) from the New Jersey Historical Commission.

Ms. Lynn Berry, granddaughter of Cdr. Fred T. Berry, granted access to family documents, as did Mr. Thomas Berry to letters written by his grandfather. Additional information came courtesy Rear Adm. Calvin M. Bolster, USN (Ret.), Capt. Howard N. Coulter, USN (Ret.), Rear Adm. D. Ward Harrigan, USN (Ret.), Rear Adm. Harold B. Miller, USN (Ret.), Rear Adm. Charles A. Nicholson, USN (CC) (Ret.), Mrs. Louise T. Rixey, Mr. Owen F. Tyler, Capt. George F. Watson, USN (Ret.), and Capt. Gordon S. Wiley, USN (Ret.). Thanks are due to archivist Rick Peuser, National Archives, for a salient suggestion early in the project.

I am grateful to Mr. Kevin Pace, who made available documents as well as images held by the Navy Lakehurst Historical Society. Through Mr. Eric Brothers, editor of the *Bulletin* of the Lighter-Than-Air Society, Mr. John V. Miller, Director of Archival Services, sent images held by the University of Akron. Thanks to Mr. Langdon H. Fulton, materials pertaining to his late father, Cdr. Garland Fulton, USN (Ret.), were made available; on this count I am especially grateful for insights into an exceptional officer.

The National Air and Space Museum, Smithsonian Institution, tendered inestimable support through a Ramsey Fellowship in Naval Aviation History. Tom Crouch, Senior Curator, proffered counsel and encouragement. Dom Pisano, Chairman, Aeronautics Division, granted precious weeks. At Garber, Marilyn Graskowiak and her staff tendered welcome as well as assistance. Mr. Allan Janus, Museum Specialist, ably navigated the writer through the Garland Fulton and George H. Mills Collections. Special mention is due Mr. Phil Edwards, Technical Information Specialist; he produced materials that improved the book more than he knows.

Mr. David Gant designed the jacket and helped guide this production—and this writer—through many frustrations.

The images of *LZ-126* under construction came courtesy of the late Lt. Gordon M. Cousins, USN (Ret.). Further photographs are from the San Diego Aerospace Museum (thanks to Mr. Ray Wagner), the Archives Division of the National Air and Space Museum (thanks to Ms. Kristine Kaske particularly), the still-picture holdings of the National Archives, and from private papers and collections. Among the latter, Warren M. Anderson, Mr. Henry J. Applegate, Mr. Clark L. Bunnell, Mr. Joseph Billiams, the late Rear Adm. Calvin M. Bolster, Mr. David Collier, CPO Moody Erwin, USN (Ret.), Mr. Thomas G. Raub, Lt. Herbert R. Rowe, USN (Ret.), Mrs. Charles M.

Ruth, Mrs. Elizabeth A. Tobin, Lt. Cdr. Leonard E. Schellberg, USN (Ret.), Mr. Hepburn Walker, and Capt. Gordon S. Wiley, USN (Ret.) must be acknowledged. ADC John A. Iannaccone, USN (Ret.) was particularly gracious with respect to images. Mr. Marvin Krieger supplied images held in the Charles E. Rosendahl Collection, University of Texas.

The late Vice Adm. T. G. W. Settle, USN (Ret.) remains an inspiration to the writer, the work of Dr. Richard K. Smith a model of historiography.

The hospitality of Michael and Marci Woodgerd meant more than they can know.

I am inexpressibly grateful to Penny for her support, patience, and understanding.

Preface

Looking back, the rigid airship seems baroque, an early-twentieth-century anachronism predestined for failure and oblivion. Such dismissal explains nothing.

Aeronautics is the product of experimentation and research. In its young decades, the basic technology of powered flight experienced a revolution, largely abandoning empiricism for engineering research and analysis. The science of aerodynamics developed with special reference to heavier-than-air craft; the airplane, indeed, was a triumph of the engineering method. During the 1920s and the early thirties as well, experimental and theoretical work on problems pertaining to lighter-than-air was conducted for the U.S. Army and Navy. Many flight researchers at the Langley Aeronautical Laboratory, the research center of the National Advisory Committee for Aeronautics, became outspoken advocates of large dirigibles.

Engineering scientist Theodore von Karman, pioneer in aerodynamics, rocketry, and astronautics, considered airships "one of the great products of early aeronautical engineering. . . . From a technical point of view I thought they were highly efficient vehicles and I believed at the time they were unnecessarily losing out to the airplane, which was faster but not as efficient or as comfortable for long journeys." Large structures composed of many elements, rigid airships required the most advanced knowledge of structural analysis and aerodynamics. "There's a lot to be said for airship technology," one engineer reflected. "It was really a leader in the field of aeronautics for most of its history."[1]

USS *Los Angeles* (ZR-3) became a flying laboratory for full-scale experimental research as well as the working out of airship and naval-scouting doctrine. Crews for the next generation of aerial strategic scouts were trained aboard "*L.A.*" The ship's service life (1924–39) exceeded that of any of her kind, even the renowned *Graf Zeppelin*. Although operated for the most part as a commissioned vessel of the U.S. Navy, *Los Angeles* was designed and built by the Zeppelin Company, then flown from Central Europe to America—the fourth aircraft to leap the North Atlantic.

Born of emerging technologies, lighter-than-air was deemed a resource by some, an obstacle by many. Its historical significance notwithstanding, little evidence of *Los Angeles* survives. This work will refresh the public memory as to a briefly viable, banished technology.

Significant Chronology

LZ-126: In American Service, USS *Los Angeles* (ZR-3)

June 1919	Seven wartime zeppelins scuttled "illegally" in their sheds
July 1919	Naval Appropriations bill for 1920 passed by Congress
28 June 1921	Naval Air Station (NAS) Lakehurst commissioned
21 August 1921	British *R-38 (ZR-2)* founders during 4th trial
1 September 1921	Bureau of Aeronautics (BuAer) commences
October–November 1921	Joint Army-Navy Board assigns responsibility for acquisition of compensation Zeppelin to U.S. Navy
16 December 1921	157th Meeting of Conference of Ambassadors, at which U.S. request for compensation airship is approved
23 June 1922	Contracts signed with Luftschiffbau-Zeppelin
1922	Design for *LZ-126* (factory number) prepared
7 November 1922	Keel of *LZ-126* laid
4 September 1923	First flight by helium-inflated rigid *(ZR-1)*
August 1924	*LZ-126* completed (hydrogen-inflated)
$713,332	Cost to German government
$150,000	Cost to U.S. government
27 August 1924	First flight trial *LZ-126 (ZR-3)*
6–25 September 1924	Second through fifth trials
12–15 October 1924	Transatlantic delivery (5,066 miles in 81 hours, 17 minutes, Friedrichshafen, Germany, to NAS Lakehurst, New Jersey)
25 November 1924	Christened/Commissioned, NAS Anacostia
8 January 1925	First moor to Lakehurst high mast
15 January 1925	First floating moor to USS *Patoka* (AV-6)
5 October 1927	First moor to improvised stick mast, Lakehurst
22 June 1929	First use of self-propelled "stub" mast

3 July 1929   First airplane hook-on (Vought UO-1)
17 June 1931   First NZY-1 hook-on
23 October 1931   First XF9C-1 hook-on
27 October 1931   Last hook-on (300th flight)

4 February–2 March 1931   Participation in Fleet Problem XII, Canal Zone

30 June 1932   Decommissioned

17 August 1934   Project order to recondition for experiments
18 December 1934   First moor-out (nonflight status), Lakehurst
8 November 1937   Last moor-out (nonflight status)
17 December 1938   Fourth material inspection ordered by CNO
27 May 1939   Report of material inspection, BIS to CNO
24 October 1939   Formally stricken
October–December 1939   Dismantled/disassembled
4–5 December 1939   Final strength tests
29 February 1940   Project order to dismantle closed out

363 hours, 15 minutes   Total hours at high mast (47 moors)
912 hours, 31 minutes   Total hours at low masts (185 moors)
709 hours, 47 minutes   Total hours with USS *Patoka* (44 moors)

336   Total number of flights (includes German trials)
4,126.36   Number of flight hours

Introduction

The concept of aviation in the navy had, from an early date, engaged Western Europeans. Aeronautics, indeed, was encouraged with men and money—in Germany and France particularly. America, though, was rather less air-minded.

In September 1908, observers from the U.S. Navy (and Army) witnessed the Wright brothers' official trial at Fort Myer, Virginia. Inventors were besieging Washington for funds, to complete experiments with various schemes and machines. Few military leaders, as yet, appreciated the potential value of aircraft. In the Navy Department, largely unaware and uninterested in aeronautics, the standing orders were "Slow Ahead." Who could be sure of the airplane's future role as a naval platform? Indeed, could aircraft be conveyed to sea? Could they be launched from a deck? Or retrieved? For pioneering men of vision, such questions cried out for development attention—and for funds.

The Wrights met the Army's requirements in 1909; the War Department bought an airplane. In the naval service, some officers embraced the possibilities. A purchase was urged. In Germany, already, large airships were national symbols. Nine years after the first flight of a zeppelin, the world's first commercial aerial transport company was founded by Count Ferdinand von Zeppelin. Within months, a passenger service was inaugurated.

The navy established an aeronautic office in 1910. It held little status. That October, the General Board—senior flag officers who advised the navy secretary on policy—recommended that, in light of "the great advances" in aviation science and its potential, "the problem of providing space for airplanes and dirigibles be considered in all new [scouting vessel] designs." Six weeks later, a biplane took to the air from the cruiser USS *Birmingham*, using a platform installed on her bow. Before the month was out, Glenn Curtiss was offering flight instruction for one officer. Accordingly, Lt. T. G. Ellyson was ordered to report to Curtiss's flying school and experimental station at North Island, San Diego.

In March 1911, the first aeronautical appropriation was approved by legislation—twenty-five thousand dollars to the Bureau of Navigation for "experimental work in the development of aviation for naval purposes." So began the naval aviation establishment. In consequence, the Wright Company offered to train *one* pilot upon the purchase of *one* aeroplane. Though hardly a real program, Lt. John Rodgers—Naval Aviator No. 2—was ordered to Dayton. That July, Glen Curtiss, then Ellyson, test-flew aircraft serial number A-1 (a Curtiss hydroaeroplane)—the first naval aircraft. Additional navy and Marine Corp officers were gradually assigned to flight instruction and to duty involving actual flying.

Tests to determine the ability of aviators to sight, scout and track submarines were inaugurated in the fall of 1911. And the possibilities of kite-borne observers led to a series of experiments culminating in the kite balloon.

By 1913, the development of dirigibles (powered balloons) abroad could be ignored no longer. Interest in aeronautics was widening. Before that year was out, a comprehensive plan for the orderly development of a U.S. naval air service was drawn up. This emphasized expansion as well as the integration of aviation with the fleet.

These initiatives notwithstanding, an American aeronautical industry was embryonic. Few U.S. aircraft were fit to fly, the laws that held them aloft but dimly understood. "The importance of the development of aviation from a military standpoint was not fully appreciated before the war," one report notes, "with the consequent lack of encouragement of the development of the art."[1] As for foreign progress, France, still, had the lead in the air. Marching briskly, Germany was building a huge military air force. In the lighter-than-air field, commercial passenger service was established fact. The *Deutsche Luftschiffart A.G.*—DELAG as it was known—demonstrated the practicability of airship transport by compiling an enviable (and propitious) prewar record: more than thirty-four thousand persons transported without injury. Berlin was developing its airships, realizing a nineteenth-century

Schwaben (LZ-10), *a pre-war passenger zeppelin, August 1911. Though not a regularly scheduled service, an excursion by rigid airship was immensely popular with* das Volk *and taken rather for granted. (Kevin Pace)*

dream: aerial travel.[2] Commercially, German superiority stood unmatched. Proud of their zeppelins, *das Volk* took this well-patronized, first-of-its-kind excursion service rather for granted.

In 1913, the Massachusetts Institute of Technology requested the navy to detail an officer to organize a graduate course in aeronautical engineering for officers from its Construction Corps (CC) and the Aviation Section of the Army Signal Corps. Aerodynamics using wind tunnels was, as yet, a novel science. Assistant Naval Constructor Jerome C. Hunsaker, a full commander, was assigned. A 1908 graduate from the U.S. Naval Academy (first in his class), keen, scholarly, and detached, Hunsaker was to devote more than six decades to aeronautical science. Since the field stood more advanced in Europe, he toured research establishments in France, England, and Germany to observe firsthand—then report upon—European work and its application. Advanced in his view of aviation and drawing on several departments back at MIT, Hunsaker developed a self-styled "course of lectures and experiments on the design of aeroplanes and dirigibles." Presented in 1914–15, this was the first formal academic aeronautic course in the United States.[3]

The navy at this time had eight airplanes and thirteen pilots, the army seventeen and thirteen, respectively.

Pursuing an elective in aerodynamic physics, one student was a recent Academy graduate (1912). Lt. Garland Fulton, USN (CC), was doing postgraduate work in

The Grand Duke of Hesse alights from Hanza (LZ-13), *Frankfurt, Germany, 13 September 1913. (National Air and Space Museum, Smithsonian Institution, 2000-9229)*

naval architecture and marine engineering. (His MIT roommate and assistant to Hunsaker: Donald W. Douglas.) As a midshipman, Fulton, Richard E. Byrd, and other floormates had been indoctrinated informally to this new thing—aviation—by Lt. Ernest J. King, an ardent disciple of aeronautics.[4] Analytical, reticent, much respected, Fulton was to be on aeronautical engineering duties from 1918 all the way to 1940.

In the spring of 1914, Congress appropriated $250,000 for army aviation. That July, the world war commenced.

Preeminence in dirigibles lay with Germany. Luftschiffbau-Zeppelin and its know-how were promptly integrated into the national cause. It is worth recording that, by midyear 1914, Washington had purchased altogether a mere handful of airplanes. Few aircraft factories were active on this side of the Atlantic, and of these, probably none was producing more than five or six units a month. Of airships, the government had none whatever. Though his prewar information had been good, Hunsaker knew little of subsequent progress overseas. "We did know, however," he wrote, "that both the French and British had developed non-rigid airships [blimps] intensively and were using them for anti-submarine patrol work."[5] In this formative period, Fulton observed, "there was a body of opinion that the dirigible airship held greater potential for Naval purposes than did airplanes or seaplanes, and that kite balloons would afford a workable observation platform for spotting the fall of shots."[6]

In March 1915, Congress passed an appropriation bill of one million dollars for Naval Aeronautics. That April, the Navy Department completed specifications for its first airship. The contract, awarded on 1 June, went to balloonist Thomas Baldwin. Among his credentials: the first American airship to return to its point of ascension (1904) and the first U.S. Army airship (1908).

Thus was the U.S. Navy launched into a near halfcentury of experimentation in LTA aeronautics. Why the interest? The department's requirement for aerial reconnaissance.

The nascent project proved ill starred. Dirigibles, critics held, were too slow, vulnerable, and expensive.[7] Notwithstanding scoffers and skeptics, by mid-1916 the General Board was proposing that three nonrigids be bought at once. Formal instruction in both free and captive balloons was instituted at the Naval Aeronautic Station (NAS) at Pensacola, Florida, where the Navy had established a flight school and experimental center. In November, the board recommended a long-range program of nineteen kite balloons, one nonrigid per naval district, and one "experimental Zeppelin" or rigid-type airship.[8]

With an eye toward the crisis "over there," the General Board updated its war plans—addressing such matters as mobilization, command, strategy, bases, logistics. Reports were required from all bureaus outlining their preparedness. Infused with serious funds, aeronautics was one of these. Though the status of naval aviation remained low, by 1916—halfway through the American neutrality period—Hunsaker was recalled from Cambridge to head the navy's new Aircraft Division within its Bureau of Construction and Repair (C&R). By the time of his posting, rigid airships of the zeppelin type (and seaplanes) were appearing with alarming frequency over coastal Britain, Paris, and other targets. The design and construction of such awesome craft—a wonder of wartime technology—tendered an alluring, abiding secret.

"American, English and French ingenuity," one newspaper mused, "have combined to perfect the biplane, the monoplane and other varieties of flying machine. But, the Zeppelin has been the creature of Germany, and it bids fair to be the first airship ever to cross the Atlantic, realizing another dream of centuries."[9] As the fighting traced its tragic course, and with every ship brought down by gun, weather, or accident, the zeppelin "secret" was shaved closer to full revelation.

The first victim of these raids, a Londoner, was killed in May 1915. (Thirty years would separate this casualty from the fire raids of the Blitz.) Aerial attacks upon unarmed civilians, it was thought, would undermine morale and, possibly, force termination of hostilities. Although actual damage was slight, the indirect results proved significant. Each raid (or false alarm) caused a shutdown in the threatened area, disrupting production. Predictably, citizens demanded an augmented home defense, thereby diverting flying squadrons, antiaircraft guns, and crews from the western front.

All in all, however, the zeppelin raids realized a military disappointment.

The Allied powers were nonetheless obliged to take note. In the print trade, these strategic bombings were translated into artful copy—thereby advertising zeppelin-type rigid airships. Deployments to sea, reconnoitering the scowling waters of the Baltic and North Seas for the High Seas Fleet, garnered scant public notice but were reported back to Washington. From naval attaches came also samples of wreckage, photographs, gossip.[10] While America was yet neutral, the professionals of its War and Navy Departments were fixed upon England and the continent. From abroad, observers reported that the zeppelin held great possibilities. For its part, the Admiralty (under Winston Churchill) pushed its own program of airship-building. "It seemed evident," Hunsaker remarked, "that a new weapon had appeared and that the United States should consider its possibilities carefully."[11]

Stimulated by zeppelin sorties, Rear Adm. David W. Taylor, the navy's air-minded Chief Constructor, requested that the Aluminum Company of America apply its resources to the development of an advanced alloy for aeronautical application. Aluminum was yet precious; when construction of a U.S. Navy rigid (or ZR) was undertaken in 1919, one of the most difficult problems was the production of duralumin, the wondrous lightweight alloy used in German rigids.[12] In October 1916, the General Board reported that zeppelins were "a matter of great importance to the Navy and demand immediate attention." In order to settle questions of fundamental policy, the Secretary of War proposed a Joint Board of officers. Intended to foster interservice cooperation in aeronautics, it was to consider the requirements for developing a lighter-than-air service in the army or navy or both.[13]

Chaired by Taylor, early in 1917 this Joint Army and Navy Airship Board (commonly called the "Zeppelin Board") submitted a tentative program of policy in development work, design work, and procurement. With funds provided equally by both the services, the board initiated design (under the Chief Constructor) while continuing efforts to obtain a craft of the zeppelin type. As well, a proposal by the Bureau of Mines for the experimental extraction of helium from midwestern natural gas was endorsed. (The first serious suggestion to use helium for the inflation of airships had been made in England, in 1914.) Funds were allotted, and three experimental plants built immediately. This action initiated government support of helium work and led to large-scale production in the United States.[14]

That February, a campaign of unrestricted submarine warfare began against all shipping, including neutrals. Washington's declaration came in April—a fateful moment for the war at sea. Operating with unprecedented intensity, that month U-boats sent 849,000 tons to the bottom. At such rates, and unless checked, Britain's resistance might well be broken. In short, German submarines might win the war.

Antisubmarine measures were manifestly inadequate. As for the naval air service, one aviation station was operational—at Pensacola. In terms of personnel, the department boasted thirty-eight qualified officer pilots and 163 enlisted men serving with aviation units. Students under instruction: none. Fifty-five aircraft were to hand to assist U.S. fleet forces: seaplanes, flying boats, and three landplanes—plus one nonrigid abuilding. By way of contrast, Britain *alone* would deploy nearly 250 assorted nonrigids

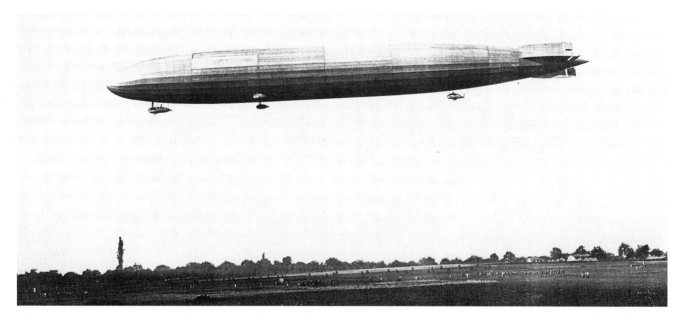

The German Navy's LZ-102 (builder's number), September 1917. Intended to resist aerodynamic bending moments smaller than those calculated for postwar designs, zeppelins fighting the Great War recorded no major structural failure. One explanation: heedful and expert handling. (University of Akron Archives Courtesy Eric Brothers)

Rear Admiral David W. Taylor, the Navy's Chief Constructor. U.S. leadership in aeronautics would depend largely on research. Among the early giants in support of aviation, Taylor was a member of the National Advisory Committee for Aeronautics (NACA) Main Committee from 1917 to 1938. (National Air and Space Museum, Smithsonian Institution (SI Neg. No. 2000-9227))

for coastal work. In German-held territory, more than three dozen army and navy war rigids were operational.

Due largely to President Woodrow Wilson's policy of neutrality, which forbade war preparations, the U.S. Navy found itself unready. At sea, the chief menace was the U-boat. To protect sea lines of communication, the navy had no antisubmarine defense plan, let alone sufficient surface vessels and aircraft equipped to fight submarines. For a fighting air arm, the department "could have done much more than it did to foster naval aviation, particularly to develop a viable torpedo plane, and to foster the technology required to destroy the U-boat."[15]

American engineers knew little about designing and building dirigibles, nothing of flying them. All concerned with the *DN-1* (Dirigible, NonRigid No. 1) project were learning by doing: contractor, inspectors, engineers. Design intricacies proved thorny indeed, not least the matter of deadweight. Completed late in 1916, the result was shipped to Pensacola. Not until spring were flight tests held; three were logged—all in April 1917. Ens. Scott E. Peck was a crewman. On antisubmarine patrol, "Scotty" went on to fly nonrigids off the U.S. coast and, during the twenties and thirties, held billets aboard each of the navy's great rigids.

In the meanwhile, it was orders to *DN-1*. Inflated with 110,000 cubic feet of hydrogen, static lift was a supposed 1,604 pounds. But the craft could barely lift itself. A clanking speed fell below specification—thirty-five miles per hour (about thirty knots), the envelope leaked hydrogen, its engines proved temperamental. Consigned to its floating shed, this "terrible disappointment" was stricken.

"The lessons from the *DN-1* fiasco were taken to heart," Hunsaker records. "We learned, first, that the design of non-rigid airships was an art not to be taken lightly, and that there was no profit or credit to the Department in awarding a contract to an incompetent builder, even though the builder took full responsibility for design. We learned, secondly, that a seaplane or balloon pilot was not at all competent to fly a non-rigid airship without training by an experienced instructor."[16]

Admiral Taylor took charge of procurement personally.[17] Hunsaker was recalled from academic duties to head the newly created Aircraft Division in C&R. As Taylor's Assistant for Aircraft, he was directed to prepare himself on the theory of airship design. Various experimental studies were executed, conjoined with the knowledge and advocacy of assistant naval attache to England, Lt. John H. Towers (Naval Aviator No. 3). Struck by the sea-patrol capabilities of zeppelins, Towers also had flown in a British nonrigid—called a "blimp" by derisive heavier-than-air men. The design was under way when, in December, the Chief of Naval Operations (CNO) requested a coastal patrol ship of twelve hours' endurance and specified performance. Designated the B-type, a reworked, enlarged design was approved in January 1917. With Berlin-Washington relations deteriorating and the design opportune, the immediate procurement of sixteen was ordered.

Contracts were placed that March. So as to pool expertise, several firms were engaged—Goodyear Tire and Rubber, B. F. Goodrich, Connecticut Aircraft—plus cars and engines to Curtiss. In a separate contract, Goodyear undertook to train additional naval personnel at a field and facilities of its own.

Model testing for the B-class (a wholly theoretical design) plus heedful calculation realized a craft that met its

Lieut. Comdr. Jerome C. Hunsaker, USN (CC), March 1919. As Chief of the Aircraft Division of the Bureau of Construction and Repair, Hunsaker supervised the design and calculations for the NC flying boats of 1918–1919 and the first U.S.-built rigid airship—ZR in Navy parlance. He would serve as Chairman of the NACA Main Committee from 1941 to 1956. (National Air and Space Museum, Smithsonian Institution, 2000-9238)

mission: coastal defense. Envelope volume: eighty-four thousand cubic feet of hydrogen. Powered by a hundred-horsepower Curtiss engine, cruising speed was forty-one knots. Since no hangar was (yet) available, *B-1* was erected in Chicago, then flown cross-country to Ohio, into Goodyear hands. The pilot: Ralph Upson, a brilliant engineer and an accomplished balloonist. Upson, though, had never flown a dirigible, there being no one in this country to instruct him.

The first B delivery was logged in August 1917. Unlike the frightful problems in cost and lead time attending modern-day procurement, all sixteen were in navy hands inside eleven months.

Aloft, antisubmarine duties held precedence. To help protect seaborne commerce, bases for naval air units were hurried to completion along the eastern seaboard and in coastal Western Europe. Near U.S. beaches, a necklace of ten patrol stations was commissioned from Halifax south to Key West (and in the Canal Zone)—two B-class blimps assigned to each. At strategic points along the British Isles and on the continent, the erection of patrol and training bases was expedited; in some instances, existing facilities were taken over by U.S. commands.

Thanks to cooperative measures, notably the convoy system plus intensified air surveillance and escort, the U-boat assault would be blunted.

A B-class nonrigid at the United States Navy's Aviator's School, Pensacola, Florida, 2 December 1918. The design and calculations were prepared under the supervision of Naval Constructor Jerome C. Hunsaker. Envelope volume: 84,000 cubic feet of hydrogen. For their type and time, these blimps were equal to anything abroad. (David Collier)

View of the Goodyear Tire and Rubber Company's flying field near Akron, Ohio, November 1918. The United States had little or no experience in lighter-than-air work prior to the war. Goodyear agreed to put up a hangar and facilities of its own to train Naval personnel. Ground was broken in April 1917, the first balloon inflated that June. (National Archives, 072-DW-41)

In Ohio, where the firm was manufacturing balloons, Goodyear graded a flying field near Akron and erected a hangar, hydrogen plant, and ancillary structures. In June 1917, the first inflation was made—a free balloon.[18] Following MIT ground school, pilot-trainees received orders to Ohio for instruction in ballooning. Students then advanced to powered craft—that is, to nonrigids. Commanding the Naval Airship Detachment at Akron: Lt. Lewis H. Maxfield, USN, Naval Aviator No. 17.

In September 1917, Maxfield reported the qualification of the first group trained by Goodyear: eleven student-pilots plus himself. Assigned Naval Aviator (Dirigible) numbers 94 to 104, the men sailed for France, some to the U.S. Navy Command at Centre de Dirigible, Paimboeuf, an LTA patrol base near the mouth of the Loire.

Paimboeuf had been commissioned a naval air station in March 1918, Maxfield commanding—the first dirigible station operated by the U.S. Navy. From that shore, Americans flew French airships over the Bay of Biscay supplementing escorts, giving protection, searching out the undersea foe. "I remember," Hunsaker said, "how Maxfield's eyes shone with pride as he explained how his American personnel took charge of the French equipment as if they had been brought up with it, and proceeded to beat all French patrol records for time in the air in all weathers."[19]

From defense bases fronting the North Atlantic basin, aircraft were detailed to accompany all important convoys. U.S. naval aviators conducted patrols in American-built airplanes, not only in home waters but along the coasts of England, Ireland, France, and Belgium. Lighter-than-air craft were deployed on a large scale. Notwithstanding, sinkings persisted, confined almost wholly to narrow—and seemingly defendable—coastal zones. Franklin D. Roosevelt, Assistant Secretary of the Navy, remarked upon this contradiction. "You would think," he said, addressing the work of the navy in European waters, "that the Allies would be able to control this area, but we have not one-tenth the amount of equipment necessary to patrol all the waters close to shore, let alone further out. The reason submarines go close to shore is because they find a concentration of ships, as almost day and night a continuous procession of merchant ships go up and out of the English Channel, Irish Channel, etc."[20]

Successful patrol or escort is a function of vigilance. The flying demanded absolute concentration. "They just had to keep their eyes glued on the water," a British officer-airman recalled, "hour after hour, in an effort to detect a periscope two or three inches in diameter among the motley collection of debris that perpetually floats on the water, particularly in war-time."[21] A U-boat, dived to avoid attack, could be expected to run at slow speed until the captain thought it safe again to surface—an interval ranging from minutes to hours. The airship—courtesy of its ability to loiter and heave to—tendered a singular asset for the hunt.[22]

Straining away, the navy had chosen the flying boat as its prime air weapon, augmented by blimps. Consigned to East Coast patrol and convoy work and used primarily as trainers, one or two B-ships may have sighted a submarine and dropped bombs. None were deployed overseas. In the fast-growing Aircraft Division of C&R, Hunsaker had the responsibility for design, construction, and procurement of naval aircraft: airplanes, seaplanes, dirigibles. Assisted by Fulton, who had requested aeronautical engineering duties,[23] and by others, the division undertook a wholly new design. The first nonrigid of the C-class reached Naval Air Station (NAS) Rockaway (New York) in October 1918.

By the Armistice, work inside naval aeronautics lay divided almost equally between airplanes and seaplanes, balloons and airships. This parity could not endure; the built-in fiscal tension soon soured. In the play of politics and opinion and chance, airships commanded neither a consensus nor the resources to secure survival as a naval instrument. Relegated to a lower tier, the large rigid type receded to marginal relevance until, on the eve of World War II, it vanished altogether.

In a final session, the Airship Board met to consider the report of its own mission—a technical committee dispatched early in 1918 to investigate the airship art in

World War I served to develop naval aviation for anti-submarine warfare. Front view of the C-class design—a larger, faster blimp for coastal patrol. To advance speed, a new form of envelope and car were developed from wind-tunnel experiments. Volume: 181,000 cubic feet. The C-type resulted from practical experience gained at home and from information gleaned aboard. The scoops (out of view) fed a ballonet—an air-filled chamber used to maintain a constant pressure within the "bag." (National Air and Space Museum, Smithsonian Institution (SI Neg. No. 2000-9494))

England, France, and Germany. This urged (among other recommendations) that any rigid airships needed in the United States be built here, to establish the art. In view of the close analogy to *ships* and their primary naval applications, the navy should undertake the work of design and construction, furnishing "full information" to the War Department.

Displeased, certain Army Air Service factions were to scheme hard for a zeppelin.

On 11 November 1918, all aerial action ceased. U.S. forces at that moment controlled three British lighter-than-air bases and had taken over "practically all" French operations. Now came demobilization. Save for those left to represent U.S. interests, officer and enlisted personnel were gradually detached and ordered home. That bitter November, German airships were immobilized—suspended in their sheds, deflated of their hydrogen.

From almost nothing, U.S. naval air forces had swelled to seven thousand navy and Marine Corps officers, with 30,700 ratings supporting. Equipment: more than two thousand planes, fifteen blimps, 215 free and kite balloons. In home waters, nonrigids had logged little direct war experience. Though discussion had reached the halls of Congress, fighting ended before our Navy Department, starting from nothing, could build or procure a dirigible of the zeppelin type.

In the years 1919–22, accordingly, the U.S. Navy sought to obtain a German-built rigid. The procurement redeemed the entire zeppelin venture in Europe and realized a genuine original for aeronautics—the most successful aircraft of its type flown anywhere. The operational career of this one machine altered U.S. naval aviation in the interwar period and, as well, influenced transoceanic commercial air-transport.

This book recounts that story.

CHAPTER 1

LZ-126: *Birthplace Friedrichshafen*

The principal value of a Zeppelin airship is for scouting at sea in cooperation with the fleet. In order to be effective, the airship must have great endurance and, hence, must be large.

—Secretary of the Navy (September 1921)

Unrestricted surrender. For its role in the great slaughter, Germany would be made to suffer. The great powers set about stripping the country of war material and squeezing its industrial establishment for information. The intent: to disarm a former adversary and competitor in the air, dirigibles as well as airplanes.

Amid humiliation and crippling indemnity, German aeronautics sank to low ebb. An air force forbidden, the future—assuming it was to have one—likely lay in the commercial field. Elsewhere, airmen turned from war-emergency to competitive peacetime pursuits.[1] International civil traffic blossomed. As for aerial heroics and commercial promise, there remained that prize of prizes—the North Atlantic.

Though it is now well forgotten, in 1918 the concept of practicable transoceanic air transport—aircraft plus payload—meant airships, not airplanes.[2] The Zeppelin concern, DELAG, had turned out ninety wartime dirigibles. Recalling prewar glories, DELAG naively assumed that commercial service could resume, not reckoning with Berlin's crushing war-reparations burden. Amid profound social disruption, DELAG's veteran personnel assembled at Friedrichshafen to await a revival, and more—zeppelins as transoceanic carriers.

In the meantime, from victor nations enforcing the peace came intelligence agents of all stripes and experts in every specialty to dine on German spoils.

Unknown to the Allied and Associated Powers, a renegade "pathfinding flight" was being plotted. The naval ship *L-72* (completed just at the armistice) would trailblaze transatlantic air service to America. In March 1919, Luftschiffbau-Zeppelin agreed to this striking—and wholly clandestine—venture. Word leaked. Without explanation, the Inter-Allied Control Commission ordered the project terminated. Motive? Beyond Germany, various commercial schemes were in embryo. That July, indeed, within weeks of the commission's order, the British rigid *R-34* made the first east-to-west crossing and the first transatlantic round trip. In contrast to their heavier-than-air colleagues in their spare, open cockpits, the airship men vaulted out and back in comparative comfort.[3]

In the face of national deterioration, a smallish but refined German design—Luftschiffbau's 120th numbered project—was constructed in six months. From late August through November 1919, DELAG operated *Bodensee* between Friedrichshafen and Berlin. Aloft, excursionists enjoyed a spacious salon in the after part of the main car.

R-34 moored at Mineola, New York, on the Army's flying field, 6–10 July 1919. The first transatlantic flight by airship, R-34 was the third aircraft to cross, its return leg the first round-trip overflight. In the realm of aerodynamics, R-32, R-33, R-34, and R-38 were the first rigids outside Germany to be used for full-scale pressure-distribution investigations. (Thomas Raub)

Though modest in displacement (706,200 cubic feet), *LZ-120* boasted a useful load of twenty-two thousand pounds—enough for thirty passengers plus freight, mail, crew, and fuel.

Throughout Germany, however, conditions held ripe for upheaval. When surrendered German warships were scuttled at Scapa Flow, airmen loyal to the old regime sabotaged seven rigids in their sheds. During June–July 1919, their shores were pulled away and suspension tackles slacked off, plunging the craft to the ground—useless hulks.

Bodensee (LZ-120) landing at Friedrichshafen, home of Luftschiffbau-Zeppelin, in southern Germany. The first postwar commercial zeppelin, Bodensee operated from August through November 1919 before being confined to shed by the Inter-Allied Control Commission. Though modest in displacement, useful load for this advanced design was 22,000 lbs— enough for thirty passengers, crew, and fuel. (Luftschiffbau-Zeppelin)

In reprisal, the Allies demanded replacements either in kind or in cash. The remaining naval airships were taken over, and that December the commission ordered operations stopped and *Bodensee*—along with the just-finished *Nordstern*—confined to their sheds. In a 1920 resolution, the Council of Ambassadors mandated the surrender of two of the lost zeppelins. Further, the German government was to pay the value of the others—and given the right to replace the payments to any power by the surrender of a *civilian* dirigible to be built. Nor did the indignities end there. The London Protocol (enforced by the commission) was to limit the volume of commercial types to a displacement of one million cubic feet—too small to leap the Atlantic.

So *Nordstern* and *Bodensee* were seized and their sheds razed. British and particularly French policy was plain: to exterminate German aeronautics. In 1920–21, reparation zeppelins were parceled out among signatories Britain, France, Italy, Belgium, and Japan. Little came of these trophies, and none flew commercially.[4] Not having ratified the Treaty of Versailles, the United States was absent from this distribution—the only leading power with no rigid.

During 1918–20, Cdr. Jerome C. Hunsaker twice visited Europe to gather information and technical data. In Friedrichshafen, he inspected the available machines. "The opportunity to inspect the German zeppelins under the cloak of the Armistice Commission," Hunsaker wrote, "was of the greatest value, although only a superficial examination could be made."[5] Construction was at a standstill; struggling to remain viable, the Zeppelin concern had turned its four hundred employees to the making kitchenware and other aluminum articles.

Luftschiffbau-Zeppelin was a parent corporation, controlling plants for building airplanes, seaplanes, engines, gears, gas bags, and hangars as well as airships. Airship know-how remained, ready for application. "The moving spirit" of the Zeppelin venture: Dr. Hugo Eckener—Count Zeppelin's collaborator and successor. A man of colossal credential, he was *the* authority on the operation of large airships. Grounded by *diktat,* Eckener toiled manfully behind the scenes—in the conference rooms and offices abroad of businessmen, diplomats, politicians—to keep the works whole. The airman's cheerless chore was a transfer of expertise and personnel so as to sustain the dream of commercial air transport by zeppelin. A strong manager and shrewd at the negotiating table, his dogged resolve was to culminate in a joint venture with the Americans.

Unlike Europe, the state of the airship-design art in America stood untried. As Garland Fulton observed, "We were definitely going along with British thinking and were receiving information and help through various liaison officers." Lewis Maxfield and Zachary Lansdowne, among others, had spent a great deal of time in England. Their nation having no big ships of its own, it was "only natural" that U.S. airmen would parrot British thinking. "Up until the Armistice" (Fulton continues), "the only sources of design information open to us were Allied sources—mainly Britain and France, plus individuals here and there."[6]

In July 1919, the naval appropriation bill for 1920 was passed by the Congress. It allocated funds for a construction shed, the purchase of one rigid (or ZR) abroad, and monies for construction of one at home. First authorized was *R-38;* procured from England while under construction at the Royal Airship Works, Bedford, this airship was redesignated *ZR-2.*[7] The other, *ZR-1,* was christened (in October 1923) USS *Shenandoah*.

Great innovators are great absorbers. Between the wars, the Americans would learn from everyone. Much data was gleaned by Hunsaker, including a full set of plans of *L-49,* a zeppelin forced down intact in 1917—"the most complete and reliable data that we ever obtained." With all available intelligence digested, a new design based on *L-49* was recommended. Fleet Airship No. 1 was approved by the secretary of the navy in August 1919.

Airships demanded a weave of (immature) technologies: theoretical and experimental aerodynamics; the study of weather; the extraction, storage, and repurification of helium; the behavior of aluminum and its alloys; fabrics, dopes, and coatings (for gas cells and outer covers); and, certainly, the training of commanders, officers, crew. For engineering and design, the naval Bureau of Construction and Repair was to occupy center stage until, in 1921, its functions were transferred to a new bureau. Hunsaker and his team, meanwhile, had designed the first practical U.S. dirigible. Now he was made directly responsible for designing the first zeppelin-type craft built in the United States.

The United States is both isolated and defended by water. The Airship Board had reaffirmed—in 1921—the navy's exclusive responsibility for development. This mandate called for the navy to make its data available to the army's lighter-than-air branch. Further, it was to assist private groups interested in developing the rigid for commercial uses. Some Air Service elements were to clamor for revisions in this policy. Among other actions, they made an abortive try for *L-72* before it was turned over. Then came the cloak-and-dagger Hensley episode.

R-38 (ZR-2) at Bedford, England in 1921. Useful lift: 100,000 lbs. Theoretical and experimental aerodynamicists had much yet to learn. One requirement for the airship engineer lay beyond the technology of the time: the "structure" of gusts and wind shifts, so as to compute the wind forces in flight and during ground handling. Still in the builder's hands, ZR-2 failed structurally during its fourth performance trial. (James H. Collier)

The army's William N. Hensley, Jr., was the chosen instrument of Brig. Gen. William "Billy" Mitchell. Having reached Europe in *R-34,* Hensley had orders from Maj. Gen. Mason M. Patrick, Director, Army Air Service, to acquire as much information as practicable concerning airships in England then Germany, France, and Italy. In contradiction of the 1918 Navy/War Departments agreement, however, and authorized in confidence to secure an airship, the colonel (a qualified nonrigid pilot) proceeded to negotiate an option for a new zeppelin. A contract was executed on 27 November 1919 with the Luftschiffbau. Within days, the secretary of war had to repudiate the project, directing cancelation of the contract. As this machination attests and the navy well knew, if it abandoned its responsibilities toward rigids, the army would take them over, thus granting its separate Air Service clique an instrument by which it could press well to seaward. So the navy could not afford to treat the big airship lightly.[8]

Implicit to the Hensley affair was the fact that the Zeppelin works was as desperate to get its postwar plant re-operating as the U.S. Army was hungry to acquire its renowned product.

As the Army Air Service entered the blimp business, its sister service curtailed its own nonrigid operations. The Navy would concentrate instead on *Shenandoah*. Though it never acquired a rigid, the army was a leading operator, retaining its interest to 1937. It had a hand in the (early) negotiations for the navy's own zeppelin and, in mid-1922, ordered an observer to Friedrichshafen during construction. Further, the Air Service persisted in a campaign to snatch control of *ZR-3*. When training commenced, at Lakehurst, its officers received orders to the station; as passenger-observers, Army airmen accompanied *Shenandoah* and *ZR-3* on operations.

In 1919, the sale to the navy of Camp Kendrick, near the village of Lakehurst, New Jersey—a naval nowhere—had been recommended. Acting Secretary of the Navy Franklin D. Roosevelt authorized purchase of about 1,700 acres. Three million dollars was appropriated for a hangar (in Europe, "shed"). Its design, based on plans prepared by the joint board, fell to the Bureau of Yards and Docks, which drew up specifications. Contracts were signed early in September. Site work commenced two weeks later.

U.S. airships were intended for hydrogen. Nonetheless, ground was being prepared for operations using helium, a heavier but inert gas. In mid-1919, the director of naval aviation requested special study for two officers in its production and usage. Lieutenant Commanders T. S.

Boyd and Zeno Wicks were selected. Wicks had commanded the naval air station at Brest during the war. "We took the regular mechanical engineering courses at Annapolis in 1919–1920," he would write, "and Columbia University in 1920–21, receiving M.S. Degrees. We spent the summer of 1921 at the Linde Air Products Laboratory and equipment manufacturing plant—a subsidiary of Union Carbide and Carbon Corp.—at Buffalo, N.Y. There we were familiarized with the laboratory methods and procedures in cryogenic testing and the determination of the percentage of Helium in a gas mixture."[9]

Helium is present in certain deposits of natural gas. Hence its production is a byproduct of natural-gas extraction. To manage the helium project, a three-way collaboration was arranged in July 1925. The Bureau of Mines took over control of all matters affecting conservation and production, including (in place of the navy) operation of the production plant at Fort Worth, coordinating with the two services; the first money was allocated from army and navy appropriations. Workable helium-bearing gases would be located, optimum processes for its separation chosen, the transport problem addressed. The element was rare, hence expensive. From around $2,500 per thousand cubic feet in 1914, helium costs would slide until, at mid-decade, commercial quantities reached "acceptable prices."[10]

Late in 1921, high-pressure cylinders reached Naval Air Station Hampton Roads, a training and experimental station and base for *C-7*, a navy nonrigid. This gas had been extracted at the Government Helium Plant at Fort Worth and accumulated over several months.[11] Its fabric hull enclosing about 181,000 cubic feet, *C-7* lifted off on 2 December—the first flight anywhere by a helium-filled airship. The command pilot was Lt. Cdr. R. F. Wood, with Zachary Lansdowne and Lt. Charles E. Bauch accompanying. Fifteen sorties logged, the nonrigid was deflated, and the precious gas (recompressed) was shipped to Langley Field, Virginia, for repurification.

In Washington, the Bureau of Aeronautics, established by act of Congress in July 1921, became a going entity that September. Taking over most of the aeronautical functions of Construction and Repair (C&R), and under the direction of the secretary of the navy, "BuAer" was made responsible for all matters relating to the design, construction, fitting out, testing, repair, and alterations of the aircraft of the Navy and the Marine Corps, and for providing all planes for the U.S. fleet. Capt. (later rear admiral) William A. Moffett became its first chief. Moffett also received appointment to the National Advisory Committee for Aeronautics, an independent governmental body. Established in 1915 by Congress and in cooperation with the army and navy and industry, NACA was charged with the supervision and coordination of U.S. aeronautical research: aerodynamics, aircraft structures, power plants, aircraft materials (e.g., duralumin and lightweight alloys and protective coatings), air navigation, and meteorology.[12] The admiral would serve with distinction as a member of its executive committee for a dozen years.[13] Moffett was an unusual and impressive officer—tall, straight, dignified, dynamic. Ambitious for aviation and a strong supporter of the naval airship program, he had an easy way with reporters and politicians.

This new office headquartered the lighter-than-air (LTA) organization of naval aviation—the intersection of political and aeronautical power within the Navy Department. Shifted in from C&R, Hunsaker was anointed chief of BuAer's Material Branch, where he brought to bear his wartime intelligence data concerning airships. In 1922, he joined the NACA, serving until detached to London as assistant naval attache. Wicks, who reported in 1923, was made responsible for helium; that July, he received orders to Lakehurst for duty in charge of its helium purification plant.

The semiannual meeting of the full NACA committee in April 1920 had considered the question of the development of rigid airships, deeming them "essential" for national defense. "The proper development of this type of aircraft for military purposes will unquestionably lead to the development of commercial types, but it is felt that the Government must take the lead by first developing rigid airships for military purposes."[14] In England, meanwhile, instruction/familiarization of U.S. personnel continued; when completed, a suitably trained crew would deliver the British-built *ZR-2*. Officer in charge of the Rigid Airship Detachment and ship's prospective skipper was Cdr. Lewis H. Maxfield, the navy's most experienced LTA officer. Back home, however, the engineering staff at C&R had developed misgivings as to British calculations of design strength.

The Allies had held Berlin responsible for the scuttled wartime zeppelins. As noted, Germany would pay compensation, either in gold marks or by constructing a nonmilitary ship in lieu of cash. The American share was eight hundred thousand dollars. At the 116th meeting of the Conference of Ambassadors at Paris, the question came up of replacement in kind or payment. During the deliberations (one account has it), Lloyd George turned to the American ambassador (sitting as an observer) and

The British Air Ministry's director of airships, Air Commodore Edward Maitland (l.), with Cdr. Lewis H. Maxfield, USN. Maitland was responsible for the trials program for the British R-38 (American ZR-2); upon turnover, Maxfield, Naval Aviator (Airship) No. 17, was slated to be ship's commanding officer. (Navy Lakehurst Historical Society)

said, "Possibly the United States, not having received an airship, might be interested in procuring one of these new Zeppelins." The ambassador so reported.[15]

Months of diplomatic fencing ensued, confused by opposition, rivalries, and setbacks.[16] At first, the U.S. Navy seemed absent from talks of a compensation zeppelin, while the army figured prominently. Quite properly, such talks were under the State Department's purview. In matters involving the armistice, State looked mainly to the War Department, with the secretaries of war and the navy conferring—and agreeing—as the matter evolved.

Washington had not signed the hated peace treaty and not been provided for when the Allied powers disposed the surrendered zeppelins. Replacement compelled new construction. In desperate straits, Luftschiffbau representatives approached the U.S. military commission as early as 1920 about possible construction. Another feeler dangled in mid-1921, when the *Direktor* (president and general manager), Alfred Colsman, wrote to Maj. Benjamin Foulois, an aviation assistant to the U.S. military observer in Berlin. Zeppelin, Colsman advised, could provide a large passenger ship for 3,560,00 gold marks ($837,000)—"an extraordinary price."

> The government of the USA being entitled to ask from the German government the delivery of two airships or the respective value in gold marks we beg to request you to call attention of your government to the fact that our firm would be ready to supply the government of the USA with an airship of the type as has been ordered from ourselves on the 26th of November 1919 by Colonel Hensley in the name of the War Board of the USA.[17]

The intent was plain. Washington should use its influence to save the big shed at Friedrichshafen, the only one large enough to construct a ship of transatlantic capability. Through channels, the chief of the Air Service expressed the army's hankering for a zeppelin. On 1 July, via the secretary of state, Ambassador Myron T. Herrick, acting on instructions, advised the Conference of Ambassadors of his country's desire to obtain a no-cost airship in lieu of cash payment. Its size was to be 3,500,000 cubic feet.

President Warren G. Harding favored the procurement, agreeing immediately to press a claim—hence the request for a big rigid, far larger than any yet doled out. Within the council and in the Allied Military Committee of Versailles, resistance proved stiff. The Allies wanted the ship's size pared to a million cubic feet—incapable of a leap to America. During this haggling ZR-2 foundered, during its fourth trial. On 24 August 1921, still in the hands of the builders, the ship failed structurally; the ensuing fuel and hydrogen fire took all but five of the forty-nine aboard. Among the fatalities were all six naval officers, including Maxfield, and ten enlisted men. The fledgling U.S. project had lost half its aircraft, not to mention the talents and experience of its best trained airship men.

The breakup represented an embarrassment for the department and for a time imperiled airship development in America.[18] In terms of engineering design, the loss occasioned intensive investigation of hull strength, and, inevitably, special NACA regard for the design construction of ZR-1—the calculation of stresses particularly. At the

Paris talks, Washington's resolve stiffened. Anticipating receipt, the navy had trained a large force and invested heavily in hangars and gas plants. Judicious pressure was therefore applied—exploiting the British who, after all, were responsible for this rank failure.[19]

The navy knew what it wanted all along: an airship of a hundred thousand cubic meters at least. As the secretary framed it,

> The principal value of a Zeppelin airship is for scouting at sea in cooperation with the fleet. In order to be effective, the airship must have great endurance and, hence, must be large. The Department would prefer an airship of 100,000 cu. m. the absolute minimum. This is smaller than *ZR-2*, but the size of the Zeppelins delivered by Germany to England and France. Any suggestion that the United States accept two airships which together total approximately 70,000 cu. m. is not acceptable as such small ships are of no value as scouts and, furthermore, could not safely fly across the Atlantic to reach this country.

Nor was military application uppermost. "The Navy Department," he argued, continuing, "is primarily interested in an airship of zeppelin construction as an example of applied science. The possible armament is of secondary interest and such equipment need not be provided by Germany. Furthermore, there are obvious possibilities in the largest of airships, as a means of commercial transport, which possibilities can be demonstrated by the possession of an airship of suitable characteristics."[20]

The matter was taken before the Joint Board, and army-navy friction was stopped, at least officially. As approved by the secretaries of war and the navy, acquisition responsibility was bestowed exclusively upon the navy, which accordingly would act as the agent of the United States. Washington would exercise its rights to compensation, demanding a nonmilitary craft of at least the *L-70* class size—"a minimum for safe transatlantic flight." In a confidential dispatch to the U.S. ambassador in Paris that November, the secretary of state advised that

> The Navy Department will be delegated by this Government to act on its behalf in all matters pertaining to the construction, acceptance and maintenance of the airship. For that reason, the airship will be incorporated into the naval establishment of the United States. The airship is to be used for experimental purposes in determining the feasibility for commercial use of rigid airships.[21]

Overmatched, the army was out—for the moment. On 16 December, at the 157th meeting of the conference, the American request was approved (with some sarcasm)—thus concluding a half-year of diplomatic effort. To mollify the British and Japanese, an assurance was added: though acquired by the navy, this would be a "civil airship *[un type civil de dirigeable]* to be devoted to purely civil purposes." Thus empowered, the navy commenced the difficult negotiations with the Luftschiffbau in earnest.

A price for the allocated dirigible required settlement. Fearing loss of the contract plus loss of its sheds and plant, the Germans were at a disadvantage. Initially, four million gold marks had been estimated for a modern craft without spare parts—a figure deemed excessive. (Three million, Moffett had cabled, was the maximum available.) At a March meeting with Eckener, negotiators requested a reduction to three million, *with* replacement spares and delivery. It would be possible, Eckener replied, to build an exact copy of the *L-70* for about 2,500,000 gold marks, but the firm was contemplating "a much superior ship" embodying its war experience and recent studies, and with new type engines. Further conversations outlined a position, as Captain Upham recorded, "In order to save the airship industry, they were willing to build this ship without profit and would do everything possible to reduce their figures." At a charge of 3,500,000 marks, a modern ship and spares could be delivered to Friedrichshafen or, taking a loss, 3,200,000 for their best ship, delivered to the United States. Then arose the matter of insurance. Continuing discussions had Eckener suggesting a separate contract for "turning over" a three-million-mark ship for another four hundred thousand after its arrival in the United States. Ultimately, Upham accepted the separate contract to cover training of an American crew.[22]

Early in 1922, the orders were cut; the builder could proceed. The Paris end of the negotiations accomplished, the focus shifted to Berlin, and to Friedrichshafen. On 23 June, the contracts were signed, outlining in four sections Berlin's obligation to pay the United States approximately three million gold marks in reparations. Section B fixed details of design and procedural matters during actual construction. To the firm's displeasure, a thirty-thousand-cubic-meter ship was the largest permissible under the Allied rules. Still, this would realize a gas capacity surpassing that of any rigid yet built. Section C outlined other obligations, notably transport of the ship to America and three months' training of U.S. crewmen by German personnel. The last section covered spare parts.

So a bit of employment was created, and the Luftschiffbau was kept from closing down. Of this godsend a Zeppelin engineer observed: "They were lucky to get that contract to keep things alive."[23]

The design of rigid airships is mainly a structural problem. The hulls, essentially large frameworks, consist of latticed girders running longitudinally and transversely, braced by steel-wire tension members; shape is independent of gas pressure. Unfortunately for Washington, the design literature was meager. Despite captured and other intelligence, the engineering calculations attending design were rather enigmatic. Given its skill and preeminence, Zeppelin held the key—an exploitable asset. For its part, BuAer was determined to gain full access to Zeppelin expertise and to retain the power of approval throughout the project.

In light of Germany's industrial and political tribulations as well as the charged atmosphere of distrust at Zeppelin, the position of the American representatives at the works called for the utmost tact. The inspector of naval aircraft (INA)—the point of contact with Zeppelin—would be faced with an exquisitely delicate duty. Lt. Cdr. Garland Fulton and Lt. Ralph G. Pennoyer moved to the building works, taking up residence in the Kurgarten Hotel. Retiring yet analytical by nature, calm and deliberate under stress, Fulton was to prove ideal for his INA duty. Pennoyer, for his part, would advise on operational matters. Both LTA and heavier-than-air (HTA) qualified, in 1917 and 1918, respectively, the lieutenant was senior surviving officer from *ZR-2*. A student *and* staff member with the first airship class at Akron, Pennoyer was Naval Aviator (Airship) No. 100. "Our best men for the job," Moffett wrote the naval attache in reference to the inspection force. At the plant, the inspector's office was opened on 6 July 1922. The assignment—enforcement of the contracts and the navy's technical requirements—would hold Fulton in Europe for twenty-nine months. For some matters, the INA would be under the naval attache; Fulton therefore was to keep Berlin informed. Most correspondence pertaining to technical matters would be with Washington, with the Moffett and his advisors.

The contract had no provision for army observers, but by informal agreement the U.S. Army was also represented on site. The admiral had asked General Patrick, chief of the Air Service, to send but a few liaison personnel, "as we have reason to believe that the Zeppelin Company will prove difficult to get information from if they receive the impression that their works are being combed by a large staff of so-called observers."[24] In January 1923, Maj. Frank M. Kennedy reported to the officer in charge.

Kennedy proved a huge irritant, barging in just as information began to flow. Defying diplomatic niceties, he insisted that the army was on scene by right and, save for technical responsibility, on an equal footing with naval personnel. His reports to superiors were laced with complaints. Some remarks reached print, imperiling his own position and further straining army-navy relations. Obliged to work through Fulton and excluded from navy-Zeppelin discussions and conferences, Kennedy lobbied for authority to go to Zeppelin direct. (By contract, Fulton, in charge, had access to "any place" where the ship, engines, equipment or parts were being manufactured.) Lt. Col. B. D. Foulois and Maj. Harold Geiger, chief observer, were Kennedy's immediate superiors in the military attache's office in Berlin. Each had participated in the early procurement conferences. With the ship abuilding, both submitted confidential reports—and "consistently and repeatedly" urged Patrick to do all that he could to secure *ZR-3* for the army.[25]

The principal actors had assembled. The INA "digs" were a pair of rooms on the second floor of the main administration building. Eckener and Ernst Lehmann, his first assistant, had offices nearby, with Kennedy (when he arrived) across the hall. Officials and Zeppelin employees were to recall the INA as a gentleman—taciturn, impassive, unexcitable. Meantime, a low profile was desired, and no publicity, none. Special precautions were installed to hold all matters connected with *ZR-3* either confidential or secret.

Deft negotiators, Fulton and Pennoyer soon were in conference concerning the design and pending construction. Eckener was to paint a picture of a happy and equal collaboration—the conventional account. Bauch recalled warm relations. In truth, the relationship was less than this, during the first half-year particularly. Zeppelin was thankful for the contract, and anxious to build. Yet it had agreed with reluctance. The firm was wary of the department's motives—nervous that the navy would grab its trade secrets and then proceed to build airships itself, thus becoming a direct competitor. The contract for *ZR-3* entitled the navy to certain data before delivery. Eckener and Lehmann, however, preferred to consider this a courtesy rather than a right to be demanded or forced.[26] For its part, the department felt entitled to design data and methods as the work proceeded; it was to insist also on being convinced, step by step, as to structural strength. In his role as inspector, supervisor, and facilitator, Fulton was to assert, firmly but with tact—and "almost endless questions"—his authority to gain what he wanted: unfettered access.[27] Holding the better hand—the works was

LZ-126: BIRTHPLACE FRIEDRICHSHAFEN

living by the grace and favor of the Americans—and authorized to force a showdown (contract cancelation) if needed, he prevailed. Steering his hosts round gradually, the INA began to solve his largest problem, the holding back of design calculation data, specifically a complete stress analysis including all assumptions of loading and all conditions for which calculations or tests were made.

> As pointed out in the Bureau's letter [Hunsaker wrote Fulton], the allowable stresses and factors of safety used are extremely interesting to know, but mean nothing to us until we have the external forces which the ship is expected to withstand with the factors of safety indicated. Naturally, the distribution of weights and buoyancy forces we can calculate from the plans and the results will not differ very much no matter who does it, but there is enormous room for discrepancy between the assumed aerodynamic forces acting on the ship. I think you ought to take the position that these external forces should be agreed upon at a very early date as all calculations depend on them and it seems unreasonable to go ahead making elaborate calculations when the fundamental assumptions have not been cleared up.[28]

As weeks added to months, summer 1922 gave way to fall and winter. In the course of innumerable conferences and evasions and memoranda Fulton extracted what he wanted. Complicating matters was Air Service meddling—including visitors to the plant and publicity as to the allotment of *ZR-3*—that exacerbated Zeppelin's (and the navy's) very tender sensibilities.[29] Within months of the keel laying, in a strong message Fulton was to recommend that the best interests of Washington would be served if army personnel were withdrawn, at least until ship's trials.

The keel of *LZ-126* (builder's number) was laid on 7 November 1922—in Germany, in a time of hyperinflation. In charge of both design and erection was Dr. Ludwig Dürr, a provincial especially affronted by the INA's insistence on inspections and approval. Responsibility for structural design rested with Dr. Karl Arnstein. Cordial and "exceptionally capable" (though at times evasive), Arnstein was head of the mathematical and structural section. *Herr Doktor* would apply his matchless knowledge of large structures (such as bridges) to the project now taking shape.

"Real shop work" underway in the big shed at Friedrichshafen, 18 January 1923. Here a fully-wired five-meter midships section—main ring No. 115 and intermediate No. 110—lie ready for hoisting. Ordered to the construction as Chief Inspector of Naval Aircraft (INA), Friedrichshafen: Lieut. Comdr. Garland Fulton with Lieut. Ralph G. Pennoyer. (INA photo courtesy Lieut. Gordon M. Cousins)

The keel corridor of LZ-126/ZR-3 at main ring No. 100, 9 February 1923. The arterial life of the aircraft would center here, along ship's keelway. (INA/Lieut. Gordon M. Cousins)

This 126th design was no ad-hoc affair. The contract envisioned "the latest ideas and experiences in airship building" and specified a commercial type emphasizing speed and endurance. When complete, moreover, it would be largest aircraft, anywhere. Why the great size? Useful buoyancy requires vast displacement. A hull 658 feet long and ninety-one feet at its maximum diameter would, in this instance, enclose an unprecedented lift-volume, 2,599,110 feet—*Bodensee* writ large.

General specifications approved, the works lumbered into action. The ship's hull would be built according to airship construction practice as it had evolved in Germany, which numbered framing from aft to forward. This was a structural arrangement of space defined by longitudinal and transverse girders, known as "frames." All framing was numbered according to metric distance forward or aft of the rudder post, at frame 5. The transverse girders formed rings that were spaced at mostly uniform distances from one another, with their apexes joined by the longitudinals. Longitudinals were the chief strength elements against aerodynamic forces and were numbered 0 through 6, port and starboard, from the bottom of the ship to its top. There were two kinds of transverse frames—main and intermediate—and wire performed an essential function in both. Brace wires afforded further stiffening in the plane of the main transverse frames and in the panels formed by the longitudinal-transverse girderwork. The main frames, intricately cross-wired, imparted transverse rigidity and also formed bulkheads between adjacent (gas) compartments. Certain transverse frames had no brace-wires, to avoid the gas-holding voids; they were called "intermediate frames." These assisted the longitudinals in handling the gas-cell loads as well as the compressive forces imposed by bending. Finally, to transmit (and distribute) lift loads to the structure, a system of netting wiring and ramie cord was adopted (net result, for laymen: an incomprehensible maze).

Hidden within but vital, a lengthwise keel ran along the bottom centerline of the ship's circumference. As well as forming a through corridor, a tube of space, this local system of girders afforded support for the concentrated weights—fuel tanks, ballast bags, the crew spaces, cargo, and so on.

Of chambers for the lifting gas there were fourteen, numbered from aft (cell 0) forward. Each was made of high-strength, low-weight cloth to which gas-tight material was secured. The cloth afforded most of the cell strength; goldbeater's skins—from the outer surface of the large intestine of oxen—constituted its impermeable element. (Packed in salt, the skins came from meat-packing concerns.) Gas-cell materials were both costly and temperamental—burdensome to store, apply, maintain, repair. Even careful handling shortened fabric life, and tears were hard to avoid. So every installation and gassing (and repair work) proved onerous. Following the transfer, the operators of *ZR-3* searched for cheaper yet satisfactory substitutes, trying various adhesives, materials, and experimental cells.

Luftschiffbau's Stress Department—Kurt Bauch and his *Meister*—analyzed the structure, the basic design of which had been laid down elsewhere.[30] Assigned an outpost office (so as to work undisturbed), the pair checked out the drawings and calculated the design sizes of structural components, the type of girder as well as the gauge of the metal. In the arcane calculus of design, so large a structure of so many members was mathematically inde-

Wiring at Joint No. 1 of a main ring. Note the net cordage. The design of a large high-speed airship is primarily a structural problem in which the most important stresses are those due to forces on the surface of the hull. (INA/Lieut. Gordon M. Cousins)

terminate—that is, the number of unknowns exceeded the available equations. As one engineer-designer wrote:

> The hull structures of rigid airships are typical of a large class of space frameworks . . . in which the shearing and bending forces are distributed between such highly complex and redundant systems of members that mathematically exact methods of primary stress calculations are so difficult as to be rarely or never attempted. Several approximate methods have been developed and used by engineers and designers, but the known methods are not very satisfactory.[31]

Of necessity, then, Zeppelin's analysis incorporated certain assumptions. The actual computations were daunting and almost endless—and all by hand. The analysis was good enough to make the ship safe; as yet, however, designers lacked full knowledge of loads.[32] Once derived, Stress's solutions were presented to Design. "We two, together," Bauch recalled with pride, "we did the entire stress analysis for the *Los Angeles*." To accommodate its client, most of Luftschiffbau's calculations—confirmed by actual tests—were written up longhand.[33]

At the Metal Shop, structural lattices for the aircraft-to-be were stamped from thin sheet then fluted for additional strength and stiffness.

In brisk, efficient steps, a hull structure took form. By early November frames 100 and 115 and the intermediates 105 and 110 were progressing satisfactorily. The first five-meter section (frames 100–115)—the airship's center—was soon under assembly. When fully wired, it was hoisted into the vertical, to be built upon both fore and aft. The date was 19 January. Nearby, further segments were fabricated and, when complete and ready for installation, hoisted into position. Using sixteen men, it was found that a five-meter section required about twenty-four working hours.

"Work is going along quietly and satisfactorily," Fulton reported as February closed. "Considered over all the job is about 35% done. . . . The engines are the only likely source of delay and their production is still up to schedule." If relations with Zeppelin had eased, the army was another matter.[34] Back home, the Air Service was boasting that it was to get the zeppelin. Finding their way to print, these statements kept the navy rubbed raw and furious. "I can't tell you," Hunsaker wrote, "how the Admiral feels with regard to *ZR-3* and the Army as his words are not fit to send through the mails."[35]

On 12 March 1923 an air-inflation test was conducted on cell number 8 (100–115). By mid-April, the structure stood integrated from frames 40 forward to 160—about two-thirds of the overall length. Work on the combined passenger and control car *(Führergondel)* was under way as well.

At this stage of the work, the after power car was about 50 percent along. Hung outside the hull, five power cars *(Maschinengondel)* would house the ship's engines—Maybach Type VL-1s, the very latest design from Maybach Motorenbau, a Luftschiffbau subsidiary. Identical structurally, each pod was light (about six hundred pounds) and streamlined, providing a firm bed for engine, transmission, and pusher propeller, plus space for an attendant. (Gasoline and oil and spare parts would be carried in the main keelway.) The centerline pod was secured aft by struts at the main frame immediately above, and the wing cars were arranged in pairs amidships. Connected by struts and cable suspensions, the four wing pods were spaced so as to reduce bending moments on the hull as well as stagger the slipstreams from the propellers. Each was accessible from the hull via a trap door, collapsible ladder, and slide door—closeable when not in use, so as to reduce drag. Since the forward and centerline cars served as "feet," each was fitted with a shock absorber or pneumatic bumpers plus handling rails.

Fabrics were used throughout, for outer cover, gas cells, ballast bags. Weights and strengths varied, depending upon their use. The framework was covered by strips of cotton cloth tailored to its lines, drawn smooth and taut. (A slack cover was undesirable, since it added to air resistance, costing speed and fuel.) Doped to make it reasonably impervious to moisture, heat, and light, this swaddle protected the cells from abrupt changes of temperature and from the sun, by forming a layer of air between them and the outer cover. To lend smoothness over the joints, removable sealing strips of cloth were stuck on with dope over adjacent lacing edges. All in all, this skin realized a smooth surface to the hull that, together with the streamlined shape, helped to minimize drag.

Although the erection work proved smooth running, the powerplants generated unending headaches. Maybach Motorenbau, under the autocratic Dr. Carl Maybach, enjoyed a fine reputation.[36] But engineering for the VL-1, a radically new (reversible) design, had yet to be worked through. Weeks passed, and no engine data appeared. Fulton had yet to learn that preliminary shop tests were disclosing a series of defects: crankcase cracking, thrust-bearing heating, connecting-rod breaks, carburetor troubles. Rebuffed in his inquiries, the INA was assured by Eckener and Lehmann that Maybach always met

LZ-126: BIRTHPLACE FRIEDRICHSHAFEN

Cell No. 8, air-inflated, at main ring No. 100, 12 March 1923. Note the valves. (INA/Lieut. Gordon M. Cousins)

his promises and that an engine would be ready for testing by May 1923. In fact, the first VL-1 did not arrive at the factory for mocking-up till July. Following repeated requests, Fulton received an installation drawing only that January, with the ship due for completion in August.

Nonetheless, spring 1923 found Fulton satisfied. About three-quarters of the ensemble hung in place. The remainder lay fabricated, with all questionable points regarding the design "now about settled." Much detail work awaited—gas cells, outer cover, ship's apparatus of various kinds. The INA confidently predicted completion between mid-August and the close of September—"the date being adjustable to suit the engines. Although progress on the engines is good, it is still too early to estimate their completion date accurately. Plans are being laid for trials in October and the delivery in November."[37]

On 12 May, BuAer nominated Capt. George W. Steele, Jr., as prospective commanding officer of *ZR-3*. In 1918 Steele had been assistant to the director of naval aviation and in 1923–24 a member of Lakehurst's first student-officer class. In Moffett's view, the command required a man of the highest caliber, in the rank of captain and a qualified naval aviator. (Steele was Naval Aviator [Airship] No. 3172.) Shortly after Steele's nomination, Lt. Cdr. Sidney M. Kraus was designated engineering officer. Anticipating a late-summer completion date, both sailed for Europe at mid-year.

The bow—the section from frame 170 forward—was

Mockups of one of six passenger coups in the main car, for day (left) and night flying. Note the portable wood table (extended). The windows had two hinged sections and could be opened. At right the middle curtain has been dropped, the passageway curtains partially drawn, the upper berth slung. (INA/Lieut. Gordon M. Cousins)

hoisted. In terms of specifications, the mooring attachment resembled that for *Shenandoah* (the cones were identical). Farther aft, the main car had been attached. A single structural unit streamlined in shape, this control-cum-passenger space was installed integrally, close up to the hull. *ZR-3* was a commercial ship; this main car could accommodate from twenty to thirty passengers abaft the navigation and radio spaces. When furnished, its saloon was flanked on both sides by large windows and subdivided into five compartments, each of which had two sofas accommodating four persons. At night, with curtains dividing the space, the sofa backs formed the upper berths, the seats the lower. During the day, tables set between the sofas accommodated hot meals—and the pleasures or routines of air travel. Two lavatories—men's and women's—were located to the rear, as well as a storeroom for supplies and baggage. "The equipment and accommodations for passengers," *Aviation* magazine observed, "will be as complete and as comfortable as the Zeppelin Co. is capable of designing, with all their years of experience behind them."[38]

The aftmost car was nearly done; the side pods were under assembly. Along the keel, installation of gasoline tanks—cylindrical aluminum casks—had commenced. By mid-June, the ensemble hung complete from frame 30 forward. (The tail had been delayed.) The keel itself now extended from frame 30 to 160.

Eight of the gas cells had maneuvering valves, at their tops. These were operated manually from the bridge by means of cables. In addition, automatic valves—for pressure relief—were installed on all but the sternmost cell (numbered 0). Attached a short distance above the keel corridor, these valves were interconnected by rubberized fabric sleeves and grouped so that a single exhaust trunk served the automatics of each two cells. At this stage, all gas shafts had been installed, plus maneuvering valves. The ordinary water-ballast bags had been fitted as well.

Mid-July found the hull integrated from frame 10 (near the fins) all the way to the bow, and the stern was ready for hoisting. Constructed of duralumin girders covered with fabric, the fins were thick and rigid, hence accessible. Within the vast, enlarging skeleton, sixty gasoline tanks had been fitted, along with most of the ballast system. Outside, two power cars hung.

On 10 July, the first VL-1 was received. Used as a mockup to fit connections and to lay out power-car equipment, the original design proved—still—full of defects ruinous to navy timetables. Not for another year would the Motorenbau realize (and install) an effective airship engine.

The engines were fitted well aft of the bridge. Control signals were mechanical, via telegraphs. These transmitted the movement of a handle on the bridge to a corre-

Typical power car—bottom half, 13 April 1923. (INA/Lieut. Gordon M. Cousins)

sponding handle in the cars, where it was acted upon by mechanics on watch. The telegraph itself consisted of the handle, a large sheave, a dial, and a tripping mechanism for operating a bell-alarm signal—a warning of an impending engine order.

Meantime, at the Balloonhullenfabrik at Tempelhof, near Berlin, eight of the fourteen cells had been fabricated, though production had slowed. On 18 August, save for the fixed control surfaces, *der Luftschiffe 126* stood essentially whole. The upper fin was undergoing attachment, the others assembly. All were internally braced, to reduce drag, as were their movable surfaces—elevators and rudders. Operating control cables led from the bridge aft over a series of sheaves and rollers. An auxiliary steering station was designed into the lower fin: Should the control system suffer damage, the ship could be maneuvered from aft, using equipment duplicating that in the main car.

The control and passenger car of LZ-126—a three-quarter view from aft, 24 April 1923. Five types of girder construction was used here. Note the attachment-joints, for securing car to hull. (INA/Lieut. Gordon M. Cousins)

Ring No. 186 clamped in ring-assembly jigs, 5 May 1923. The whole-numbered longitudinals butted to the cylindrical ring of the nose. (INA/Lieut. Gordon M. Cousins)

As for externals, the outer cover lay in place from the bow aft to frame 55; over approximately half of it two coats of dope had been applied. By month's end, observers for the transocean delivery were settled: Captain Steele, Lieutenant Commander Kraus, and one army observer (Kennedy).[39] Kraus, the prospective engineering officer, had been assigned to the inspection staff. Steele, the prospective skipper, would not arrive until trials were imminent, the following spring.

If all broke right with the VL-1s, Pennoyer advised Berlin, *ZR-3* might be ready for trials about the first week in November 1923 and for the crossing late in the month—"but I am not that optimistic." He advocated thorough engine tests and trials; "Do everything," he urged, to prevent delivery during the winter months, "say from 15 December until 1st April."[40]

A gas-tank installation fitted at main ring No. 100. Note the water ballast bag. (INA/Lieut. Gordon M. Cousins)

At Lakehurst, the first U.S.-designed-and-built rigid was undocked. (Design work had included wind-tunnel model tests.) On 4 September, Lakehurst heard these commands: "Stand by"—"Up ship." *ZR-1* lifted off and sailed to seaward; this first performance trial was the first-ever flight by a helium-inflated rigid airship. As the craft climbed away, two airplanes (representing the army and navy) rose to accompany it. In the control car, Cdr. Ralph Weyerbacher (CC) (i.e., Construction Corps), stood in

Outer cover being applied to the hull aft to ring No. 100, 22 August 1923. Note the staging, foreground. Aerial ladders and drop platforms also were employed. (INA/Lieut. Gordon M. Cousins)

View aft from the bow, 24 August 1923. The outer cover has been laced onto the hull framework to ring No. 145. Note the two emergency water-ballast bags, at ring No. 175. Cutouts in the cover-swaddle (poorly visible) at the propeller areas were fitted with heavier fabric. (INA/Lieut. Gordon M. Cousins)

LZ-126: BIRTHPLACE FRIEDRICHSHAFEN

State of completion in der Halle, *11 October 1923. Insisting on full access to German design methods, theoretical calculations and stressing procedures, the Navy Department sought, step by step, assurance as to structural strength. Of particular interest: a complete stress analysis including all assumptions of loading and all conditions for which calculations or tests were made. (INA/Lieut. Gordon M. Cousins)*

charge. Designated Naval Aviator (Airship) No. 2126, he was assisted by Capt. Anton Heinen, a former Zeppelin employee—the only man aboard who knew rigids firsthand. His presence was emblematic. As events were to underscore, this U.S. machine was unknown, expensive, untried. Also aboard were nonrecording strain gauges, with four observers.

That October, *ZR-1* was christened USS *Shenandoah*.

Controversy between the Navy and War Departments still menaced the erection and delivery of *ZR-3* by the Luftschiffbau. Thanks to the Air Service and its schemes, the navy continued to seethe, with Moffett near to boiling. By international agreement, the new zeppelin was slated for the naval service. Any change in allocation prior to final acceptance, the admiral worried, would injure the good faith of the government.

> Furthermore, the acquisition of the ship, together with the design, have been attended by the greatest difficulty due to German suspicion, economic disorganization, and labor troubles in Germany. The Navy has, it is believed, finally convinced the German builders of our good faith, friendly attitude, and technical and professional competence, so that the ship has made good progress, inspection troubles are minor, and, unless, unforeseen complications develop, eventual delivery is assured. To turn the uncompleted ship over to the army to-day would destroy a great deal of the satisfactory position built up and make eventual delivery of a perfect ship very doubtful.

The tail section covered aft to main ring No. 25, at the forward fin attachment. Large aerodynamic forces occur on the tail surfaces, so very strong transverse frames were designed to transmit these loads into the hull. Forces on the rudders and elevators of ZR-3 had been calculated from wind-tunnel data, helping to realize the thick cantilever design and location of two cruciform structures at Frames 5 and 15. (INA/Lieut. Gordon M. Cousins)

The most promising field for commercial application lay in transoceanic passenger and mail transport. Accordingly, the navy (Moffett continued) was better positioned to carry out the further development, to prove whether or not ZRs were "practicable not only for Naval purposes but also for commercial uses." As bureau chief, it was his conviction that "nothing whatever, including the desires of the army, be allowed to interfere in any way whatsoever with the program of the Navy to carry out its mission of the development of rigid airships."[41]

In mid-Europe, the object of this intraservice fracas neared finish. The gas cells were slated for shipment, after which installation and inflation would require at least nine days. Excepting her troublesome Maybachs, the dirigible could be ready by 10 October. This same week, Paul W. Litchfield and William Young, the president of Goodyear Tire and Rubber and one of its manager-assistants, respectively, visited the shed to view the work.

From his vantage, Moffett had developed reservations as to a late-season crossing. "The Admiral would rather delay until spring if any delay at all is involved—because a) don't like the idea of new engines coming across before face tests, etc., b) weather is windy at Lakehurst in winter and risk of putting in shed is great, and c) training contract can't operate in windy winter weather. . . . We can't under the contract refuse to take delivery but we might intimate we are in no sweat about it." There was another factor. "Helium enough for two ships is not available," he continued. "Army is hot after ZR-2. If one ship is laid up they can howl more. They control 50% of helium and

Power car in cradle, 19 October 1923. All five nacelles were alike except for rear center-line car (No. 1), which was fitted for a bumper bag. (INA/Lieut. Gordon M. Cousins)

will withhold their half." Nor was the political situation reassuring: "But if Germany blows up we may never get the ship."[42]

At mid-month, *LZ-126* stood very nearly whole. But officials had abandoned all thought of a transatlantic leap before spring, courtesy of teething problems attending the VL-1s. Further, about a hundred employees had been dismissed, so work was progressing slowly.

On 20 November 1923, and save for the troublesome Maybachs, the aircraft was ready. Fall shaded into winter. Engine tests spilled into the new year. Not until August did the VL-1 pass muster—a ninety-hour test. That spring (1924), the bureau, in its frustration, considered the substitution of U.S. powerplants, then thought better of it:

> After canvassing the situation here with all concerned, you are advised that the Bureau's decision is to insist on the Zeppelin Company delivering *ZR-3* with German engines for this German trip. There are a great many reasons why American engines in a German ship would not be satisfactory, the most pertinent of which is that, no matter what might happen to the ship, the fault would undoubtedly be laid by German personnel at the feet of American engines.[43]

Mere days before this note, Moffett had vented his underlying reservations. "[The Germans] were told by the Admiral that the whole reputation of the Zeppelin Company, the Goodyear-Zeppelin Company and the reputation of Zeppelin rigid airships depended entirely on results obtained with *ZR-3*; that the whole world interested in lighter-than-air was watching this ship with the keenest interest and it was up to them to come through

clean no matter what the difficulties encountered were. Captain Lehmann agreed to this in its entirety and we therefore feel that it is absolutely essential for the Zeppelin Company to properly engine the ship and deliver it to Lakehurst."[44] The navy wanted its zeppelin during July or August at the latest, so as to exploit good weather for training. But even a 1925 delivery would be tolerated—as long as satisfactory engines were furnished.

In anticipation, various plans were put forth. The contract provided for three months' training immediately following acceptance. The first U.S. flights, then, would be purely training, during which the ship could be shown off to Philadelphia, New York, and Washington. Later, longer flights could be made—Chicago and St. Louis—followed by a transcontinental demonstration. Emphasis then would shift to confirming commercial possibilities—mail and passengers. A Lakehurst-to-London flight was suggested or, with the weather's blessing, a San Diego–Honolulu run. For the latter, *ZR-3* "could be run with the regularity of the steamship service and reduce the time by two days. If the ship be correctly operated, there is no reason why it can not operate on schedule for months on end."[45]

By August, Fulton seemed satisfied. "Today the engine situation looks hopeful for a first trial around 1 September. Maybach has a world's record on engine troubles." On 6 August, the stipulated tests were declared satisfac-

A power car, aft, portside, at ring No. 85, 29 April 1924. The water-filling line is just forward of the hull opening, oil line just aft. The gasoline line runs along a strut to rear end of the car's ridge girder; the distribution line is immediately over the engine. Clutch coupling is just aft the Maybach, the thrust bearing just forward the propeller. (INA/Lieut. Gordon M. Cousins)

One-quarter front (port) view of the 126th project, 29 April 1924. Note the handling-line trap doors, at the keel. (INA/Lieut. Gordon M. Cousins)

In civilian gear (l. to r.), Major Frank M. Kennedy, USA, Lieut. Cdr. Sidney M. Kraus, and Lieut. Cdr. Garland Fulton, Inspector of Naval Aircraft, Friedrichshafen, at the chart table of the Zeppelin Company's latest creation—the world's largest aircraft. (National Air and Space Museum, Smithsonian Institution (SI Neg. No. 2000-9242))

tory. In the space of twenty-one months, a complex of systems had come together, a great aircraft realized.

Per the contract, Luftschiffbau-Zeppelin was to determine—via "sufficient trial flights"—airworthiness and reliability. A program of trials had therefore been prescribed. The ship would fly! Now the publicity lid was off: anticipating flight, full cooperation was extended to newspapers and news agencies, picture services, and other channels, including representatives of virtually every American news agency in Germany.[46] Deeply excited *Deutscher Volk* streamed toward Friedrichshafen. This was a German story, after all. *Der Amerikaluftschiff*, the largest aircraft yet built anywhere, embodied a long-eminent (and envied) prowess in aeronautics. If the *126* succeeded in crossing the Atlantic, Zeppelin would be credited with the longest flight yet logged by an airship. Its release to the Americans, moreover, would mark the end of the construction works—slated for dismantling in fulfillment of the Versailles Treaty.

Following shed tests of all mechanical gear, the inau-

Under the glare of shed lights, LZ-126 is made ready for its first undocking, 27 August 1924. (National Air and Space Museum, Smithsonian Institution (SI Neg. No. 2000-9234))

Dr. Hugo Eckener and Capt. George W. Steele, Jr., USN, on board LZ-126—Luftschiffbau-Zeppelin's 126th design. Eckener would command ship's delivery to the United States. As the prospective commanding officer of ZR-3, Steele—Naval Aviator (Airship) No. 3172—would accompany the transatlantic delivery to Lakehurst. (National Air and Space Museum, Smithsonian Institution (SI Neg. No. 2000-9231))

gural *Probefahrt* (trial trip) was set. In the forenoon calm of 27 August, the hangar shook to life. Scaffolding was removed from beneath the cars, builder's gear was shunted ashore, last touches were put to ship's outer cover—left open near the keel to admit the cells. Once the docking rails were cleared, one by one, the Maybachs barked to life and were tested.

Preflight complete, the observers joined the ship—Fulton and Steele forward, Kraus in the after car, Schmidt to the keel corridor. As well, Jack Fulton (Construction Superintendent, Goodrich Aeronautical Department) and Kennedy joined the test crew. Before the choreography of ground handling commenced, Eckener proceeded to "weigh-off" (to define buoyancy). This consisted of determining the airship's static condition and discharging water fore and aft to bring it into a state of equilibrium, in which weight was literally zero. With two hundred workers grouped on the handling lines and car rails on both sides, *der Luftschiffe* began to move stern first. Time: 0935. Hauled clear of the shed and held to the ground, the great ship veered until bow-on to the morning wind. Eckener waited till satisfied that all was in readiness, then boarded. Word was passed on his command, and guy lines were disconnected. Alongside the passenger and after cars, helping arms pushed up, lending a skyward force. Cast off, the craft rose statically. Ballast water poured out. Telegraphs jangled, the Maybachs obeyed, and propellers bit, pushing *126* off under its own power.

Ballast released forward, the ship lifts away from ground handlers for the first of its trials (probefahrten). *In command: Herr Doktor* Eckener. *(National Air and Space Museum, Smithsonian Institution (SI Neg. No. 2000-9230))*

The Zeppelin rises off the Friedrichshafen field on even keel. Despite their size, under certain conditions airships are, literally, lighter than the surrounding atmosphere. (National Air and Space Museum, Smithsonian Institution, (SI Neg. 2000-9228))

The aircraft edged forward. Answering its controls well, the ship steered to the neighboring Bodensee. Over water, helm was ordered to port, course southeast—tracing the north shore, toward Lindau, near the Austrian corner of the lake. A squall struck. Beneath thick, gloomy skies and in occasional rain (that confirmed where cabin walls and windows leaked), the dirigible kept offshore.

The objective of this trial was to test systems and structures as well as familiarize crew, technicians, constructors, and observers. Luftschiffbau stood responsible for the craft until its delivery-landing at Lakehurst. Hence only seasoned staff were aboard. Eckener, beyond debate the reigning airman, was chief pilot. Hans C. Flemming was second in command, assisted by Hans von Schiller and Lehmann and other experienced hands.

As men came forward or disappeared aft, *Herr Doktor* had his helmsman steer east, then make for the Swiss shore. Turning north, he returned to Friedrichshafen before driving west, past Meersburg, then back over Konstanz to the vicinity of the shed—for another weigh-off. Operational practice called for a weigh-off in the air above the field before landing. The engines were stopped or idled and the ship's rise or fall measured. Gas was then valved or ballast released to bring the aircraft into equilibrium.

Like sailors, airmen must know the wind and respect its power. Eckener was masterful, superb in the air. Though cautious, he was confident; for the launch, Eckener had deliberately chosen threatening conditions. Above a roiled Bodensee, and passing through "a very rough line squall," the ship had handled well. At full

Herr Doktor *Eckener and his senior officers flanked by the U.S. Navy's Kraus (l.) and Steele. (National Air and Space Museum, Smithsonian Institution (SI Neg. No. 2000-9244))*

power, it developed 79.5 miles per hour (sixty-nine knots), and seventy miles per hour at cruising speed, with three hundred horsepower from each Maybach—a gratifying performance. (A knot is a unit of speed equal to one nautical mile, or about 1.15 statute miles, per hour.) "It appears," Steele said, "that the *ZR-3* will be a handy ship." Engine gremlins, though, had yet to be banished. Shortly after revving up to full-speed revolutions, the mechanic at number-two engine had signaled "Stop"—a thrown crankshaft counterweight had punctured the crankcase. Back on the ground, engineers (Eckener reported on the 29th) would decide to strengthen the bolts in the crankshafts and counterweights of all five engines.

Two hours and forty-seven minutes had been spent aloft. Maximum altitude: 2,600 feet. All in all, the audition proved successful—"a bully trip," Steele pronounced. "Just like riding in the *Leviathan*."[47]

On 6 September, *LZ-126* was unhoused for a high-profile test that included speed trials. Eighty-three were along for *die zweite Probefahrt;* its morning takeoff was attended by a military band and crowds of onlookers, who cheered and waved flags. "It was a beautiful flight, we had excellent weather, and [it] extended over most of southern Germany," Bauch later recalled. "It was also, again, a test flight but also at the same time a demonstration flight—to show the people the ship."[48] The trials held a subtext: Eckener intended to parade German technological prowess.

With American observers on board, der Amerikaluftschiff shows itself to the motherland during flight trials.

LZ-126: BIRTHPLACE FRIEDRICHSHAFEN

Immediately after touchdown, following the first trial sortie. (National Air and Space Museum, Smithsonian Institution (SI Neg. No. 9235))

He made no secret of this. A superb publicist, Eckener had set out to generate international interest and American backing for commercial air travel by airship.

On board for the United States were Fulton, Kraus, and BuAer's John Towers, as well as Kennedy and Geiger. Nodding shrewdly to public relations, Eckener hosted two movie cameramen plus a cast of correspondents. The salvos of copy and the reactions of city crowds to this latest zeppelin were to be all that he could have wanted.

As a fairy-tale countryside slid beneath, engineers monitored and probed their newest creation. Bauch, for his part, measured structural deflections during certain maneuvers, by sighting (via transit) down the gangway along markers attached to the main frames. With zero defects evident, Eckener, pleased, had taken departure from the wide, lovely lake on a northerly course. Ravensburg was below at 0945. A turn to starboard (eastward), and the zeppelin overflew Kempton at 1020. Pressing northeast, she was cruising the environs of Munich in less than an hour. Eckener now ordered left rudder, to steer almost due north—destination Regensburg. From there, a dogleg to northwest brought her to Nürnberg, thence southwest, to the suburbs enfolding Stuttgart. In midafternoon, 1530, Überlingen and the Bodensee were again visible in the forward windows. En route homeward, *LZ-126* overflew lakeside Meersburg before ending its trials and, at 1815, landing. Average height for the sally: one thousand feet; maximum altitude: 3,300. Four hundred seventy-five miles and nine hours, thirty-two minutes had been added to ship's log.

This portentous September, *Aviation* reported that the new ship would not leave for America until a thirty-five-hour trial had been completed.[49] As for pressure to effect the crossing, Fulton was unequivocal: "There was never any idea nor any pressure to start the flight prematurely or before the engines were proved reliable to Eckener's satisfaction, Dürr's satisfaction and Maybach's satisfaction. There were many frustrations and much talk from both sides and from the press. But, basically, the Zeppelin people could not afford to 'take a chance.'"[50]

Meantime, Cdr. Jacob H. Klein, skipper of the Lakehurst naval station and Naval Aviator (Airship) No. 3185, set sail on the liner *Mauritania*. To Moffett's smoldering displeasure (and the irritation of Steele), Klein had inserted himself into the project, conniving temporary-duty orders to Friedrichshafen. There, at the eleventh hour, he joined his fellow officer-observers.

The third trial, on 11 September, logged eight hours and twenty-five minutes. Again, reporters were along; in all, seventy people were on board. The star passenger was Countess Hella von Bradenstein-Zeppelin, daughter of the late Count Zeppelin and mother of three. "The trip was somewhat tiresome to me," she said of the experience, "but a great delight to the children, who were up for the first time."[51] Responding to a flood of requests, *der Luftschiffe* this time graced Swiss airspace. First, though, a series of tests were flown over the relative calm of water—measuring dynamic lift and calibrating the radio compass. Eckener then took a westward course. Overflying Konstanz, on the southwestern shore, the helmsman followed the Rhine River Valley to Schaffhausen, thence to Basel. A turn to port, and the *126* pushed deep into Switzerland. From Lucerne, Eckener turned northeast; four and a half hours after takeoff, Zürich spread beneath ship's keel, then Winterthur, then St. Gallen. Beyond arced the lovely Swiss shore of the Bodensee and, opposite, Friedrichshafen, with its zeppelin airfield.

Though a joyride for passengers, this sortie brought disappointment. One of the Maybachs—the after-car engine—had to be stopped, due to a defective thrust bearing in the crankshaft. Back at the berth, a postflight autopsy was conducted on all five powerplants, the defect corrected.

The next sortie took place on the 13th, another turn over water. In its two hours, two minutes' duration, the radio compass was calibrated and a bit of blind flying practiced—by taking bearings from Friedrichshafen to the radio station at Konstanz. An announcement followed the trial: the transatlantic flight had been deferred until late in September or early October. Still, in the aeronautical world the commercial possibilities glittered. At the outset, *ZR-3* would be exploited for training. "It is quite probable," *Aviation* intoned, "that one or more experimental flights may be made with a view to demonstrating the commercial utility of airships of this type." The ship, it further speculated, might be turned over to the Air Mail Service, for the transcontinental route, or possibly sold to a U.S. commercial firm.[52]

The fifth and longest *Probfahrt* was not logged until the 25th—the delay owing to the powerplants and, one day, to unpromising weather. This *Deutschlandfahrt* was attended by an immense outpouring of public enthusiasm. Rumors had the new zeppelin running as far as the Baltic port of Königsberg in the Soviet Union, or to Stockholm or Copenhagen. Requests for flyovers had flooded in. As many as practicable, Eckener decreed, would be met.

Takeoff was at 0900 with seventy-one on board, thirty of them crew. Eckener expended an hour circling, testing the radio direction finder and, possibly, conducting deceleration tests.[53] Satisfied, he ordered a northerly course for Frankfurt-am-Main. This track, her captain intended, would thread urban skylines and rural landscapes—the full length of the Reich, "to show," as Anton Wittemann, a seasoned airman put it, "the last Zeppelin to the German Fatherland, and to say farewell." Its luggage racks piled high, the main car hosted a spirited company; in short order, Steele observed, the "whole crowd were eating. Wine and beer also appeared." A newsman began typing, photographers darted from one side to the other, and two motion-picture cameras pointed out the windows.

Among the first points was the university town of Tübingen, well to the north of the Zeppelin works. Then on to Heidelberg, paralleling the Rhine; she was over the city at 1151. Darmstadt was below at 1218, the metropolis of Frankfurt but minutes later. Course due north, the ship passed over Geissen (1300), then Marburg (1315). Eckener gained Kassel, in the very heart of Germany, at 1357. There a plane met the airship.

As the big airship droned over the countryside, the peopled roofscapes of villages, towns, and cities unrolled beneath: the blocks of Hannover at 1506, then, approaching the Reich's northwestern corner, the wharved harbor-metropolis of Bremen. Steaming northeast, Eckener reached the sprawl of Hamburg at 1700—precisely the hour advertised. Here six planes joined. Below, canals leading to the city were filled with freight barges; the streets were seas of waving handkerchiefs. Flying over as much of the Fatherland chart as possible, Eckener now pressed into extreme North Germany: Rendsburg and the Kaiser Wilhelm Canal (at 1750), Schleswig (1802), Flensburg (1817), the Baltic—virtually the Danish frontier. Eckener circled his boyhood home. On board, a hum (the kitchen generator) announced the prospect of supper—sausages, cold ham, bread, and hot tea. The airship jogged south, to overfly Kiel (where a searchlight kept on her), then the river city of Lübeck (1846), then the Ostsee. For the night, Eckener chose to loiter offshore. The crew practiced overwater navigation via radio and compass bearings (the latter taken on lighthouses ashore) and by drift measurements using smoke bombs dropped to

The 126th design over Dresden, 26 September, 1300 hours—the Zeppelin's fifth and longest probefahrt. (Author)

the sea. As for sleep, there were more men than berths, obliging some to sleep two to a bunk. Others resorted to the deck or slept sitting up.

Though Eckener had intended to penetrate east to Königsberg, heavy squalls forced reconsideration. Having touched the Swedish coast, he came ashore near the German-Polish frontier at dawn; residents of Stettin, on the Oder River, heard him at 0740. It now was but a short run for Berlin over dead-flat terrain. Appearing at 0930 in low cloud, *der Amerikaluftschiff* loitered forty-five minutes over the capital. People had massed in thousands, including schoolchildren let out to see it pass. The effect was stupendous, a frenzy of jubilation. Then home—southward—via the suburb of Potsdam. En route to Dresden, sunlight played between detached masses of cloud, producing bumpy air and a slow pitching motion. Dresden, though, was circled in bright sunshine. This mealtime, bowls of hot soup were followed by a goulash and dessert. Ulm, then Augsburg (1622) underlay the keel and then, at 1710, the Friedrichshafen shed—in rain. Sailing overhead, Eckener circled offshore, the lake smeared in whitecaps. Having hung aloft thirty-three hours thirty-two minutes, *LZ-126* was brought to earth at 1832, in semidarkness, "as easy as a bubble."

Ship, captain, and crew had performed admirably.

So much for sprints. Now the marathon, the big test. Eckener's team felt ready for the North Atlantic. The U.S. Navy concurred. "In view of the fact that the rigid airship constructed in Germany for the Government of the United States in accordance with an agreement signed in Berlin June 26, 1922, has completed in Germany trials satisfactory to the Navy Department of the United States of America, and has been declared by the Luftschiffbau Zeppelin G.m.b.h. ready for delivery by them to the United States, I, as the representative of the United States of America, hereby certify, in accordance with Paragraph 7 of said agreement, that the general characteristics of the airship are satisfactory, and that subject to its delivery with its customary spare parts to the Naval Air Station Lakehurst, New Jersey, in good condition, the airship will be accepted by the United States."[54]

But the calendar read October—a time of autumnal storms and fog. Expressions of "complete confidence" aside, privately Eckener viewed the crossing with some apprehension. No zeppelin had yet flown the North Atlantic. Further, given its displacement, his ship's range was marginal for so great a leap.

Today's air travelers are blasé; distance has lost its wonder. In the 1920s, vaulting oceans was a bold, even heady notion. Airplanes operated but short distances out to sea. Further, there existed no reliable ocean-weather service—timely as well as accurate data, and plenty of it. Changing storm positions demanded synoptic ocean maps, so operators could choose the most favorable course. In 1924, the system in place (such as it was) served only coastwise traffic.

Speed is essential in weather services. Observations must be promptly made, reports transmitted without delay—and free of error. The 1924 network relied upon that marvel of the age: wireless communications. But the number of stations were few, and their electronic reach to seaward covered a pitiably small zone. Further, ocean reports were obtainable only from shipboard observers. These were far less numerous than over land and tended to concentrate along regular sea-lanes. This left vast sweeps underreported. In terms of Europe-America traffic, the most notable blank lay between Greenland and Newfoundland, and southward to latitude fifty degrees north.[55] Moreover, at-sea reports were often in error, and no standardization had been imposed. The relationship between weather and changes in pressure is intimate; storm systems deliver low pressures. Yet the most common errors were barometric pressure and wind force and direction. Cloud forms—especially important for frontal identification—went unreported. Finally, U.S. ships were making regular reports only every twelve hours—far too infrequent for transoceanic service.

The net result was forecasting so meager that it was impossible to identify any but well-marked fronts. Indeed, the efficacy of air-mass (or frontal) methods of analysis had yet to be recognized fully.[56] Nonetheless, the navy was interested; the history of frontal analysis in this country reflects the Navy Department's concern for the subject. Why? The method gave promise of improved advisories for naval aircraft operating over the open sea, where, plainly, the airways network system could not be applied.[57]

Still, the seagoing airman had aids. Broadcasts from the powerful naval station near Washington—Radio Arlington (call sign NAA)—was one. In 1921, the U.S. Weather Bureau had begun forecasts for aviators. Already, naval aerological officers were applying what frontal principles they could in their work. Their (daily) maps and forecasts were of particular value at stations like Lakehurst, cursed by the passage of frequent and often severe fronts.[58]

For this crossing, however, Eckener and his fellow airmen had no choice but to rely upon installed systems—augmented by their own instincts, experience, and judgment.

As liftoff neared, interest soared. The emotional attachment was abiding; in this Teutonic romance, the "last" zeppelin represented a salve for the national psyche. Acres of overheated prose spilled forth as, from throughout the Reich, expectant citizens and tourists descended on Friedrichshafen. In the town, imaginative plans for a transoceanic berth were contrived—and refused. Denied official sanction, would-be stowaways plotted.[59] In the United States, airship men knew that immediate development of the rigid type hung upon how well the ship performed. (The commercial aspects of the delivery, indeed, were creating the most comment.) If it was to secure contracts for transcontinental or transatlantic mail or passenger service, Goodyear, for one, needed to see success.

In an atmosphere of nervous exhilaration, preparatory work closed out. Eckener and his inner circle, along with engineers and Zeppelin officials, reviewed plans and settled a great deal of detail. The final tasks included a precise, minutely itemized loading plan specifying the weight and distribution of the useful load—gasoline, water ballast, tools, provisions, cargo. Fuel was crucial. So as to haul every possible gallon, some nonessentials—luggage racks, crews' quarters and furnishings, ballast bags, winches—had been removed and extra tanks fitted along the keel at frames 50, 70, 130, and 145.

Came the predawn of Saturday, 11 October. The aircraft stood loaded, ready. Despite the hour, the crowd was such that Steele (en route from his hotel) doubted he would reach the shed in time—0700 hours.[60] Inside, he found much bustle and found that "everything was not going well." The ship would not rise. Elevated temperatures had combined with wet fog to soak its outer cover enough to offset lift—too many pounds heavy. Sand bags were removed, three gas tanks emptied to ease static condition. Still it refused. Rather than jettison more fuel to gain equilibrium, and given contrary winds reported off coastal France, Eckener elected to reset departure—to 0600, the following day. Ceremonial pomp thus embarrassed, discomfited dignitaries and the swarm of early risers decamped the shed area.

Sunday brought haze, as well as dripping fog at eighty feet. Yet the temperature was better—that is, colder. To the westward, a ridge of high pressure extended from mid-France to the Bay of Biscay and the Azores. Shortly after 0600, beneath lights, the crew was ordered on board. A weigh-off was taken—the craft was heavy at the bow. Shunted out at 0625, the great hull drilled through a clammy, gloomy dark, bow into a light breeze. Final particulars were completed on the wet grass; at 0635 (Berlin time), handlers cast off lines and handholds on Eckener's shout of *"Hoch!" Der Amerikaluftschiff* rose slowly, statically, in perfect quiet.

The good-by journey had begun.

The transatlantic team comprised twenty-seven, five of whom possessed their master's tickets.[61] Free spaces (and disposable lift) allowed four Americans home passage by air: Steele, Klein, Kraus, Kennedy.[62] In the control cabin, Eckener would not stand regular watches; for the rest, a port-and-starboard rotation of four-hour watches would extend throughout the crossing, under two officers, Lehmann and Flemming. (Lehmann's was on deck first.) Manning the elevators was Max Pruss (*Hohensteuer*); to his right, forward, was Ludwig Marx,

Encased in early-morning damp, 0635 local time, LZ-126/ ZR-3 rises from the Friedrichshafen airfield, 12 October 1924. Destination: America! There, in the United States, ZR-3 would change the trajectory of the Zeppelin story. (National Air and Space Museum, Smithsonian Institution (SI Neg. No. 2000-9237))

Seitensteuer, on the rudders. Hard by stood von Schiller; he and Wittemann would assist in navigation and weather observation. A trio of radiomen were along—two always on duty. As for the motors, an engineer, one head motorman *(Maschinenmeister)*, and a dozen machinists were distributed aft. En route, the Maybachs would be stopped one at a time, in succession, twice a day, in order to examine spark plugs, fastenings, and fittings. Ludwig Knorr, *Zellenpfleger,* nursed the cells.

Below, the home field lay awash in wavering foggy light. And emotion: as a band puffed "Deutschland uber Alles," thousands of well-wishers (Countess von Bradenstein-Zeppelin and Bauch among them) roared cheers and sad farewells. The show was brief. *Der Luftschiffe* melted from view on the ascent and, at 0637, vanished. On the ship's bridge it was the same: All was lost to the dank suspended moisture. "We could only hear our motors and a faint murmur from the earth," von Schiller later recorded.[63] As for Bauch, on the ground, "I stood about in the middle of the ship," he remembered. "It was so foggy, you hardly could see the bow or the tail. And when the ship really took off, the bow disappeared in the fog and, after a little while, the tail disappeared. And she was gone."[64]

As sound faded on her wake, reporters scrambled to wire word: liftoff for America! Editors printed headlines hung heavily with background, bulletins, and assorted hullabaloo. Europe was agog. In Berlin, Frankfurt, Munich, news copy merited hurried extra editions. Two or three daily printings—even special feature editions distributed gratis—all trumpeted the airship's progress and the implications of its journey.[65] From liftoff to landing, for three days, this singular aerial push to the westward would transfix two continents. "ZR-3 STARTS FLIGHT TO CROSS ATLANTIC," the *New York Times* announced in bold, page-one type.

At the airfield, knots of people, in no hurry, milled about in anticlimax before shifting homeward. At the Postal Station, as planned, time of departure was reported by radio.

Eckener had gotten off with a useful load of slightly more than forty-three tons. In addition to crew and passengers (2.7 tons), he had managed to loft thirty-three tons of fuel; at full power that granted seventy hours, or a range of 5,400 nautical miles (6,210 statute miles).[66] Along the keel hung three tons of water ballast. Also stowed along the corridor and elsewhere were oil (2.2 tons), provisions (837 pounds), luggage (1,411 pounds), and mail (330 pounds) for addressees in the United States, Canada, and Central and South America.[67] With an empty weight of forty-four tons, *der Amerikaluftschiff* now displaced 87.3 tons altogether. (With helium, lift would have been about one-seventh less.) Though not on any manifest, the promise of zeppelin transport was very much on board. "It was clear," Eckener reflected, "that not only the lives of the participants, but also the fate of the zeppelin idea, would depend on our success. A failure would, in these circumstances, have meant the end of the Company in Friedrichshafen and with it the only place where the rigid airship tradition and the belief in its future were still alive."[68]

The elevatorman had her climb, to top the fog—cleared at 2,500 feet. A westerly heading was ordered—along the course of the Rhine—and the engines were set to cruising speed. To port and starboard, raw mountain walls punched a solid undercast. The traveling knot of humanity found itself "floating above a shimmering ocean of mist lit by the rising sun, while the chain of the Swiss Alps rose through the mist to the south. We were happy that the 'great day,' for which we had hoped so long, was here at last. The Zeppelin was going to show what it could do!"[69] Eckener pressed westward above fog for about two hours, shedding altitude as the fog's depth decreased. At 0808, to the north, a silver thread lay in view—the Rhine. German soil lay astern and Basel beneath—the place where Germany, Switzerland, and France conjoin. Obligingly, the background of fog dissolved, cleared by a gusty wind. With cameras snapping and whirring away, the gardens of France—for the next 450 miles—spread in bright sunshine. Average altitude: about 980 feet (three hundred meters.)

The pass near Belfort was negotiated. At 1047 Autim spread below. Less than an hour later, at 1138, Eckener wirelessed, "We are descending the Loire Valley southwestward aiming toward Bordeaux. Everybody's spirits are excellent."

At 1200—noon—the watch changed. In the control cabin, Flemming assumed officer-of-the-watch responsibilities. Wittemann relieved navigator von Schiller; at the wheels, Albert Sammt replaced elevatorman Pruss, as Walter Scherz relieved Marx on the rudders. Just aft, a fresh radioman came on. Well aft, five refreshed *Motorenwarte* took station. Already, the clock had been set back an hour; the craft was moving westward. For off-duty spells, the options were few: sleep, enjoy the electrophone, eat, maintain diaries. St. Armand lay before the airmen within minutes. At 2,550 feet, the midday meal brought oxtail soup, beef ribs, fresh peas and carrots, apple sauce, caramel pudding, coffee. The anxiety attending the first hours aloft had vanished; despite the ship's rocking and a shirtsleeve warmth, all hands were "quite at home." The vineyards of Burgundy met the plodding dirigible, then the level fields of Bordeaux. From here signals from her Telefunken radio equipment were heard in the United States—an extreme distance for the time.

The radio gear was mounted portside, in a specially ventilated, soundproofed cabin adjoining the bridge. A twenty-watt radiotelegraphic sender—wavelengths from five hundred to 3,000 meters—had a maximum range of about 1,500 miles; in terms of radiophone, though, exchanges were limited to three hundred miles or so. (Weighted antenna wires could be lowered, separately or together.) For navigation, a radiotelegraphic bearing finder was fitted, equipped with a rotatable loop antenna and a network of wires passed transversely about the hull. (In its essentials, navigation here was comparable to that used by seagoing vessels.) Outboard was mounted a combined continuous-current and single-phase alternating-current generator, driven by propeller-like windmills. Its capacity was 1,200 watts for continuous and 1,500 watts for alternating current. It fed a twenty-four-volt current to the ship's electric light plant. The gyroscopic compass was equipped with a generator of its own.

For the crossing, *LZ-126* (call sign NERM) would be in general radio communication with the station at Norddeich (KAV), on Germany's North Sea coast, until that thread (wave length 2,300 meters) was interrupted or the aircraft passed approximately forty-five degrees west longitude—roughly halfway across the ocean. Thereafter, NERM would communicate (on 3,700 meters) with Bar Harbor, Maine (NBD). Throughout day one and then all the way across, the radiomen worked direct as well as relayed traffic: *Standort* (position), *Wetter* (weather), and private messages.[70] Weather reports from Eckener would be transmitted daily at 0100, 0700, 1300, and 1800 Greenwich mean time. Back at Friedrichshafen, reports of progress to the westward were faithfully posted, and a U.S. operator advised the Navy Department.

Transiting southern France, Confolens, northeast of Cognac, scrolled past the windows at 1356. At 1529, nine hours out, the ship hung over the Bay of Biscay, having crossed the French coast at the mouth of the Gironde River. Clocks were set back. Now south, over water. At 1950, Cape Panas was sighted—the Spanish seacoast. Filling the crew's views forward and to starboard, the usually stormy bay reflected a glittering sheen, quiet and peaceful.

The position at midnight was forty-two degrees, thirty-six minutes north; longitude eleven degrees, thirty minutes west—a point off northwestern Spain. Leave had been taken of Europe. Along with Eckener's report of favorable, auspicious weather (short-lived, as it happened), officers and crew wirelessed greetings home.

The great circle route—over England and Ireland, across to Newfoundland, thence southward—represented the shortest track. However, having studied wind and pressure distribution, Eckener had elected to shift his track southward.[71] The northern North Atlantic was now dominated by low-pressure troughs offering nothing but strong headwinds. German meteorologists saw a "high" prevailing from the Spanish coast to the Azores. From wind direction and rising temperature, Eckener induced the presence of a "low" to the northward. Conditions deteriorated somewhat and, as expected, the wind shifted to the west—a stiff southwest wind arising about 0700, spotting the skyline with rain clouds. At 1950 the northwestern corner of the Iberian Peninsula, Cape Ortegal, was passed. The wheel was spun to port; *Herr Doktor* continued to cling to the shore. At sunset the disturbance passed. A full moon played on the sea "like a searchlight," Lehmann reflected, "over the picturesque rocky coast and enchanted us with magical effects."[72] Course was taken for the Azores. Position at midnight: 42.6 degrees north latitude, 12.2 degrees west longitude—a vacant point well out to sea. "We are over the open sea and making at present ninety miles an hour [about seventy-eight knots], making for the Azores," Eckener radioed at 0015. "All the airship's machinery is in perfect order, working splendidly. All aboard are well and in excellent spirits."

Beyond Portugal's subtropical isles, Eckener would await developments in the weather. On Spain's northwestern tip, the light at Cape Finisterre (so important to Spanish navigators) flashed a final greeting from Europe. Now the Atlantic: threatening and exhilarating. "Before us," the captain reflected, "lay the boundless plain on which we were to gain our first evidence to decide the question whether airships could safely be navigated across it. Our dreams and our hopes were on the verge of fulfillment."[73] On this startoff day, the *Berliner Tageblatt*'s second *Extra-Blatt* flashed a late word: "ZR-3 OVER THE OCEAN." On a far shore, with three-column morning headlines, the *New York Times* fairly roared, "ZEPPELIN *ZR-3* OVER OCEAN ON WAY HERE; MAKES 65 MILES AN HOUR TOWARD AZORES AFTER CROSSING FRANCE AND TIP OF SPAIN."

Mile by maritime air mile, the horizon ahead pushed westward.

An intuitive, pioneer airman, Eckener was flying by the weather map—and *shortening* his transit time. On transoceanic nonstops, detours can convert storms from obstacles to aids, by exploiting their quadrants of favorable winds. By virtue of their cruising radii, the great rigids could fly around these tropospheric systems, profiting from their air-circulation patterns. "An airship is so slow that one never butts into a head wind if it can possibly be avoided," an *R-100* passenger remarked of the England-to-Canada crossing, in 1930, "and we altered course repeatedly to go the right way round the depressions and find a favouring wind."[74] The operators of *Graf Zeppelin* and *Hindenburg* were to hold to their schedules—and conserve fuel—by detouring around delaying sky conditions in this way.

Through the night the wind blew from the northwest and the north, boosting the airship's speed. Checking drift, Wittemann and von Schiller found that to maintain course *ZR-3* had to be held twenty degrees into the wind. At breakfast on Monday a freighter was in view: *City of Boston*. Astonished, her captain radioed, "Who are you? Where are you going?" An inquiry back as to position showed the aircraft to be south of assumed coordinates. Course was altered. Shortly after noon, land in sight! Off the starboard bow was the island of San Miguel, in the eastern arc of the Azores. *ZR-3* had leapt nearly halfway to New York—2,100 miles flown, 2,350 yet to cover. Weather signs are scarce in this neighborhood, but the faint clues looked good. The sighting was marked by dinner: bouillon, ham with Burgundy sauce, beans, peaches, and pudding. Kudos were dispatched to the galley. The Maybachs, hungry also, had made the ship "light," so altitude was gained—to five thousand feet, to blow off hydrogen and thereby offset the weight of fuel consumed. (This was the second of three such maneuvers during this voyage, in addition to manual valving.) Upon descent to cruising altitude, a two-foot tear was discovered near the lower part of number 8 cell. A seam had parted. Lift was escaping. No one wanted an explosive mixture in the keel or cell. The tear was held together by hand, and a patch was prepared in about five minutes.

Of the islands Steele wrote, "They were all lovely." At midafternoon, to starboard, the cliffs of Terceira paraded—a verdant monotone of vineyards, olive groves, and rich vegetation. At Angara, over midcity, two bags of mail were thrown off (parachutes attached). The strait between San Jorge and Pico, with its great peak, shifted astern. Then came the island of Faial and, at about noon, Flores. The watch again was set back an hour. One calamity for comfort: wash water gave out. "We have joined the ranks of the unwashed and unshaven," Steele jotted. Outside, the temperature was rising, and the humidity as well.

New conditions now interposed—headwinds. Sea-surface sightings confirmed a stiff west wind, driving whitecaps before it. The island of Flores was glimpsed—the last of Europe. By sunset the breeze had freshened, halving ground speed. In the midst of featureless wastes, such an unhurried pace was cause for concern but not for alarm. "Naturally," Eckener dryly remarked later, "we would not expect this wind to persist all the way to the American coast, but it could increase further and was a little disturbing."

The map of air pressures before them was incomplete, leaving much to the instincts of Eckener, Lehmann, and aides. Pressure centers were suggested, certainly, but only a very general idea of the direction and force of the wind. "It was evident that some disturbance must be on our starboard bow," Eckener wrote, but its intensity was unclear. A pressure synopsis suggested highs and lows, but no detailed reports were to hand. Given the data before them, it was not possible to lay out a definite course most likely to grant the best flying.

Finding the right wind at the right time could be vital, airmen know. No one's experience inspired more weather tales than Dr. Eckener's. Augmenting his talents were all vessels within radio range, but in particular U.S. cruisers. USS *Milwaukee* lay on station at forty degrees, about 250 miles south of Halifax, and *Detroit* at forty-five degrees, about 150 miles southeast of Cape Race. The oiler USS *Patoka* lay well to the north.[75] Following fruitless tries (foiled by static), contact with these ships was established. *Milwaukee* reported fresh winds from the west, its partner warship a light southeast breeze—"and from this we could conclude with safety and certainty that the ship was at that time at the northern edge of a low or storm center." Integrating his data, Eckener, with seemingly impossible confidence, ordered a heading to the northwest, toward Cape Race, to go around the north side of the storm astride his track. Maneuvering skillfully and to great effect, he placed the stiff southwest wind on his beam, thus minimizing its drain on airspeed. The reward: its gradual dissipation until, toward noon, conditions were quite calm—evidence that the low's center was nigh. By deliberately exploiting the southeasterly and northeasterly winds near the low's center, Eckener accelerated his progress.

The airmen bore down on the American continent. At a quarter-hour past midnight, Tuesday, the 14th, Lakehurst heard *ZR-3* asking a vessel for temperature, wind, and barometric readings. By 0300, the advancing zeppelin was in direct wireless touch with RCA's station at Chatham, Massachusetts. Position: thirty-eight degrees north, 41.1 degrees west—about six hundred miles southwest of the Azores. The thermometer held remarkably steady: von Schiller "could hardly sleep from the heat. This morning I roamed the forward gondola in pajamas." Lunch proved a hurried affair. That afternoon, southeast of Cape Race, a helping wind sped the airmen due west—a direct compass course for Boston. Ground speed was now seventy-eight miles per hour.

By late afternoon, the bridge watch faced a screen of fog—a lapping, glistening blanket ceding no glimpses of the sea and rising to meet *ZR-3*. Altitude was added until the fliers hung at five thousand feet. No surface sightings meant no estimates—and no checks—of drift. (On U.S. Navy airships at sea, the drift angle was determined at least once every fifteen minutes.) The wind was rising; a zone of low pressure had been penetrated, its airflow racing from the northeast; and the temperature was plunging, from seventy-seven degrees to forty-one. (Officer von Schiller was soon eying Kennedy's lambskin-lined coat and felt boots.) By 2000 the storm was upon them, so strong indeed, per von Schiller, "that the ship trembles and throws one slightly back and forth. One feels that she can withstand it; she feels well built."

Convinced that the fog and racing cloud forms denoted the Gulf Stream south of Nova Scotia, Eckener ordered a descent—to five hundred feet by barometer—into the colder air to ascertain the wind. Within about two hours, conditions were slowly beginning to clear.

At 2218, America blinked into visibility—a light from Sable Island. In terms of navigation (and weather), risk had ended; piloting aids would be in sight more or less continuously now. The 550 air miles to New York would be child's play. Small-scale ocean-transit charts were exchanged for larger-scale approach and pilotage charts. Halifax lay abeam at 2300. At about 2325 hours, Lakehurst picked up a message: "Now off South Nova Scotia; everything well." About midnight, an illuminated coast and Halifax stood out clearly. With Atlantic coast stations now so near, congratulations and inquiries crowded the weather signals. Eckener ignored most; save for occasional factual signals—chiefly calls for compass bearings—his operators gave no reply. On they droned through the chill dark. Based on radioed bearings, Boston Navy Yard had the craft 120 miles off shore at 0300, about opposite that city. In New York, the Associated Press reported "strong and clear signals." Lakehurst, though, was fretful; no direct contact had been established. (*ZR-3* did not know Lakehurst's call sign and had to get it from Bar Harbor.) Relayed to the zeppelin by land wire to New York's Naval Communications Station, a message from Lakehurst implored: "What is your position and what is probable hour of arrival at Lakehurst?"

Rural Cape Ann hove in view, followed by the suburbs of Boston at 0400. *ZR-3* swept in at a thousand feet or thereabouts, her lights flashing clearly in a moon-lit dome of sky. "The city was sound asleep," Eckener records, "but it awakened for us. Sirens and steam whistles began howling while we were still far off; but they gave us some idea of what was in store for us." Von Schiller thought the welcome "flattering."[76] "Will lay off operating" Eckener radioed Lakehurst at 0420, "till I get a look at beautiful Boston."

Throughout the Fatherland, the nationalist press roused itself. Headlines bannered this giddy triumph of engineering and airmanship. America! As the *New York Times* had it, Germany was "a nation drunk with enthusiasm." Its showpiece creation, its zeppelin airship, had gained the far continent.

Steaming south, the visitor vanished from over Boston in mere minutes. Eckener, in no rush, rung up reduced speed, so as to meet New York in the daylight—a canny decision. That city's mood—expectant, alerted for the big event—now was electric. "*ZR-3* IS EXPECTED OVER CITY THIS MORNING," the *Times* had announced; "WAS ABOUT 120 MILES OFF BOSTON AT 3 A.M.; MOVING FAST AFTER PASSING NOVA SCOTIA." Steering southwest, the airship traversed Providence. Now plainly visible, the pathfinder overflew Newport around 0500. Forty minutes later, Westerly, Rhode Island, lay blinking below. Then New London. At this hour Washington intercepted a message meant for Lakehurst. New York, it confirmed, would see the great airship: "We will land about 9 or 10 o'clock." Meantime, the rhythmic throb of Maybachs became audible over New Haven, then Bridgeport.

The moment of presentation: daybreak. Seventy-seven hours out of mid-Europe, in fog and light haze, *ZR-3* hung off Sandy Hook and lower New York Bay, slightly to seaward of the skyline and incomparable harbor of New York. Planes from Mitchell Field buzzed up as escorts. Although the shore lay shrouded, "the fantastic shapes of the skyscrapers towered above it and gleamed in the rising sun. The imposing picture, celebrated in all languages, which this overpowering giant metropolis of daring and enterprising spirit offers to the arriving stranger, appeared to us with a double beauty, a fairy-tale city to which we had abruptly come out of the dark, empty sea."[77]

Eckener's instincts were minutely calibrated. He had chosen to loiter; this press-agentry would work to his advantage. At 0745, as the city awoke, his command could be seen circling Ellis Island. Pounding uptown over Manhattan, it steamed toward the Palisades at about 125th Street, then toward the Bronx, thence back to the center of Manhattan. Next, the helmsman steered southeast, to sail high over Brooklyn. This Old World envoy was a genuine wonder; enthusiasm proved irrepressible. Street crowds, workbound, halted and were swept up—"oblivious to all save the glistening shape drumming its way above. More thousands, knowing that the big craft was due, were encamped on roof tops." As the zeppelin passed, "folk rushed to windows, verified the arrival and then made for roof or street. Others were summoned to see by neighbors who shouted across airshafts or courts or whose excited conversation in corridors, en route to a vantage point, spread the news. Some dashed to house tops with only an overcoat concealing night garments."[78]

The metropolis was bedazzled.

Circuiting five times, dawdling, Eckener cut across the Battery to the Hudson River, granting its Jersey and Staten Island shores a close-up. Over the harbor, von Schiller closed out an extra-duty special-correspondent assignment with a fresh dollop of news. His log of the crossing was stuffed into an airtight bottle, dispatched overboard, retrieved by a waiting boat, and then delivered to the *New York World* and the North American Newspaper Alliance.[79] Ashore or afloat, denizens of New York's public places savored the spectacle: a huge silver form, a zeppelin, pounding past their streets, their homes, their town. "We gave the New Yorkers," its commander observed, "the pleasure of watching us cruise a long time over the city, while we slowly spiraled up to an altitude of 10,500 feet," to blow off lift and equalize pressure in the cells.[80]

Having thrilled all who might witness the sixty-five-minute visit, Eckener ordered course set south-southwest, for the naval airfield amid the scrub and pine of South Jersey.

CHAPTER 2

ZR-3: *Homeport Lakehurst*

A good ship makes its own luck.

—Capt. George F. Watson, USN (Ret.)

At 1100 on 14 October, unofficial word reached air station Lakehurst: *ZR-3* was making "excellent progress." A further report at 1500 had Eckener braving headwinds. On base and round about, anticipation had soared. Personnel were rehearsed, paraphernalia ready. As yet, station operators had not established direct radio communication: they could intercept its signals but had yet to be acknowledged by the advancing zeppelin.

Boston had had the ship in sight by 0400, so unless he chose to loiter at New York, it appeared, Eckener would be landing at dawn. The base snapped awake. "Immediately," the *New York Times* had it, "the air station was aroused." Navy men dressed. An army of commentators (many asleep on desks) stretched into wakefulness. Outside, the tall mooring mast burned a homeport welcome on the West Field. The canteen was opened, breakfasts put on the fire. In gathering light, moon and stars paled. Just off the landing field, at Aerology, increasingly distinct dot-dash traffic from *ZR-3*—position reports, statements of weather—was read and excitedly plotted.[1]

The Radio and Aerological Building now was headquarters. In charge of communications was Naval Aviator (Airship) No. 3350, Lt. T. G. W. "Tex" Settle and his staff.

Hovering near telephones, the acting station skipper, Lt. Cdr. Maurice R. Pierce (No. 3176), and Lt. Cdr. Joseph M. Deem, acting executive officer (or "exec"), were kept advised. Handling the weather aspects at this end of the crossing was Lt. Francis W. Reichelderfer (No. 2422), ordered up from Washington.[2] "I can remember very well," he said long after, "what a thrill we got on the first report we got directly from *[ZR-3]*, by radio. We plotted, of course, the ship's position right on the weather map and followed it from then on just where it was. We were up almost day and night. . . . It was an extremely crucial, important operation. And we just wanted to do everything we could."[3]

At 0800, following hangar muster, sailors and marines were marched out in V formation, divided into groups, arrayed as the ground crew. Eckener would alight bow-on. Given the morning's wind (from the north), Aerology advised that the East Field receive him. Lines and ladders and landing signals were made ready, the east shed doors opened. Special-duty planes took off and climbed to await and to photograph the arrival.

No less than the public spectacle at New York, this welcome would be ecstatic. It was the midtwenties—an exuberant, noisy, optimistic time. Aeronautics (and aeronauts)

Table 2-1 Log of the world's fourth air-journey across the North Atlantic, Central Europe (Germany) to the United States, 12–15 October 1924. All times Eastern Standard Time. (U.S. Air Services)

12 OCTOBER

0025	Ground crew commences undocking.
0035	Departs Friedrichshafen, Germany [0635 local time].
0240	Ship over Belfort, France.
1536	Ship over Cape Ortegal, Spain.
2030	Twenty miles off Portuguese coast.
2230	Midway between Spanish coast and the Azores.

13 OCTOBER

0900	Ship sighted from Azores.
1130	Airship passes Fayal, Azores.
1435	Position 135 miles west of Fayal.

14 OCTOBER

0120	Position 960 miles west of Azores, approximately 640 miles from Bermuda.
0800	Position 1,500 miles east of NAS Lakehurst, New Jersey.
NOON	Position 1,300 miles east of Lakehurst.
1600	Position 1,000 miles east of Cape Sable, Nova Scotia.
2200	Position 500 miles due east of New York City.

15 OCTOBER

0300	*LZ-126/ZR-3* sighted Cape Cod lighthouse.
0415	Ship passed over Boston, Massachusetts.
0448	Ship passed over Providence, Rhode Island.
0500	Sighted at Newport, R.I.
0540	Ship at Waverly, R.I.
0630	Ship at New Haven, Connecticut.
0640	Bridgeport, Connecticut, overflown.
0710	Ship sighted at Mitchel Field, Long Island, New York.
0800	Ship reached New York City.
0840	Airship departed New York for Lakehurst.
0916	Ship arrived over the Lakehurst landing field.
0955	Airship safely in hands of U.S. Navy ground crew.

held a special allure. *ZR-1* had primed the popular mind for airships. Now *ZR-3*. The nation was becoming air minded—a mood not lost on Admiral Moffett and his young bureau. Civilian America had crowded onto the reservation, pouring from Fords and Packards and Hudsons. Hydrogen—a "treacherous gas"—demands respect. For safety, the smoking lamp was out everywhere on station—an edict enforced by patrolling military policemen.[4]

As marines stood watch and the radio chattered, Lakehurst waited. At 0900 came the report: ship in sight—over the scrub pines, well away. Preferring to inspect the field first, Eckener delayed his final descent. On approach from the southeast at last, he pushed in toward the ground party. Fifty feet off a trapdoor opened, a line uncoiled. Under the eye and orders of the ground handling officer, "spiders"—blocks with smaller ropes attached—were fastened, then group strength was brought to bear. A second line fell. Pulled down by the bow, the dirigible's stern angled up. But all forward drift was checked. "It was plain that the soldiers and ground crew had had plenty of practice," Lehmann recalled. "The landing maneuver was executed as precisely as it was at Friedrichshafen." Aviation Rigger Third Class Leonard Schellberg was on the field. "[Eckener] came in," he marveled, recalling the German shiphandling, "and instead of driving it into the ground like we did our airships—full of helium, he brought it in by valving hydrogen and dropping a little water ballast. And he made *beautiful* landing, right into our hands."[5]

Arms aloft, sailors seized the control-car railing. At 0956, ship's bumper kissed the sand, thus ending the first

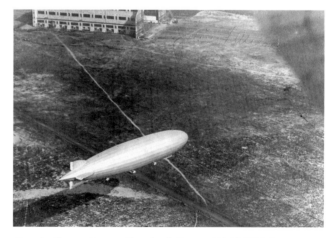

Concluding a transatlantic delivery, LZ-126/ZR-3 noses down onto Lakehurst's east landing field, 15 October 1924—81½ hours after lift-off from Central Europe. The zeppelin was the fourth aircraft to leap the North Atlantic. (National Archives)

The dirigible is struggled to a landing as Navy and Marine personnel reach for the handrails of No. 1 power car. Time: 0955. (National Air and Space Museum, Smithsonian Institution (SI Neg. No. 2000-9085))

Moments after touchdown, Lakehurst greets the airship and its returned U.S. officers, including air-station skipper Cdr. Jacob H. Klein, Jr. (National Air and Space Museum, Smithsonian Institution (SI Neg. No. 2000-11195))

mainland-to-mainland transoceanic flight. Eckener had logged the fourth nonstop leap of the North Atlantic—and the longest nonstop journey yet made by man in any craft. The ability of lighter-than-air to realize long-distance, intercontinental commerce with freight[6] and passengers had been verified.

Klein was first off. Climbing through a window, he resumed command of the air station.[7] "In general [Klein was to report] the operation of the ship through the trip was in all respects very successful, very convenient, and could be repeated without difficulty any number of times. It has undoubtedly proven the fact that a ship with sufficient lift to carry a good reserve of fuel can make the trip across the Atlantic with only minor hazards and minor inconveniences."[8]

Via ladders, fellow observers disembarked: Steele (who joined his waiting wife and mother), Kraus, then Kennedy. "It was a most interesting and successful voyage," Steel said, adding, "There was only one minor mishap. And that was when the fabric of one of the gas bags tore as much as two feet." The first *up* a ladder: F. W. von Meister. Educated in Germany as a mechanical engineer, "Willy" held the distributorship for Maybach Motorenbau in the United States and Canada. He had made it a point to be on hand—anticipating sales and service in connection with ZR-3. A customs officer pulled him back; until the airship was cleared, only officials could board.

Amid the hurly-burly came a wind shift—from north to east. A downwind docking was ill advised, so a "walk" ensued. With ground parties clutching the centerline cars and smaller handling groups fore and aft manning trail ropes and snatch-block spiders, the dirigible was manhandled round the shed until abeam its west doors. Pulled into docking position well out, it was struggled over the sill and inside. Time: 1025. Upon entering the high, arched room—*frisch und gesund* (safe and sound)—the zeppelin became the property of the U.S. Navy.

Bow to the east, the ship was secured in the north berth by 1050. Lines snaked down, berthing complete, a detail of sailors led by Chief Petty Officer William A. Buckley commenced preparations for deflation. Low cradles were wheeled beneath the rear and forward cars as, along the keel, shoring was wedged into place. For their part, U.S. officials had embarked, to clear the foreign vessel—a ritual

The zeppelin is "walked" around the ZR hangar preparatory to docking. (National Air and Space Museum, Smithsonian Institution (SI Neg. No. 2000-11192))

chore that would be repeated later for *Graf Zeppelin* and *Hindenburg*.[9] Within forty minutes, Customs and Public Health had inspected both crew and belongings. The airmen filed off, Dr. Eckener leading—a hugely emotional moment.

News bulletins ricocheted worldwide: "She has landed!" Within the echoing hall, waves of news gatherers, officials, and well-wishers crowded close. "A torrent of newspaper reporters and photographers descended on us," Eckener recounted, "and there followed an hour of being pushed around and embarrassed, until we were rescued by friendly naval officers and led away to the quarters set aside for us."[10] But first, a few remarks. A new world's record for continuous flight—5,006 miles—had been put up, ship's captain declared. No airplane yet designed could hope to depart mid-Europe and come down next in America. In a stupendous display of discipline, daring, and planning, Eckener and Company had delivered the goods.

Formalities concluded and duty done, the airmen-celebrities were escorted to lodgings—officers to the Bachelor Officer Quarters. (Each man would pay for his own subsistence.) Despite the simple facilities, every effort had been made to provide comfortable quarters. The base was more convenient than the town, a mile or so off—an unglamorous village devoid of luxury accommodations. Following a stop at Aerology (to offer thanks and share a postmortem), Eckener was driven to the commanding officer's quarters; as station skipper, Klein hosted the charismatic airman.[11] Most of the enlisted men, for their part, received housing commensurate with their rates.

It is impossible to exaggerate what this flight meant. Eckener and his crew had redeemed the zeppelin venture while inoculating the U.S. Navy's own ambitions for this oldest branch of aeronautics. Large airships, hitherto largely a German monopoly, were about to blossom as an American industry. Though fraught with risks, the commercial future—regular transoceanic air routes—seemed ripe with promise.

ZR-3 had conveyed more than itself, thirty-one men,

Via side-handling lines and ground parties, "America's Zeppelin" nears position opposite the west doors. Upon entry, the airship became the property of the United States Navy. (Charles E. Rosendahl Collection, University of Texas)

cargo, and mail to the New World. It had carried an immense emotional freight as well. The exploit was pathbreaking (and hugely appealing), its implications revolutionary. The national conversation now embraced airships. As aide and interpreter for *Herr Doktor* Eckener, von Meister assisted in discussions of zeppelin technology—newly fashionable—with the Americans. "When *ZR-3* landed at Lakehurst," he said, recalling the fervor, "the enthusiasm for the first transatlantic voyage of a commercial airship—for delivery to the U.S. Navy—had a tremendous publicity result in the newspapers."[12] Trusting that merit and performance would be rewarded, airship men everywhere knew a heady exhilaration. This machine, indeed, would compile a splendid run of accomplishment and good fortune. Though the years were finally to disappoint, the day of the airship had dawned.

Eckener was a superb publicist-marketer, the skies his podium. Elevated to international renown, lionized in the press, committing his energy, passion, and belief in zeppelins, he inhabited the headlines for the next dozen years. Meantime, as befitted a science-mad age of ballyhoo, the airmen were feted everywhere—Washington, New York, Akron, Cleveland, Detroit, Chicago. Amid a crush of photographers and reporters, interviews, and speechifying, the luncheons, banquets, and business meetings proved unceasing. The day following touchdown, Washington honored the airship's officers. First, Dr. Eckener, Captain Steele, and members of the crew appeared before the annual meeting of the NACA committee, together with members of the General Board of the Navy and other prominent army and naval officers. Through an interpreter, the German airman sketched his vision: airships in transoceanic transportation between the United States and Europe, or the West Coast and Hawaii. Overwater naviga-

ZR-3 is man-hauled under roof. Note the slot-type docking rails; in this period, the rails and attendant "trollies" offered the sole mechanical assists for handling airships on the ground. (Airship International Press)

tion, Eckener prophesied, would be more comfortable than flight over land, with its uneven heating and bumpy air; airships belonged at sea. Rigids enjoyed a great advantage over airplanes and ships: the absence of seasickness. In concluding, he expressed "perfect confidence" that such transport would develop.[13]

A luncheon at the Hotel Hamilton—attended by the chief of naval operations (CNO), the president of the General Board, and other prominent officers—opened with a toast (ice water) by Secretary of the Navy Curtis D. Wilbur "to the President of the German Republic and the President of the United States." He closed with a toast to "Our Guests." In between, the secretary announced that ZR-3 would be named *Los Angeles*—a symbol of peace and of friendship, and a reminder of the angels who sang at the birth of Christ.[14]

President Calvin Coolidge wired congratulations of his own: "It is a matter of great satisfaction to me and to the people of the United States that peaceful relations between Germany and America have been fully reestablished, and that this great airship has inaugurated the first direct air flight between Germany and America." Welcomed at the White House, Eckener met its taciturn, unsmiling chief executive.

Coolidge, elected as vice president, had become president upon the death of Warren Harding in April 1923. Within weeks, a contented nation would elect Silent Cal to a full term.

In New York City, a parade down lower Broadway advanced beneath a storm of paper. Within four years, New York was to salute Eckener with yet another hurrah, upon the arrival of *Graf Zeppelin*. At the Capitol Theater, the full crew watched a newsreel of their crossing. With a spotlight upon him, Eckener's tendered his (understated) appreciation, receiving a "stormy applause." A quarter-century later (he would note), "the orchestra rose to a man and played the German national anthem. Part of the audience, probably German, sang along. My eyes clouded over, and to this day, I cannot think of the occasion without tears coming to my eyes. As a German, I sensed how the national anthem once again paid tribute to me and my people. I left the theater as if I were in a dream."[15]

Thus were the war's old animosities thawed, a certain rapprochement gained.

The delivery represented a watershed on several levels. A new subsidiary had been worked out: Goodyear-Zeppelin Corporation. The terms of the collaboration (which took effect with the delivery) transferred patent rights and design skill to Goodyear and, further, ensured that key Zeppelin personnel would be made available. ("I wanted men as well as blueprints," Goodyear's Paul Litchfield recalled of the negotiations.) Thus was Teutonic expertise saved from a bleak future in postwar Eu-

Safe harbor: "America's Zeppelin" in the cavernous berthing space at Lakehurst. (Warren M. Anderson)

Dockside, civilian station workmen attend to insignia and lettering. From 1924 through June 1932, amid a sense of possibility, Los Angeles *would serve as a commissioned vessel of the United States Navy. (San Diego Aerospace Museum)*

rope—or so it seemed. Hitherto, commercial operations had been a German monopoly. No more. Editorials foresaw epochal results; great passenger craft were assured. Within months, indeed, the administration announced a policy fostering the development of commercial aviation, including lighter-than-air commercial service.

Ultimately, it was decided that the navy would operate *ZR-3* as a demonstration platform—to gain experience and data on costs, and to confirm feasibility. Investor capital then would take over, leasing her for commercial work. In short, the department was crucial to Goodyear's long-range plans. In turn, the navy recognized that a private industrial base was needed for future procurement of military rigids. On the success of this one machine perhaps hung the future of commercial dirigibles. The *Literary Digest* summarized the prevailing mood: "The Last Zeppelin—Ours."[16]

Before that year was out, the promised experts, headed by Dr. Karl Arnstein, were Ohio bound. A contract was thought imminent. For the next two years, however, the immigrant-engineers did little but plan; not until 1926 did design work commence in earnest, by which time Britain had adopted an airship transportation plan of its own and restrictions on German construction had been lifted.[17]

Less noticed was the return (by commercial vessel) of Cdr. Garland Fulton. The office of Inspector of Naval Aircraft Friedrichshafen had been closed out. Fulton had orders to Washington, as head of BuAer's Lighter-Than-Air Design Section. For Moffett and his successors, "Froggy" Fulton (ably supported by designer-engineer C. P. Burgess) became the right-hand man on technical matters inside the bureau.

Hydrogen gas was cheap but flammable. Inside the vaulting hangar at Lakehurst, all motors, buzzers, lights, electric leads, and equipment were secured—a safeguard against sparks, friction, and electricity. The berthing space was restricted. "No one will go aloft except wearing rubber shoes," orders read. "Men going aloft are cautioned of danger of sparks from tools." Why the climbing? Before

Long lines in hand, ground parties fight ZR-1 earthward. Not until 1922 were airships inflated with inert helium. To conserve the costly gas, commanding officers were obliged to minimize valving and to bull their landings—approaching nose-down and driving their commands to ground handlers before headway was lost. (U.S. Navy Photograph courtesy Navy Lakehurst Historical Society)

its hydrogen lift could be purged, the 656-foot hull had to be supported. Carefully. Topside, one detail worked to sling cables and tackle from the overhead as others opened the ship's cover at the top-center longitudinal (number six) and at the hull joints of the main transverses with numbers four and five. This rig would help support from above ZR-3 in its deflated state and would be used to position empty cells during reinflation. On the deck, wooden supports were set in place and wedged—the remaining support.[18] (Lakehurst recorded the airship as "moored" up to this time, after which it was logged as "shored up.")

With lines connected, the cells' ties to the structure were cut, deflation sleeves opened, maneuvering valves and control leads unshipped. The release began at 1300 on the 18th—under the watchful eye of the watch. In 1924, Calvin Bolster, Naval Aviator (Airship) No. 3930, was a young lieutenant (CC). "Everyone was concerned in and around the hangar since she was inflated with hydrogen," he remembered. "Our first job was to support the ship's structure and to deflate the gas cells, all being done with great care to avoid static sparks or an open flame anywhere in the vicinity." By 0300, the ship's cells were reinflated to about 6 percent—enough to float them. Lifting medium: expensive, nonflammable helium. Roughly 180,000 cubic feet had been transferred—all the helium on station. Production as yet was very limited. Full inflation would await *ZR-1*, and an exchange of *her* gas.)[19]

That afternoon, automobiles and excursion trains rolled onto the reservation as approximately twenty-five thousand visitors filed inside the great room to see firsthand, close up, the navy's zeppelin.

Four days later, Rear Adm. G. H. Barrage and his Board of Inspection and Survey arrived to make a inspection of the airship "in all departments," to report upon her condition and make recommendations as to official "acceptance of the vessel."[20] A joint inspection—Eckener accompanying—was held on the 24th.

The board's pleasure is transparent in its report, submitted in November. The main car (for instance) was thought "a very much better design than the one on the SHENANDOAH and is very neatly arranged and comfortable." Pronouncing the craft to be "a well-designed rigid airship of the latest design," the board recommended acceptance—when certain items had been executed as per contract, viz., when *ZR-3* was fitted in "the American condition," all equipment and spare parts had been delivered, and when Fulton had cabled from Germany that "all plans, descriptive pamphlets, instructions for operation and technical data" had been supplied.

A six-month guarantee period had been specified during which Luftschiffbau would effect certain alterations and make good on defects—using its personnel who were now standing by. Accordingly, materials, workmanship, and the functioning of parts and accessories received heedful examination. No defects in the hull, fittings, or equipment were found, though unsatisfactory items were noted—repairs to lattices, broken wires, rips in the cover, absent or sheared rivets, and the like—for

Shenandoah moored to USS Patoka (AV-6) in Newport Harbor, Rhode Island, 8 August 1924. Patoka was a Navy oiler converted to an airship tender. Based on a German prototype, the long, slender hull-shape of ZR-1 was, aerodynamically, a poor one. (Navy Lakehurst Historical Society)

High-pressure helium storage at Lakehurst. The Army-Navy Airship Board had endorsed experimental production; by the time ZR-1 was ready, the decision had been taken to inflate all Naval airships with helium. From 1920 to around 1936 the Army and Navy juggled the available supply between them, sharing costs and equipment and coordinating various helium matters through the Board. These cylinders held about one million cubic feet. (Rear Admiral Calvin M. Bolster)

which the contractor was chargeable. The candlepower of ship's running lights was deemed too low, the lighting throughout "scant." Assisted by Zeppelin workmen, removals, installations, and repairs incident to the transatlantic run were now effected, such as crew's quarters, the extra fuel tanks installed for the crossing, mooring winches and mooring gear, a water-ballast line and certain ballast bags—the latter replacing the extra tanks.[21]

In brief, for October–November, substantial work awaited.

During the midnight hour of 25–26 October, *Shenandoah* was hauled inside. Her transcontinental flight logged the longest journey ever made by an airship—a precursor perhaps of commercial air service. That afternoon, about seventy-five thousand visitors came to view the co-starring pair—and to gawk at the hangar itself. With a clear reach of 804 feet by 264 feet, its berthing space was the largest single room anywhere. A pair of rolling doors closed off each end; operated by four twenty-horsepower electric motors, each leaf—177 feet high and seventy-six feet wide at the base—weighed 1,350 tons.

As October faded, alterations were effected as a nucleus crew was organized and familiarized. Among those ordered to *ZR-3* were transferees from *Shenandoah* and from the airstation command. Seaman David F. Patzig was neither: a destroyer sailor, he was "striking" for the aviation machinist's mate rating. The Board of Inspection had requested an opened engine. So the Maybach in number-one car was unbolted from its bed and shunted to an overhaul shop. There it was assiduously disassembled. (With no spare engines yet to hand, the unit had to be reinstalled.) Meantime, navy mechanics studied the Maybach—a unique design "in that it had all roller bearings and was reversible by means of shifting a camshaft. . . . None of them were familiar with any engine similar to that."[22]

These were the introductory years, a time to acquire practical data based upon actual practice and operating experience. As the NACA summarized, "There is much to be learned about airships." Helium, however, was scarce, hence expensive. On 12 November, *Shenandoah* was shored up and suspensions on all main frames secured. (The shed's overhead was equipped with catwalks and hinged gangways at different levels to provide access, and traveling platforms and cables for suspension of docked airships.) Next morning, the transfer commenced. The *Shenandoah*'s gas was transfused gradually to the helium plant, thence into the new ship.[23] By the 14th (1610), the transfer was done.

External insignias and lettering would be applied, minor maintenance and repairs effected. To protect against corrosion, the structure was given a coat of varnish at the ballast bags, hatches, and at the valve openings topside. Such miscellanea as yellow-pine boards were procured for the keel corridor, as protection when in port.[24]

Expectations for the debut were outsized. Every experience, though, adds to the sum of a journey. Over the next half-dozen years *ZR-3* would log the longest useful life of any craft of her type; during this span personnel by the dozens would come to know her, largely during watches stood and duties performed on board, in every billet. Yet operations were saddled from the start. One factor undercut all others: the supply of helium. On her transcontinental trek, *Shenandoah* had valved, hence lost overboard, a half-million cubic feet. Production from Fort Worth approached eight million cubic feet—of which half was allotted the army.[25] (That fall, the secretary of the navy "hoped" that sufficient gas could be arranged from the Air Service.) Aggravating the matter, the German-made cells were becoming porous, losing additional cubic feet of lift—whether *ZR-3* flew or not.

Table 2-2 Helium production 1921 through 1930, Bureau of Mines, U.S. Department of the Interior. Production began at a plant near Fort Worth and continued until 1929. Due to depletion of helium-bearing natural gas, operations were transferred to a plant at Amarillo, Texas. (Bureau of Mines)

Fiscal Year	Cubic Feet Produced
1921	241,242
1922	1,739,383
1923	3,801,626
1924	7,709,924
1925	8,889,051
1926	8,805,204
1927	6,021,531
1928	6,352,654
1929	3,373,072
1930	9,801,060

The annual—and variable—supply foreclosed concurrent operations of the two rigids. Though an ample accumulation was expected by spring, in fact the two ZRs never flew in company. Instead, "helium waiting" was instituted: one ship received extended overhaul and repair while the other operated for several months. Lakehurst made the best of it. The decision for helium introduced other headaches as well. Storage was one. As *ZR-1* had neared completion, only standard commercial compressed-gas cylinders had been to hand; each held a miserly store. Conceiving the idea of large, high-pressure cylinders, Wicks prepared specifications and oversaw their installation. Total volume: about one million cubic feet. Wicks departed the station not long after; in January 1924, he relieved Lt. Cdr. A. G. Olson as officer in charge of the navy's helium production plant, at Fort Worth.[26]

That November, Eckener returned home, as did the greater part of the Luftschiffbau crew. Zeppelins, the airman was convinced, could lead a revival for his nation abroad, thus easing the humiliations of Versailles. The *126* had shown the way to redemption for a battered, chaotic Reich.[27] Its press embraced the sense of national pride and self-confidence instilled by the venture. This "Columbus of the Skies," as one patriotic observer had it, had been built "by German engineers and German workers, with German machines and German tools."[28] The question of the hour for the Allies, then: should Germany be allowed to build more zeppelins? In Paris, the fate of the Zeppelin plant hung in the balance. As it happened, the interest aroused by this exemplar of its technology seeded the hoped-for reprieve.

Klein convened a board to study and report upon the most suitable method of employing the navy's largest aircraft. Its members were Bruce Anderson, Kraus, Lt. Cdr. Maurice R. Pierce, and Lansdowne. The Navy Department was restricted by treaty to *commercial* deployment. In addition to training flights and practice moors, then, one option was to operate *ZR-3* as a model commercial platform on extended missions, testing her cruise radius at various speeds over water. Businessmen worldwide were giving aviation serious consideration. In meeting the nation's transportation needs, the railroad, the airplane, and the airship would soon be linked. The navy hoped to attract investors who would take the craft over as, in a quickening industry, further procurements were placed.[29]

But first, while key Germans were still in America, Moffett pressed for as much flying as practicable. Accordingly, ad-hoc seminars were convened and opinions traded. Much was learned.[30] For Zeppelin, the psychology was plain; the contract had saved the works from razing and (not least) had continued careers. The "American ship" bonded a liaison born overseas—a connection that

Civil employees effect repairs to USS Shenandoah *(ZR-1) after the January 1924 breakaway from the mast. In 1923 and 1924,* ZR-1 *was used to determine the stresses in girders of a rigid airship in actual flight—the only then known experimental determinations. Commencing in 1925,* ZR-3 *enlarged this database. (San Diego Aerospace Museum)*

would endure the next European war. In the commercial realm, zeppelin technology and doctrine was studied here with the keenest professional interest. On the operational side, moreover, U.S. officers accompanied *Graf Zeppelin*'s pioneering flights and later the transits of *Hindenburg*. For its part, the Friedrichshafen organization exploited navy facilities for its transoceanic jumps and was not too proud to adopt its handling methods and equipment at home.

On 17 November, BuAer notified the Bureau of Navigation (BuNav) that *ZR-3* would soon be ready for christening. In his endorsement, the secretary of the navy noted that she was to be christened near Washington during the week of November 23—the exact date to be announced—by Mrs. Calvin Coolidge. Five days thereafter, an examining board (senior member, Lansdowne) met to examine candidates for the designation "Naval Aviator (Airship)." Examined and found qualified were Lt. (j.g.) Edgar W. Sheppard (Naval Aviator [Airship] No. 3173), Lt. Charles E. Rosendahl (No. 3174), and Lt. Cdr. Lewis J. Hancock, Jr. (No. 3175). Within hours, Rosendahl was dispatched south—temporary duty at the naval air station at Anacostia, outside Washington, to help prepare for the christening and commissioning.

Takeoffs tended to be at odd hours—late evening or early morning. "What with hanging around aerology last night until about midnight and then getting the *"L.A."* into the air about 3:30 A.M. [I] was damn tired after a strenuous day," one airman wrote of such musters.[31] At 0530 on 25 November, the call for a ground crew sounded. All flight stations were manned, engines warmed—ready to answer the telegraphs.[32] Within forty minutes, a ground party had been mustered and *ZR-3* shunted out. At 0903 the dirigible lifted away—"Up ship!" The jump to Washington was the first with helium and the first under U.S. Navy command. Though Steele was aboard, Klein, as second in command, was actually commanding because Steele was ailing. A total of thirty-nine were on board, among them eleven of the German crew.

By midafternoon, having overflown the District of Columbia, the zeppelin circled preparatory to landing.

Operating funds were problematic. Under the iron laws of helium economics, valving had to be minimized. The result of such hoarding was hobbled operations. On a flight of any duration, tons of fuel were burned; this left the aircraft light. Reluctant to valve and so reduce lift, skippers bullied their landings—approaching nose-down and driving in with speed sufficient to hold the ship down aerodynamically. If enough men seized the lines before headway was lost, the maneuver could be completed by pulling her down and loading more men on. The practice was slow and uncertain—an iffy business. "Frequently, we were unable to hold the ship down," Bolster observed, "and on command everyone would release the lines and the ship would rise and go around for another approach. This was a very dangerous maneuver for the men on the ground, who had to be trained not to become entangled in the lines and to let go promptly on command. Once off the ground, a man could be carried to dangerous heights in seconds."[33]

Commissioning day was an emblematic event. As *ZR-3* stood in to Bolling Field at Anacostia, in southeast Washington, the general public (massed to one side) looked on as the invitees—including the president, his wife, and their party—were ushered to reserved seats. Conditions proved gusty, the aircraft light. Declining to valve, Klein pressed in. He could not gain the ground. All present watched and waited as Rosendahl, in charge of a bluejacket crew, strained to haul the floating bulk earthward. A handling line snapped. Jaws tensed, smiles tightened. On ship's bridge, the maneuvering-valve controls were (reluctantly) ordered pulled until seventy thousand cubic feet of lift bled overboard. After several failed tries, upheld hands gripped and held. To ensure *ZR-3*'s obedience, men scrambled on board as ballast while the rest held her down.

Stirred by the Navy Yard band, ceremonies commenced. At 1614, as reporters scribbled and cameras recorded the

Commissioning Day, 15 November 1924. Outbound from Bolling Field, USS Los Angeles *steams north by the Potomac River, west of the Capitol. The Navy had accumulated helium sufficient to keep one ZR operational. Helium-waiting was therefore instituted: shored and deflated in October,* ZR-1 *had had its lift transferred into "L.A." (James R. Shock Collection)*

Capt. Steele, President Calvin Coolidge with Mrs. Coolidge, "sponsor mother" for ZR-3. The airship is being held to the ground at Bolling Field, Washington, D.C., for christening. (National Archives, 04026)

scene, this "American instrument of peace and progress" was christened USS *Los Angeles* by the First Lady, using a bottle of water from the River Jordan. Carrier pigeons were released, each bearing a message of good will. Then a portrait was presented, inscribed by Mrs. Coolidge, "To the good ship *Los Angeles* from her sponsor mother. 'Go forth under the open sky and may the winds of Heaven deal gently with thee.'" Six minutes thereafter, the vessel was placed in commission by the commandant of the Navy Yard and delivered to Captain Steele, who read his orders assigning him to the command. (Klein, as per *his* BuNav orders, temporarily relieved Steele immediately upon commissioning.)

At 1625, Coolidge and his party were escorted up the ladder for a brief inspection tour, during which the president's flag hung displayed. The line of VIP traffic consumed fifteen minutes, after which Admiral Moffett boarded and *his* flag was displayed. Seven more men embarked, among them Lt. Cdr. Joseph P. Norfleet, Aviation Chief Rigger Frederick J. Tobin, and Rosendahl. Weigh-off done, all ready, the "Up ship" command was heard. The ground crew released their lines, the men along the car rails shoved upward. The dirigible was under way at 1704, four of her Maybachs hammering at two-thirds speed.

The bridge had the "barn" in sight in less than three hours, bearing north. Klein hove to, weighed off, circled preparatory to landing. Among those standing to the airfield on this proud, expectant evening were officers and men who were to figure large in naval lighter-than-air. Rosendahl was one. Lt. Raymond F. Tyler had received his naval aviator (dirigible) designation in 1920, with duty in Akron and at Pensacola before being assigned to *Shenandoah*. "Ty" would be attached to *ZR-3* for several tours.[34] Lt. Cdr. Maurice R. Pierce (No. 3176) was to command the air base now spreading away below, as would Klein, Steele, and "Rosie." Heading ship's engineer's force was Chief Machinist Emmett C. "Casey" Thurman, Naval Aviator (Airship) No. 3351. At their re-

Commander Jacob H. Klein, Jr. in the control cabin of ZR-3. Disgusted with its politics, Klein, Naval Aviator (Airship) No. 3185, resigned from the program in 1926. Note the inscribed portrait of Grace Coolidge, wife of the president, a commissioning gift from the "her sponsor mother." (Navy Lakehurst Historical Society)

spective duty stations stood Lieutenants Jack C. Richardson (No. 3354), Herbert V. Wiley (No. 3183), and "Tex" Settle (No. 3350). Also riding along, observing, were three U.S. Army personnel and one marine.[35]

Watch upon watch, scores of names would be recorded in the ship's log. Upon its earliest pages, aviator-graduates from Lakehurst Classes I and II dominate. A few fine men like Buckley, Thurman, Tyler, and Lt. (j.g.) Charles E. Bauch (No. 3377) had experience predating Class I, which had convened in 1923. Others, such as Steele and Klein, Rosendahl and Wiley, Richardson and Settle, Lt. Cdr. Joseph M. Deem, Lt. (CC) Roland G. Mayer (No. 3244), and Lieutenants Joseph C. Arnold (No. 3184), James B. Carter (No. 3355), and John M. Thornton (No. 3356), were to cut their professional teeth over the pine woods enfolding the naval air station.

Officially, *Los Angeles*'s mission (in the words of the secretary of the navy) was "to further the development of this type of vessel and to demonstrate its practicability for commercial use." The existing, approved policy had been recommended by the General Board, namely: "To complete rigid airships now under construction and to determine from their performance in service the desirability of further construction." Klein's "Board on Employment of ZR-3" had recommended long flights to test cruise radius over water, and extended sorties to evaluate commercial value, among other proposals.

This, then, was to be a time of learning—of training and experimentation and doing. Most immediately, the Germans billeted at Lakehurst had to be exploited. (Their contract would expire 10 February.) The department planned to order *ZR-3* to seaward, to rendezvous with the tender (former fleet oiler) USS *Patoka*, to practice mooring to its stern mast. Thereafter, the tender would steam to Panama, to act as a floating base for a series of protracted overwater sorties.

By the beginning of the 2000–2400 watch on 25 November, arriving from Washington, Klein had begun his approach. At the dock, a door leaf stood aside to receive him. The log's next entry is illustrative of helium's bedevilment: "Stood into landing field . . . and attempted to land; ship too light; valved helium for one minute. Made several unsuccessful attempts to land. Valved helium total of 6 minutes and 20 seconds."[36] Success was achieved at 2100, whereupon Moffett and passengers disembarked. Walked to the southeast end, *Los Angeles* was secured at 2200 in the north-side berth, bow to the west. A quarter-hour later, the east and west doors slid shut.

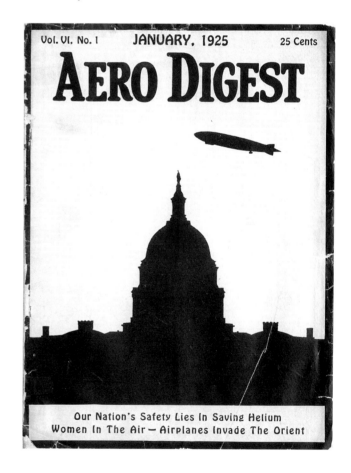

This same month, corrosion was detected throughout the aircraft. To combat its effects, a program of cleaning off the coatings (steel brushes) then varnishing or replacing was stepped up. Orders were issued also to clean and grease all car-suspension cables and, along the walkway, to renew broken lattices. During the ship's service life, countless lattices ("spoons," in crew jargon) would be broken or distorted—largely the result of inspection-force traffic in the keel and along girders as far as ship's equator. Proper upkeep meant frequent replacement.

December saw installation of a water-recovery apparatus. This system—air-cooled condensers and separators—was peculiar to helium-filled airships. The Germans did not hesitate to valve in order to compensate for burned fuel, to maintain a state of near equilibrium under way. With helium, the most practical method of avoiding valving was to recover water from engine exhaust gases. The vapor was there; within every hundred pounds of gasoline burned hid 145 pounds of water. Retrieval might seem simple, but in fact cooling this vapor out introduced back pressure, head resistance, and the weight and bulk of the condensers themselves. Freezing of the ballast collectors vexed winter operations. Heat-transfer efficiency proved most nettlesome, due to oil carried over from the cylinders, plus soot (from imperfect combustion). Even a thin deposit quickly fouled the apparatus. This "baffling problem" demanded nearly continuous research, spurred by the imperative of helium conservation, to seek ways to enhance recovery and render the condenser scheme both lighter and more efficient, for less drag.[37] Recovery would be tried on a J-class nonrigid as well.

On 15 December, the CNO approved BuAer's recommendation for employment of ZR-3 for the first half of 1925, including practice moors to *Patoka*. Launched and commissioned in 1919, the tender had served as a fleet oiler (AO-9) until 1924, after which she had been refitted to serve aeronautics—as an aircraft tender (AV 6). This conversion included installation of a mast equipped with yaw booms over the poop and, below decks, facilities for aviation gas, helium, and spares, plus accommodations for flight personnel. *Patoka*'s original masts were shortened, to allow her charges to swing about the stern mast.

The contract with Luftschiffbau-Zeppelin estimated a minimum of three hundred flight hours and thirty flights for half the guarantee period. Desirous of hours aloft, the front office pressed for independence from the hangar, for keeping *Los Angeles* outdoors, moored to the mast whenever possible (see page 54). That December, though, two sorties only were logged—less than six hours.[38] Mostly, *L.A.* sat hostage in her berth. Trip after trip was announced, only to be postponed or aborted. Early in January, exasperated, Moffett dispatched a memo salvo to induce the base to perform. Thus far, he fumed, operations had been wanting.

> The Navy Department stands committed in the eyes of the nation to the project of determining from the performance of the LOS ANGELES the practicability of commercial airships. . . . The failure of the LOS ANGELES to operate on schedule and with reasonable frequency has reached a point where it is causing unfavorable comment and is jeopardizing not only the particular project relating to this airship, but also securing favorable action on any further program for developing rigid airships within the Navy Department.

Success, the admiral added, closing out his rebuke, would be in direct ratio "to the initiative, zeal, application and care of the personnel charged with flight operations."[39] Later aggressive commanders would in fact lay on a heavy, more flexible operating schedule. But for further weeks, the northeastern winter as well as mechanical gremlins—notably the water-recovery gear—served to foil flight.

In Steele's defense, Lakehurst is miserably located. Most storms from the eastern Gulf of Mexico and Atlantic move up the coast before departing the continent. "The result," as the *Rigid Airship Manual* (1927) laments, "is that practically every storm that passes over this country or along the Atlantic coast exercises for a

USS Los Angeles *moored to USS* Patoka *(AV-6) in Chesapeake Bay, 15 January 1925—the first use of the floating base by ZR-3. By 1932, L.A. would log more than 700 hours in company with the mast-ship. (National Air and Space Museum, Smithsonian Institution (SI Neg. No. 2000-11187))*

Captain George W. Steele, Jr. in cold-weather gear, January 1925. Steele had just assumed command of ZR-3 and the Lakehurst station. To his left is Aviation Chief Rigger (ACR) Frederick J. "Bull" Tobin, of Shenandoah's complement. (Rear Admiral Calvin M. Bolster)

greater or less time an influence over the winds at Lakehurst." Further, given the hangar's axis, the reigning winds often were cross-hangar. Pending mechanical solutions to ground handling, ZRs too often lay in berth.

The immediate schedule called for continuing demonstration flights, including, in January–February, practice with *Patoka* in Chesapeake Bay. Thereafter, steaming to southern waters, the ship would await *ZR-3* and a series of exercises. Of these, as Steele saw it, the most important sortie was to Bermuda. He faced hurdles, the highest of them helium. Conservation remained imperative; yet, absent workable water condensation, valving was unavoidable. Further, the condenser equipment (two sets mounted) wanted redesign as well as overhaul, as did his engines.

In the forenoon watch (0800–1200) of 8 January, a ground crew was called to the dock. All engines warmed,[40] *Los Angeles* was marched out.[41] When clear of the shed (1100), the men alongside struggled her to the edge of the west landing field. At 1205, the assemblage reached the three-legged, 160-foot mooring tower. The decision to construct this first-of-its-kind in America had been taken in April 1921. Experiments in England (at Pulham, in East Anglia) confirmed that big ships could be maintained at a mast in poor weather, even in high winds. U.S. personnel had been duly impressed.[42] Alighting at Mineola, New York, the British rigid *R-34* had masted in New York for four days. Drawing on British ideas, when C&R prepared its own specifications and plans, its design resembled that at Pulham.

At its base, Lakehurst's tower boasted a machinery house with three main winches driven by electric motor. (The variable-speed gears had been salvaged from battleships.) Each was equipped with a ship's telegraph and speaking tube, for communicating with the masthead. The machinery house also held pumps for water and gasoline. The telegraph controls were to prove the mast's bête noire. Without direct control of the winches, delay in the transmission of orders from topside—from the mooring officer, who was watching the cables—was built in. Many a cable was destined to break or foul.

Further problems were foreordained. A mast watch had to remain on board the airship—in the forward car, keelway, and power cars, "flying" to the tower—that is, maintaining trim. Temperature effects (tail lightness, especially) and the tendency of gusts to rebound against the hull had to be counteracted promptly. Men and or water were therefore shifted. The use of "galloping kilos" (men) was slow, awkward and, at severe angles, hazardous. "It was a regular flight job," one officer remembered, "to keep the ship either at a high mast or at the *Patoka*. You had to be quite alert as to the weather conditions to keep her from going to one extreme or the other, either tail-up or tail-down."[43] As for repairs, no major work was possible at the mast.

With the advent of the "stub" mast as standard equipment (see chapter 4), the tower was shunned; by 1928, it was quit altogether. Meanwhile, the flaws attending the "high" scheme became clear only with experience—a commodity still sorely lacking at Lakehurst.

For her inaugural moor-out, *L.A.* was weighed off and her main cable secured to another led out from the masthead. *ZR-3* then was allowed to rise. Her drag ropes—lowered from the cone and taken at angles away from ship's nose, to keep it more or less steady en route—were used as yaw guys as she lifted to the mast cup. The mooring cone was secured at 1232. Held by the bow, *ZR-3* could now swing to the wind, much like a ship at anchor.

After nearly a day, and with all hands (including five Germans) at landing stations, the order was given—at 1014—and *Los Angeles* cast off. No upward thrust could be had from ground crew, so up momentum came from static lift—that is, positive buoyancy (about five hundred pounds) and a slight bow-light angle. Ballooning clear of the tower, Klein ordered his engines cut in at about two hundred feet, whereupon course was set for Lakewood and the coast.[44] The hop was local, cruising the New York–New Jersey area. Touching down then weighing off, the airship loitered with engines slow ahead before making a flying moor to the tower—a first for ship and aircrew. Promptly cast off again, ZR-3 was hauled down into the hands of the landing party (crewmen from ZR-1 assisting). By 2000, she was berthed and secured, the crew dismissed, doors closed.

Six days later, at 0630, the first floating moor entered ship's log—in the Patapsco River, below Baltimore Harbor. (As per routine, men had been sent ahead with *Patoka* to prepare—in this instance Rosendahl, Tobin, and two more.) Afloat, the mooring scheme resembled that at Lakehurst, save that distances were shorter. To compensate, the yaw lines were boomed out, so as to get a better angle into the nose. Though ZR-1 had been first to exploit the tender, ZR-3 was to be its best client, logging forty-four moorings over eight years in various bays from Rhode Island to Panama. *Akron* would moor to *Patoka* but twice, *Macon* not at all.

Failure of any part of the ground handing system at a critical moment may mean loss of the ship. This particular mission nearly ended before it began. In the dawn hours of the January 15, with the hull protruding stern-first beyond the east doors, the starboard aft trolley stuck and had to be tripped. Freed aft, the tail started to fall rapidly to port, the ground crew, standing on sheet ice, unable to hold it. On the bridge, Hans Curt Fleming sprang to the telegraphs and rang up "Cruising" on the side engines, then "Ahead, Full." ("Stop" was rung up after fifteen seconds.) ZR-3 was thrust back inside. Her discomfited handlers staggered to a halt, the aircraft resecured. Run intermittently, the engines were kept warm. *L.A.* was marched clear at 0845; by 0939, Klein was under way for Maryland.

Four days later, Steele assumed command of *Los Angeles* and the air station, thus relieving his executive officer. Nonetheless, he continued to leave shiphandling and internal administration to Klein.

Late that January, ZR-3 lofted specialists to observe a solar eclipse. By exploiting the thin, clearer air at altitude, the Naval Observatory hoped to make photographs of the "greatest clearness." Special cameras and other gear were placed on board, rehearsals for the eclipse held. (Installed also, on cars number two, three, and four: condensers removed from *Shenandoah* for overhauling.) But a stiff, bitter wind thwarted an escape to the mast. For two days, the eclipse party and ship's crew stood by at or near the hangar—shut-ins awaiting a suitable lull. "We waited day and night—it seemed like forever—for the wind to die down," Patzig recollected. "It was damn cold. And damn windy." All hands—from skipper to lowest seaman—were relieved when, at 0503 on the 24th, *Los Angeles* was unhoused out the east doors by a ground party into four-degree air and a cross-hangar blow. Before the craft was fully clear, and despite extra men clinging to its windward flank, a gust pushed it broadside, its handlers sliding on ice. A thrust from the props got her past.

Ground crew dismissed (at 0520),[45] ZR-3 crabbed away to seaward, toward Nantucket and the curving path of a totally complete eclipse of the sun's disc by the moon.

A brightening dawn painted its colors. The air was punishing, the temperature within matching the cold without. On board, ambient temperature was controllable only in the warm engine cars, where the "mechs" watched gauges, answered the telegraph, and bathed the valve springs in kerosene (to inhibit sticking). For the bridge and keel watches, winter gear was the uniform of the day. Exhaling clouds, the guests—led by Capt. Edwin T. Pollock, superintendent at the Naval Observatory—stood similarly bundled.[46]

Given the late start and a mass of cloud to eastward, a decision was taken at Long Island: observations would be recorded from just south of Block Island. Klein steadied on course 150 degrees, on the center line of totality. *The moment arrived: 0912.* (To minimize vibration, only one

Commander Klein and Captain Edwin T. Pollock, Superintendent of the Naval Observatory, in the control cabin during the solar-eclipse sortie, 24 January 1925. Ambient air temperature: 4 degrees, Fahrenheit. (National Air and Space Museum, Smithsonian Institution (SI Neg. No. 2000-11194))

Back in berth, the party of astronomers and photographers for the eclipse flight pose by Los Angeles. *The trio at left are professors; Pollock is at center. Second from right: A. R. Peterson, the Navy's Chief Photographer. (National Air and Space Museum, Smithsonian Institution (SI Neg. No. 2000-11193))*

VL-1 pair was running.) At their stations, tense yet thrilled observers began their respective tasks. "Every officer and man on board was engaged, if not at their regular stations, then at some observing," a participant recorded.

Klein had his ship back on base at 0340. Harassed still, *Los Angeles* did not gain its berth, however, until 0510. The scientists seemed pleased with their results—recompense for the protracted wait. "The use of LOS ANGELES for scientific work of this type," a department press release crowed, "is a striking demonstration of the variety of important uses to which rigid airships may be put in spheres outside of purely naval and military activity."⁴⁷

These were anguished days. Winter persisted in ravaging the naval reservation; winds, snow, and ice held *Los Angeles* shedbound. The unforeseen interposed as well. Calcium chloride and denatured alcohol were the only antifreeze agents deemed practicable. Alcohol, though better, was about three times as expensive. So the former was introduced in the ballast water (about 20 percent by weight). This proved ill advised; spillage attacked certain keel girders and adjacent frame members. "Practically all lattice work in way of water ballast bags is in such condition that replacement is necessary," Klein wrote. Replacement of corroded and mechanically damaged alloy—in the hull's lower half particularly—became standard practice, along with scratch-brushing and varnish protection. Beginning in February 1926, samples would be removed annually from the ship and tested at the Bureau of Stan-

dards, to serve as a general check on condition. (The specimens were assumed representative of the entire hull.) For this 1925 episode, the damaged alloy was removed and either replaced or, if sound, immersed in nitric-acid solution, brushed, then washed with warm water and, when dry, varnished. All in all, here was a rude lesson. "Unfortunately," Steele sighed, "we did not know all the German practice[;] . . . whenever they spilled calcium chloride solution on the duralumin framework they immediately cleaned such parts." Alcohol was in.⁴⁸

Some renewal and a general cleaning were applied at this time, while *L.A.* continued to operate. When her helium was transferred in mid-1925, however, further examination revealed the need for extensive replacement. A lengthy inoperative period ensued. Most troublesome were the water-recovery units. Under way, condensers froze. Moreover, the higher power and exhaust temperatures of the Maybachs ravaged the units inherited (as an expedient) from *ZR-1*. They badly warped the inlet headers, for example, burned and cracked the exhaust manifolds, and heated the pipes between the manifolds and condensers red hot. To counter the problem, operating personnel kept the fuel mixture as lean as possible. External fins were affixed to the exhaust pipes and headers, to increase radiating surface. These and other correctives, Steele lamented, "have not been sufficient to overcome the unsatisfactory thermal conditions existing with the present design."⁴⁹ Eventually, special condensers and additional cooling banished them.

Though the schedule called for continuing demonstrations, early in February Lakehurst notified the bureau that, given the condition of the recovery apparatus (hence excessive lift losses), another *Patoka* moor would not be attempted until a new system was complete—estimated as the 16th. Further, lack of helium foreclosed two-ship operations for six months at least. On the 5th, during preparations for repurification, a flashlight dropped from the overhead punctured cell number four. The missile continued on, exiting ship's bottom. Even a badly holed cell would not discharge itself abruptly from the *bottom*, buying time to effect repairs. (Holes were not unusual or particularly serious.) Due to pressure at the *top*, however, a satisfactory patch could not be made to hold as quickly. All in all, it was a seven-thousand-cubic-foot loss.

The next phase of deployment involved long overwater flights, to simulate—and stimulate—commercial air operations. The first objective: Bermuda.

A complete weather service had been vital from the start. Lakehurst's, though independent, was supplemen-

L.A. swings to its "high" mooring. The magnitude and distribution of forces imposed on airship bodies were, in the twenties, a largely unexplored field of investigation. As much additional data as possible was therefore extracted from each new design. In this context, Los Angeles *proffered a flying, full-scale laboratory for experimental research. (Charles E. Rosendahl Collection, University of Texas)*

tary to that of the U.S. Weather Bureau. Consulting with Aerology, the outlook was assessed, takeoff time chosen, a zero hour ordered. Aviators must live weather. "Weather was a twenty-four-hour-a-day, seven-day-a-week, thirty-one-days-a-month proposition," Reichelderfer explained. "It was never possible for the science, or the profession, to give all of the information that the airship operators needed to have." For the aerological staff, then, the demands were unremitting: to outguess the weather and anticipate what operations would need. The station's observatory was key. There the maps—drawn and analyzed four times daily—were gone over, along with reports from U.S. airways stations. (The latter flowed in constantly, via teletype, and were watched unceasingly.) A good aerologist was prized. "This really made an ideal, a marvelous position—role—for the airship meteorologist," Reichelderfer continued. "There's no question about it. You saw the most important people, they came to you to ask for information, you were junior to them, you were subordinate, but *they* were asking you for information.

Los Angeles *in position for a "high" moor. Her main wire connected to the (slack) mast wire, the aircraft is "weighed off." Here ZR-3 is rising slowly on her trail ropes. When secured to the high-mast cup, ballast will be adjusted, a mast watch set. (Airship International Press)*

Steele driving his command into the clutch of a ground crew, 8 June 1925. The store of helium was yet provisional, the supply insecure. First attempt this day: at 0905. Touchdown: 1342, concluding the 18th try. The ship as much as three tons "light," 125,000 cubic feet of helium were valved to complete the landing. (National Air and Space Museum, Smithsonian Institution (SI Neg. No. 2000-9086))

This was one case where even the captain [deferred to you].... [H]e made the last decision, but his decision was made upon the questions he asked you."[50]

Preparations under way, the Bermuda-bound guests arrived on the station and were rendered the prescribed honors. For Assistant SecNav Robinson and his aide, plus Moffett and his party, this was to be the first voyage of its kind. Its guest list ever changing, ZR-3 was to host scores of rider-observers during eight full years of operations.

Auguring clear sailing, good anticyclonic weather had been forecast. Course and altitude were chosen accordingly. At 1400, 20 February, a ground crew was called to assembly—three groups of three short blasts on the Power House whistle. Men assembled by divisions for muster, after which groups were stationed alongside ZR-3—car parties, trolleys, line groups, tail. Division officers then reported to the mooring officer. From "assembly" until "pipe down," this officer ruled the evolution.[51] Under his eyes, *Los Angeles* was hauled out past the east doors, stern first. Let-go was logged at 1520. Guest-passengers: Robinson (whose flag was displayed), Moffett and BuAer's Captain Land, and Cdr. Robert L. Ghormley. Attended by the latest charts and wind soundings, Steele rang the (idling) engines to "cruising" and set course for Barnegat Light.[52] Minutes later, the lighthouse lay on the starboard beam at four miles—the last view of the coast.

The dirigible pushed southeast.

Prior to departure, a dispatch had informed a waiting *Patoka* that *L.A.* would get under way that afternoon. (The tender had left Hampton Roads on the 11th, arriving in Bermuda on the 14th. Three days later, Rosendahl, Tobin, and Aviation Chief Machinist's Mate William Russell were dispatched south.) The two vessels maintained wireless contact throughout ZR-3's outbound passage.[53] That first evening, the men dined on roast beef and "the necessary accompaniments." The midwatch (0000 to 0400) found the airship cruising on four engines, airspeed fifty-two knots. She was a ton heavy—a load carried with ease. The stab of Bermuda's Grassy Bay lighthouse was seen at 0255. Early in the next watch, with two Maybachs idled, Steele steered various courses about the island. Ashore, "early birds astir could plainly see her, brilliantly lighted as she was, and those too disinclined to turn out and view the sight could hear the throbs of her engines."[54]

In the morning watch, at 0550, Steele had *Patoka* in view—steaming to Murray Anchorage to await her charge. *Patoka*'s skipper was hosting a gaggle of local luminaries, among these the American consul, Mrs. Robertson Honey, and J. R. Roosevelt, whose son-in-law, the assistant secretary, was aboard the advancing dirigible.

Patoka pronounced herself ready at 0745. But a moor was deferred pending improved conditions. At 0845, rain started; soon, *L.A.* met heavy squalls and took on weight. (A ship her size could gain by saturation 2.5 to 3.5 tons of water, depending on the character of the fall.) Braving the blow, Bermudians had cruised in boats to the North Shore, to obtain a better vantage. Steele, hovering vainly, decided to abort. Before taking his leave, however, he made a delivery drop. For the first time in U.S. aeronautics, overseas mail was being carried by air. Steele radioed that three bags would arrive near Government House. During the forenoon watch, at 0945, he ordered ahead-cruising on four engines; ten minutes thereafter, eighty-five pounds were surrendered—letters, postcards, and registered packages, including copies of the *New York Times*. The bags plopped down just outside the residence grounds. So arrived the first official U.S. mail via dirigible—and Bermuda's inaugural air mail.

"Impracticable to moor to *Patoka*," the log records, "on account of continued heavy rain squalls." Having no further business, Steele took his departure, course 326 PSC (per standard compass), northwest. Altitude: two thousand feet above disturbed water.

Though disappointed, Bermudians had consolation; *L.A.* would return, Meyers assured the press, as conditions permitted. The voyage, he added, "was successful in every respect" save for weather. "The water recovery outfits functioned perfectly," he continued. *ZR-3* had refrained from mooring, the skipper explained, "because she did not desire her envelope to become saturated with rain which would have formed undesirable ballast evaporating in the sun and causing a lightened condition which could be corrected only by the valving of precious helium." Steele's passenger-guests enjoyed themselves. "My experience upon this voyage," the admiral observed, "convinces me that vacationing by dirigible will be highly popular."[55]

Bermuda fourteen hours astern, a weary *Los Angeles* neared homeport in the wee hours of the 22nd.

Each return to earth is different. At his station forward of the elevatorman at the port forward window, Steele would have been observing his instruments and ship's behavior well in advance of arrival. When the word was passed—"Landing stations"—all hands proceeded to assigned locations. There each man remained, on the alert, until ordered to move or "pipe down" was passed. Descending, the airfield not yet in sight below, the keel telephones stood manned; should it become necessary to trim ship, men could be shifted quickly. Steele himself was not handling the engines, ballast, or ropes. As second in command, Klein was conning the rudderman from *his* station, at the starboard forward window. To his right, the engineering officer was at the telegraphs, while, portside, the first lieutenant (keel officer of the watch) stood at his station by the gas and ballast control board. The watch officer was keeping the log; when ordered by the skipper, he released the trail ropes. As props were cut out and speed ebbed, the dynamic factors controlling the machine vanished. (In low surface wind particularly, the operation was pure statics.) The most favorable condition for landing was a steady wind of from ten to eighteen miles an hour. The easiest: when the ship was slightly heavy, dropping ballast to check her fall once ground personnel had the ropes in hand.

At 0036, her way checked, *Los Angeles* settled to the ground party. Two hours' fuel remained. Time aloft: thirty-three hours, sixteen minutes. The dirigible was secured by 0230.

The economics of sustained overwater flights were under study. This round-voyage performance (Steele calculated) worked out to $837.31—for the sea portion, about $25.17 per flight-hour.[56]

No further flying was logged until April; until then, a new problem grounded ship and crew. The cells, made using old methods, had about exhausted their useful life. Steele deemed them unsatisfactory for the northeastern climate, and for helium. Indeed, diffusion losses from numbers two and thirteen were so severe that both were unshipped and their loose goldbeater's skins—thousands of them—relaid and revarnished. In mid-April, Steele urgently recommended a new set. Unwilling to incur the expense, the bureau could not agree. Accordingly, during the next overhaul (which commenced that June), all fourteen were unshipped and laboriously repaired, and all for nothing—when reinflated, the material was worse for the handling.[57] Gas cells (and budgets) were to plague operations, causing repeated layups for repairs or repurification.

A less porous, more durable material plainly was called for. "We realized," Fulton remembered, "that a substitute for goldbeater's skin had to be found and the Bureau of Standards was given the job and worked continuously on it for a decade."[58] A NACA subcommittee on coverings, dopes, and coatings was incorporated into the work. First try: cellophane. By 1928, a better substitute had been developed, fabric impregnated with viscose latex. The first experimental latex cell (built by the Naval Aircraft Factory) was delivered that April. Within a year BuStandards devised a gelatin latex compound that was far more heavy-duty than viscose latex. By late in her career, *L.A.* would be fitted with several.

In the meanwhile, operations persisted on a fiscal shoestring. Lakehurst was beggared—desperately improvising as it importuned for funds. *Los Angeles*'s cells are illustrative. By the close of 1926, the airship was flying with gas cells from five separate sources, including German originals, retailored *Shenandoah* spares, and experimental units contracted for her.[59]

During the overhaul, recording strain gauges and angular accelerometers had been fitted, adding to those from German tests. On 16 April, a (delayed) postrepair offshore flight was made. Alighting in the afternoon watch, *ZR-3* became abruptly heavy. Emergency ballast streamed out. Though the lines were dropped some distance from them, the ground party managed to grab on before she rose away. *Los Angeles* assumed a steep angle during the haul-down. When maneuvering, her engines exhausted direct to the atmosphere—that is, the condensers were cut out. Lacking an accessible condenser cutout, the Maybachs were generally stopped one at a time and their the exhaust plugs removed. On approach, the engineering officer obtained permission to do this. But for this landing, no opportunity was afforded the engineers in three of the cars to remove their plugs. The condensers became very hot, and two of them warped.[60] Upon hangaring, repairs to the unit on number-four car commenced immediately.

In the morning watch of 21 April, following a general-assembly call, engine warmup, and a weigh-off, preparations were effected for walk-out. But *ZR-3* was resecured: wind force and direction were unfavorable. At 0840, another weigh-off was taken, after which cells two and thirteen were gassed to 87 percent. By 1130, the call for ground crew again sounded. The day's third weigh-off logged, *Los Angeles* undocked in the afternoon watch. (The crew of *ZR-1* had been turned over to assist the mooring officer.) Near the mast almost a ton of fuel was taken on. At 1435, *L.A.* was under way, her VL-1s set to cruising speed. Total on board: fifty, including three army men as regular crew, three army officers, and five from the navy in training.

Safely away, Steele ordered a compass course of 148 degrees true. He was making for Bermuda—the airship's eleventh mission under U.S. command. Gibbs Hill Light, bearing 140 degrees, hove in view at 0100 on the 22nd. The skipper chose to hove to, keeping to the northwest of the island, delaying weigh-off until 0510.

The worst conditions for mast moorings were calm or light and variable winds, which demanded prolonged backing of engines, causing the aircraft to yaw. In a steady forty-knot wind, however, *Los Angeles* was unable to secure until 0636—the second moor to her floating base. (So held, *L.A.* was generally lightened at the bow, to ensure a good strain on her cable.) As fuel and water came on, mail was delivered to *Patoka*, anchored not far from the light. Steele loitered—adjusting ballast—for twenty-six hours, during which time the gods of chance proved moody. Late in the afternoon watch, winds became gusty, reaching forty-two knots. Gusts are worse than steady velocities. Why? A force steadily applied is less dangerous than one abruptly exerted. "Time and again during the operation," Rosendahl writes, "gusts drove her below the mast and perilously near the water. Often the mooring wires literally lifted the ship to safety."[61] Despite such antics, exchanges through the bow hatch persisted: fuel and water, men, mail. Altogether, eight bags were offloaded: three for the tender, five for shore.

At 0857, 23 April, Steele unmoored, shed 440 pounds of water, telegraphed all engines to "cruising," and ordered various courses over Hamilton. Having thus shown off his command, he then set a course for that familiar New Jersey minaret, Barnegat Light. Water-recovery systems were cut in on four Maybachs. Stowed aboard in four bags was the first mail by air from Bermuda.[62]

The lighthouse lay dead ahead, distance about twelve miles, at 0341 on the 24th. *ZR-3* stood snug in berth, secured on cradles, three hours later. Later that day, at Moffett's invitation, members of the NACA's entire committee visited the air station and inspected both ZRs in berth.

Steele's next sortie was the most ambitious to date: to Puerto Rico. In the forenoon watch on 3 May 1925, a Sunday, *ZR-3* began taking on fuel; at 0845, flight crew relieved the mast watch. (*L.A.* had secured to the masthead at 2100, 27 April, more than five and a half days before.) With four passengers plus three dozen officers and men, cast-off was ordered at 0958. Disposable weights for the outward trip included 25,383 pounds of fuel and 2,750 of water. Five motors were set to cruising, and Lakehurst slipped astern at fifty-five knots. Thirty-two minutes later, with Barnegat Light four miles on the starboard beam, the side VL-1s were put on water recovery, and the crew took departure south at two thousand feet—course 175 degrees true.

In accordance with bureau instructions, Mr. Horace D. Ashton, a magazine writer and lecturer, was in passage. After photographing the takeoff (from number-one car), he made his way forward, where Klein extended welcome. "He brought a phonograph and records which had been presented to the ship a few days before. The Filipino steward entered with cakes of chocolate and passed them around. The chocolate took the place of pipes and cigars." At luncheon (three tables arranged end to end), Ashtown sat down to ham, cheese, and bologna sandwiches, spaghetti, coffee, fresh fruit, and more cakes of sweet chocolate.[63]

Save for dawdling with the steamer *Delfina*, about which Steele maneuvered, observing the sun and taking photographs, the downwind drive above the cloud-shadowed blue of the Gulf Stream proved uneventful—save for one incident. That first forenoon, while shooting from number-five car, Ashton dropped a crank handle from his camera. Swept out by the airstream, the handle struck the prop, making deep dents, impossible to repair, in its leading edge—a two-hundred-dollar loss.[64]

In the 1600-to-2000 watch of the 4th, a radio-compass bearing from *Patoka*—lying to anchor at Mayaguez, on the island's lee side—indicated the need for a slight change in course. At 1740, land was sighted off the port bow and, twenty minutes thereafter, the tender. (Rosendahl had again been sent ahead, for temporary duty aboard the mast ship.) After weigh-off three miles off Point Jiguero, Steele maneuvered in. His command was light, so the maneuvering toggles were operated for three minutes. Trailing its wire, the dirigible was driven down. In the tropical heat and oil smoke from the stack of *Patoka*, *ZR-3* closed with the mast ship and her two boats (deployed astern) as daylight ebbed. Not until 2215 did Steele secure. Flight time: thirty-six hours, seventeen minutes. Immediately, water was taken aboard and, later, fuel—660 pounds in the 2000-to-2400 watch, during which this weight in water was shed.

L.A. swung to the mast for thirty-six hours, during which men shunted off and on as water, fuel, and helium were pushed aboard. Throughout the moor, a succession of watch sections—sixteen men each—focused on "flying" *ZR-3*. This consisted of adjusting water to meet changes in temperature so as to maintain a more or less even keel, and on running engines intermittently to check her from swinging in over the tender. Nonetheless, the watches suffered several trying moments.

On the 6th, at 1019, the dirigible let go. The rudderman steered an easterly course, thus skirting the north coast—the first leg of a leisurely ten-hour circumnavigation of Puerto Rico, including a probe toward St. Thomas. At 2035, *ZR-3* again secured off Mayaguez. The next day brought close calls. During the forenoon watch, with *Los Angeles* swung in over the tender, ballast bags at frames 140 and 55 carried away. This shed nearly four thousand pounds (the culprits: a tropical sun and hot ballast water from the tender, which opened seams on the bags). At 1215, due to gusts, *L.A.* took an "extreme angle up," then down. A second oscillation had the stern up fourteen degrees before settling—this time dipping the rudder. Between 1430 and 1440, *ZR-3* shifted heading approximately 180 degrees. This swing put her directly over *Patoka*. "The ship encountered a gust which started the stern down, at which time an emergency ballast bag was pulled to prevent collision with the cargo mast of U.S.S. *Patoka*. The stern then rose to an angle of 21 degrees after which it settled while the ship swung clear. . . . Unusually gusty weather with strong vertical currents. Adjusted water ballast and used men to trim ship to meet conditions."[65]

At 1433, 8 May, following forty-two hours more joined to *Patoka*, Steele conducted a final weigh-off and unmoored. On board for the homeward run were five sacks of mail. Next morning, his command was circling Miami. With persistent slowness, *L.A.* floated up the Atlantic seaboard; by the midwatch, the lights of Maryland's Eastern Shore gleamed beyond the windows. At 0431, 10 May, the watch relieved, *ZR-3* eased into homeport. Airtime for the return: two minutes shy of thirty-eight hours.

Next day, the station log recorded a pair of ZRs snug at berth: "U.S.S. LOS ANGELES docked on North side of hangar, bow to East. U.S.S. SHENANDOAH docked on South side of hangar, bow to West. Kite balloon number 2799 secured in kite balloon hangar. Plane DT 6592 secured at West end of hangar."[66]

Thus ended the first extended operation exploiting *Patoka*. A success, the sortie yielded astounding air mileage plus superb publicity. Yet *Los Angeles*'s swings over her seaborne base had discomforted not only the airmen but *Patoka*'s commanding officer—and BuAer. Early in 1926, to add clearance, mast height was increased by thirty feet.

Before deflation and overhaul, slated to commence on 10 June, *Los Angeles* received orders from CNO to conduct passenger flights and, in June, visit Annapolis and the Midwest. As for *ZR-1*, it was to be made ready for operations with the fleet, taking necessary supplementary helium from *L.A.* Further, after the overhaul period—and at the first opportunity after 1 August, by when ample he-

Closeup of power car No. 1, USS Shenandoah. *Note the "car cradle," sand bags, and ship's 18-foot propeller—one of five. (San Diego Aerospace Museum)*

lium might be accumulated—*L.A.* was to be made flight ready for a mission to Hawaii, supported by *Patoka*.[67]

As per BuNav dispatch, a contingent was transferred to Naval Air Station Pearl Harbor for temporary duty in charge of the mooring mast at Ewa. (No airship ever rode its circle.)

The Hawaiian chain was a logical place from which to deploy a ZR for strategic scouting. Or for commercial service, given the route's heavy traffic and favorable weather. Indeed, Hawaii asked that *Los Angeles* establish a service between Honolulu and the West Coast of the continental United States.[68] The chance vanished that September when *Shenandoah*—and her helium—were lost. Later, the CNO planned to have *Akron* (ZRS-4) participate in Fleet Problem XIII, slated for Hawaiian waters. That mission too was not executed. In December 1934, at the insistence of Adm. Ernest J. King, Moffett's successor, the CNO recommended that *Macon* (ZRS-5) prepare a schedule of flights between the West Coast and Hawaii, to exploit her capabilities in strategic problems. Two months later, *Macon* foundered off California. So it was that the field of operation in the Pacific had hardly been investigated, let alone developed.

May–June added seven sorties more to ship's log before the scheduled overhaul. For the balance of May, three were flown, all local. In an transparent publicity gesture, the guest list at midmonth included thirty-three notables, among them a trio of rear admirals, the chief of staff of the army, a secretary of war, a radio personality, seven company and university presidents, and two editors from the *Saturday Evening Post*. With Moffett as host (and three naval cooks and two mess attendants assisting), *L.A.* cruised up the Delaware River as far as Easton before looping back over eastern Pennsylvania, thence circling Wilmington.

A flight on the 26th saw the dirigible calibrating Lakehurst's radio-compass station. This involved circling at a radius of ten miles or so while the calibrating party on the ground compared actual visual bearings on her with those derived from equipment tuned to her frequency. (With interruptions only to receive instructions, the airship's radio spewed a series of continuous dashes.) ZRs were ideal for this—steady, slow, long-ranged. In 1926 alone, a dozen or so calibration missions were flown. Thanks to *Los Angeles*'s rapid work, all the Atlantic-coast stations would be opened to aircraft that year.[69]

During this trip, C. P. Burgess, a design-engineer from BuAer, and colleagues busied themselves recording the readings of strain gauges and angular accelerometers at various points within. Ten minutes out, number-one engine was stopped due to failure of a connecting-rod bearing. Time on that engine: 359 hours. A ten-hour, ten-minute run concluded with a flying moor to the mast.

Next day, 27 May, with four motors operational, *L.A.* cast off the tower mast, again with bureau representatives. Proceeding off shore, Steele cruised up the coast via New York City to the Military Academy at West Point, then on to Newburgh, a dozen newspapermen accompanying. Disembarking passengers at 1800, the aircraft was walked to a three-point mooring to take on water. Held on the field by a cross-hangar wind, *L.A.* was berthed in company with *ZR-1* at 1937. Work commenced on exchanging the damaged engine for one of the only two spares to hand.

In the last watch (2000 to 2400) of 1 June, at 2022, the east door slid aside; almost two hours later, *Los Angeles* cleared the sill en route to the high mast, to which she secured at 2305. The flight to Annapolis (a frequent destination, as was the nation's capital) took place on the 2nd, *Patoka* in the Chesapeake Bay serving as host. Off Maryland, a new procedure was tested. Devised by Rosendahl, a special grapnel was attached (via shackle) to *ZR-3*'s mooring line. Rather than having the tender's launch connect the two lines, the grapnel was dragged across a cable from *Patoka*'s mast—led astern by her boat. No less than landings shoreside, at-sea moors called for exceptional airmanship.[70] Steele's first try realized success. As the tender steamed ninety degrees off the wind with about five hundred feet of line led astern, *L.A.* headed into the wind dragging her main cable. When the grapnel caught the wire, the launch released its end, the airship rose with motors throttled, and the line reeled in as *Patoka* turned into the wind.

Time from the grapnel's catching the wire to all secure: twenty-eight minutes. Despite the hitches, further practice refined the approach procedure. Use of the grapnel was soon routine in *Patoka* operations.

Tarrying two hours, Steele cast off to hang about the Annapolis area then retest the procedure. Maneuvering for the approach, his main line was dropped and a descent made to three hundred feet. But the grapnel failed to connect properly. The launch connected up the main wire (the old method) on the fourth try, the yaw guys were joined and the ZR secured. *Patoka* stood into Annapolis Roads trailing its silver charge. Steele stayed fourteen hours before casting off for home, at 1014, 3 June. Four passengers more had joined, including Moffett and Fulton, plus Rosendahl—who had directed the couplings from *Patoka*. First, though, Steele started down Chesapeake Bay to conduct deceleration tests for "C.P." and other bureau personnel. Back over the academy, *ZR-3* steadied on course 258 degrees true, for nearby Washington. At a thousand feet, she hung over the Capitol at 1415, destination Lakehurst via Baltimore and Wilmington.

The excursion to Minneapolis remained. This was sheer salesmanship—an answer to countless requests to see the ship. ZRs belonged at sea; a jaunt into the nation's midsection (while newsworthy) was empty of *naval* significance. Rather than withdraw its big airships from the public eye, *Shenandoah* and *Los Angeles* were obliged to log publicized "barnstorming" visits to county fairs, conventions, parades, air meets, and the like. Calls for a blimp or two at various civic functions also were obliged. Why? To gain public and budgetary good will. On one level, these appearances contributed a great deal to the acceptance and popularity of the program. But it was exposure to—and experience with—units of the fleet (and fleet doctrine) that was needed. Indeed, the entire ZR venture was predicated on the notion of the seagoing aerial scout. To be assimilated, the type first had to be assessed. The operations of *Akron* and *Macon* were to reflect the pernicious effects of too much barnstorming.

The midwatch of 6 June found *Los Angeles* docked on cradles, in company with *Shenandoah*. Aboard were 17,600 pounds of fuel, 11,560 pounds of water, "and all other miscellaneous items necessary for flight." A quarter-hour past midnight assembly was sounded; twenty minutes thereafter, the northeast door rolled. At 0145, following a final weigh-off, a thousand pounds were let go—120 gallons—and the ship walked out. Another weigh-off was taken before masting. Just before 0300, the tower ablaze in a circle of light, *ZR-3* was secured.

Ground crew dismissed, the mast watch aboard, attending on station, the aircraft oscillated about its nose-hold in a light, boisterous wind—from fifteen degrees up by the stern to six degrees down. (Twelve degrees down-pitch had the tail hitting the ground.) In the forenoon watch an experimental tail drag was rigged aft. Attached to a pelican hook at frame 25 (where the fins met the hull), the drag consisted of a ninety-foot length of chain weighing 533 pounds attached to a manila line. Though crude, the device appeared to dampen all motion; despite gusts, it cut the maximum oscillation to plus six and minus five degrees from the horizontal. Pleased, Steele recommended a longer, heavier drag chain be used as standard practice.[71]

At 0130, 7 June, having hung twenty-two hours at the masthead, *Los Angeles* tripped free, destination Minneapolis. Crewmen numbered thirty-seven, among them a pair of army officers plus a trio of enlisted men. But no passengers—to Steele's delight, BuAer had decided to deny all requests for rides. (Deemed too heavy to take along, the strain-measuring apparatus had been unshipped.) Despite arrangements with the War Department to use the army's Scott Field if needed and to land at Fort Snelling for refueling, neither base saw the dirigible.

Pounding westward, *L.A.* overflew Camp Dix to near Mount Holly before crossing the Delaware. Reading sparkled below at 0415. A succession of Pennsylvania communities entered the log in turn: Shamokin, Sunbury, Lewisburg, Lock Haven, Clearfield, Rockton, Brookville, Clarion, Lamartine. At 0440, *L.A.* reached "pressure height," the altitude at which decreasing atmospheric pressure caused the lifting gas to expand inside the cells such that the automatic (spring-loaded) valves opened and helium "blew off." It was determined by the percent fullness upon leaving the ground and by atmospheric conditions. Without exceeding three thousand feet, Steele traversed the Alleghenies between Sunbury and Clearfield. At 1010, he passed over the Allegheny River on course 265. Four minutes later, a side propeller was lost—engine number five had burned out a connecting-rod bearing. The casualty was identical to that with number one twelve days before. Time on the power plant: 390 hours. The immediate concern for Steele was the endurance of two of his remaining VL-1s: number three (440 hours) and four, with 406 hours. At 1122, number four was stopped. At 1505, 7 June, the Ford Motor Company received a radio message from *Los Angeles* (call sign NZALA): "Account engine trouble abandoning remainder of trip and returning to Lakehurst from Cleveland. Expect to arrive sunrise Monday."

Steele nursed his ship via Youngstown and Akron toward Cleveland. Having circled Lakewood, Ohio, and back eastward along the lakeshore to Cleveland, he steamed for Pittsburgh thence the shed via Freeport, Pennsylvania, Lock Haven, Sunbury, Reading, Philadelphia.

A midwatch arrival: at 0315, 8 June, *ZR-3* hung over station Lakehurst, its field smothered in fog. Landing was put off. Around 0845, having loitered in the area, the skipper made ready. When air temperature dropped rapidly as he tarried, Steele had ordered all hatches in the keel opened, to cool the gas cells and help make them "heavy." (As volume decreases density increases, shedding lift. Lift is the difference between the weight of the air displaced and the weight of the gas.) But the risen sun warmed the cells; soon, his command was very light. Below, the store of helium was yet provisional, the supply insecure. "In those days," Rosendahl reminds, "one never knew where the next carload of helium was to come from, nor when, and besides, the specter of old man economy shook a warning finger at any thought of deliberately valving helium."[72] On first approach, at 0905, Steele tried to drive his trail ropes into the clutch of a ground crew without valving. For this return to earth, try succeeded unavailing try. Following two minutes' valving and five failures, more (timed) tugs were granted the valve cords—ten minutes altogether. Not until 1342—the airship as much as three tons light—was touchdown effected, on the *eighteenth* attempt.

In nearly five hours of maneuvering, 125,000 cubic feet of helium had been dispatched overboard—six thousand dollars.

Though no one could know, the year's operations had ended; this ballyhoo mission proved her last for 1925. Since the first builder's trial, *ZR-3* had logged 275 hours, nine minutes, in twenty-nine sorties. Further, 235 hours had been "flown" to the mast, and a half-dozen flying moors had been made to *Patoka*, for another 121 hours. *Los Angeles* would not be "be ready for flight" again for seven months—her longest incapacitation as a vessel of the fleet. Repairs had been scheduled and, as per instructions, a helium transfer effected.[73]

Operations concentrated now on *Shenandoah*.

Much activity filled this layup. By mid-September (Fulton had calculated) sufficient helium would reach Lakehurst to operate both rigids. Tragedy interposed. When *Shenandoah* was lost over Ohio, this idle period was necessarily extended: *ZR-3* would not, could not, be returned to flight status until March 1926. As repairs continued, with special attention to engines and cells, one crew section had duty and another was granted a bit of leave.

Prior to tragedy, *ZR-1* had operated with both *Patoka* and the battleship USS *Texas* on search, gunnery, and other tactical problems. Findings? Fulton characterized these trials as "promising but elementary and . . . never adequately evaluated—in other words[,] . . . inconclusive."[74] In the meanwhile, Lakehurst's *entire* stock of helium was lifting *ZR-1*. Valving, therefore, was out of the question if ample fuel was to be lofted. By way of counterstrategy, Lansdowne (skipper of *ZR-1* since February 1924), chose his getaways warily.[75] Accordingly, he moored out in darkness, to await the sunrise and its warming effects on the cells—known as "gaining superheat." *ZR-1* was held to the mast until all possible superheat was gained.[76] (With each degree difference between ship's helium and the atmosphere, *ZR-1* could lift a further three hundred pounds. This waiting strategy became standard operating doctrine. Inversely, landings were often at night, when the cells were relatively cool.

The impact of such delays and complications on operating commands afloat may be imagined. An alien, dubious platform subject to extravagant disdain, the rigid had yet to be worked into the evolving scheme of naval aviation. Nor was it destined to be. The ZR's primary mission (partisans stressed) was exactly that of the Scouting Force: to search for, locate, and develop the forces of the enemy. Pioneering proved slow, however, with each unit (and the attending shore establishment) consuming huge sums. Held by skeptics to be a waste of resources and organization, its contacts with surface forces little noticed and inconclusive, the ZR platform—its value undemonstrated—would fail to win the fleet's regard.

Given this hiatus, we might examine the Lakehurst base itself—homeport for USS *Los Angeles*. A naval air station offers logistic support for the fleet. In these years Lakehurst's fleet units were unique. ZRs (as we have seen) were commissioned vessels under commanding officers, with full complements of officers and men, and duty aboard them was deemed "sea duty." Accordingly, their commanding officers were autonomous from the air-station skipper. In terms of responsibilities, then, the latter had one overriding mission: support of the fleet. If *Los Angeles* required certain repairs, or needed a draft of men for a midnight evolution, the base had better have an organization to proffer them. As a port for ZRs, the station command furnished and maintained facilities in support thereof: hangars, office spaces, helium, labor, marines (security), supplies, and other vitals, such as heat and electrical power.

For station personnel, the early "zero hours" held scant appeal. The village itself was a backwater; Lakehurst held few attractions for young, aggressive sailors. Though the Pine Tree Inn hosted dances (as did the base), by evening's end a visit from the Shore Patrol was likely. Some of the chiefs owned homes in town, but the younger unmarried enlisted men tended to stay away. Officers living "ashore" seemed to favor the Toms River–Lakewood area. Navy men having cars might travel to Seaside Heights or to Lakewood or as far as Asbury Park for diversion. For those carefree yet footloose, there were liberty buses to New York and Philadelphia.

All in all, the isolation was less than splendid. Facilities were scant, primitive, or nonexistent—a source of complaint. "I have just been up to Lakehurst to look the place over," Lansdowne admitted, "and am quite satisfied to remain in Washington for the time being."⁷⁷ In 1929, a new recreation (Welfare) building opened, offering such diversions as regulation bowling alleys and billiard room. One entire end was the Ship's Service Store, "carrying on its shelves not only a full assortment of tobacco, cakes, candy, ice cream and soft drinks, but in addition a full stock of the necessities of every day life." The upper floor held the library, offering more than 1,800 volumes. Here also were a tailor shop and chaplain's office. Most of the building's space, however, was devoted to a lobby and lounge for enlisted personnel. Here dances with refreshments (and guests), and various other functions, were held. Here also the women of those lost with *Akron* were to gather, awaiting word of husbands, sons, and brothers.

Inside the perimeter fence, not all was strictly business. The social swirl was a busy one for most personnel and their families: rounds of obligatory functions, good times, and conviviality. In quarters, officers or their wives hosted dinners and parties.⁷⁸ Bridge clubs thrived. As with theaters everywhere, the movie program—five evenings weekly—was well attended, with films secured from the Navy Motion Picture Exchange. Bowling season boasted league competition and intramural teams—"Inside Civilians," "4th Division," "Marines," "L.A. Officers"—as did interstation basketball. Picnics for groups could be scheduled; a standard swimming pool would be installed near to the bachelor officers quarters.

As for navy children, there were scores of them. "All of us," the daughter of Ty Tyler recalls, "had what we now realize was a carefree and idyllic childhood. The bunch of us were permitted an incredible amount of freedom to roam the station pretty much at will. . . . One of our favorite haunts was Hangar 1, especially after the *L.A.* had been decommissioned [in 1932] and we could board it, fantasizing about being aviators ourselves and taking long flights. The free balloon baskets were convenient props, too, for letting our imaginations soar."⁷⁹

The house journal—a mimeographed weekly of changing name—served the interests of all, proffering news snippets of the station and its programs as well as league standings, announcements, promotions, personnel changes, and the like. For the ambitious, a wide range of correspondence courses from BuNav were available.

The athletic officer was a busy fellow—responsible for baseball and basketball leagues that also fielded teams representing Lakehurst. (In January 1930, for instance, its basketball team engaged the army at Fort Monmouth.) A football team was active, the occasional boxing and wrestling match was held. Given that personnel were divided into various departments and divisions, keen competition obtained. Carved from the pine woods northwest of the BOQ, a nine-hole golf course awaited; the first fairway was under construction in the spring of 1934. Though popular, it proved miserable in summer swelter. "Saturday afternoon Scotty Peck and myself braved the heat and went out for an afternoon of golf," Fred Berry recorded. "I darn near expired but enjoyed the exercise." A tennis court and athletic field bordering the West Field are evident on plans drawn early in 1932.

For church services, the Welfare Building served as temporary station chapel. For the "Catholic Church party," a motor coach left the base on Sunday mornings. When, in 1928–29, the American Legion learned of the lack of proper facilities, the idea of a nonsectarian Memorial Chapel was born. A corporation was formed, donations solicited, an architect consulted. The enabling act was signed by President Coolidge—one of his last official acts. A contractor was not procured until 1932, however. A path was cleared to a site on the base perimeter, near the old main gate. Groundbreaking ceremonies were held on 26 June; SecNav Charles F. Adams placed the cornerstone that November. Depression economies halted interior work. Not until the general military buildup in 1939 was the campaign renewed.

For those steering off base, there were the liberty buses—or a shipmate's automobile. Early in 1930, the Central Railroad Company of New Jersey inaugurated a motorcoach service between the station, Lakewood, and Toms River. Local towns and the seashore were popular destinations. "Yesterday I celebrated the completion of the [LTA] course," Berry wrote, "by going fishing with the Bolsters, their relatives and Ann Richardson [wife of Lt. Jack C. Richardson, USN]. We hired a boat at Forked River and went out on Barnegat Bay for the day. We caught enough fish for both the families and I brought a market basket full back for the mess."⁸⁰

Base-related souvenirs were on sale throughout the area. In a small shop near the main gate, Rell Clements, a civilian photographer, sold his work. An enterprising resident, Clements was allowed to roam the reservation (though not its aircraft) and was on hand seemingly everywhere. Until he folded his venture, Clements was to enjoy a brisk business. For years, his work hung throughout the air station.

Don Brandemeuhl did Lakehurst duty in 1934–35, as part of the station's marine detachment. "Recreation was available almost daily: baseball, soft ball (weather permitting), and movies nitely," he later remembered. "There were liberty buses running to Seaside Heights in the summer and to Trenton on Friday, Saturday and Sunday."

Marines at Lakehurst? The commandant is responsible to SecNav; the two services are synergistic. For marines, the primary on-board chore was guard duty. Besides regular watches, marines could be called onto the field—in the dark and cold, highly unpopular duty. "They got stuck for zero hours too," one sailor confirmed. "If they weren't on watch, they were stuck for zero hours the same as everybody else." "The daily routine," Brandemeuhl recounted, "started with breakfast, then roll call and police call (cleaning the living quarters, wash rooms, etc). Then close order drill sessions, schooling on navy laws etc., personal hygiene, military conduct, rifle and machine gun practice, and occasionally mess duty (KP). This we liked since we were paid an extra five dollars a month and were excused from guard duty and the other formations."

Base pay: twenty-one dollars per month. As for getting along, marine and naval personnel interacted "real well. It was only in the movies that they were always fighting."[81]

In sum, the naval air station was representative of any large military facility, anywhere, in those days.

The base has always hosted tenant commands. In the big-ship era, the Parachute Material School and the Aerographer's School (both Primary and Advanced) were there. For the former, the course of instruction lasted twelve weeks, during which the trainee was indoctrinated in the care and operation of navy parachutes. To complete the course, a student was required to jump one of his own chutes—in 1930, from the J-ships. The course of instruction at the Aerographer's School was of sixteen weeks' duration.

Under way, its free balloons (see chapter 3) were deprived of communications save for homing pigeons. Hence, a box was always along—several birds from the Pigeon Loft. Often, their messages were back aboard before word by telephone from absent ballooning airmen. Birds also accompanied the blimps and (early on) *Los Angeles*. By 1930, however, Lakehurst was one of but two naval stations that still raised and trained pigeons.

On 3 September 1925, *Shenandoah* was torn apart in the air—as one commentator had it, "for a few days the country supped on horror."[82] The effect was wrenching. "Of course there was great consternation and anguish, because we all lost some very good friends," Reichelderfer later reminisced, recalling the blow. Operationally, the crash was a lesson in the value of radio communications and foreknowledge of the weather.

Meanwhile, (as we have seen), her lift gone with *ZR-1*, her shedmate was dry-docked—to accumulate helium. Gas cells were overhauled and *ZR-3* inspected and re-

The Parachute Material School, a tenant command, dates from the earliest years of the Lakehurst station. Students concluded their twelve-week course by jumping with a "chute" of their own packing. (U.S. Navy Photograph courtesy Navy Lakehurst Historical Society)

Shenandoah *secured to a semi-portable or expeditionary mast at NAS North Island, San Diego, 10–16 October 1924. Her transcontinental round trip underscored the commercial potential of large airships. (San Diego Aerospace Museum)*

A modification of a 1916 Luftschiffbau *design, ZR-1 was obsolete before commissioning. Its break-up, in September 1925, signaled a need for much greater hull-strength to meet American weather and, as well, rough handling compared to German practice. Following closely upon the loss of ZR-2, the crash realized heated criticism of the program. (Harold Tookey photograph courtesy Dan Hagedorn)*

paired. The date for completion was set for late December; as the inspection progressed, however, the need for further renovations became clear.[83] This work, along with unforeseen delays, would have *Los Angeles* hangar bound until April 1926.

The loss of *Shenandoah* left the United States with no large airship for strictly military purposes. Voices urged that experimentation be abandoned, at least for the rigid type. "The committee fully appreciates the seriousness of the airship situation," the NACA reflected, "and believes that despite all that has been done in many countries to develop airships, they are still rather delicate structures. . . . In the judgment of the committee, the time has come to decide to do one of two things, viz, either to carry on with the development of airships or to stop altogether."[84] The real policy differences over the issue precipitated a crisis. One element, rushing to judgment, clamored to close Lakehurst. Leasing of the base was explored; the budget was slashed about three-quarters of a million dollars. In March 1926, Wicks was called to the bureau to confer with Moffett and Assistant Chief Capt. Emory S. Land on how to keep the air station open for fiscal year 1927. As Wicks remembers, "A very large reduction of civilian personnel was recommended in order to reduce expenditures to meet the new reduced budget and still provide all necessary services for operating the station on an active training and flying status."[85]

This approach was approved. Civilian personnel were cut and a complement of ratings established. Under orders issued by BuNav, these men were detailed to the station's power house, helium plant, and garage. July 1926 saw the base operational, its funding restricted.

Meanwhile, an hysterical press shrilled for an end to spending on "murderous airships." Moffett maneuvered against the storm. In discussions with Wilbur, the admiral took the long view, insisting that development continue, as weapons, the airship—indeed all of naval aviation—was comparatively new. A program therefore was proposed: two fleet-type ZRs, one 1,250,000-cubic-foot training ship, and a base on the West Coast. An artful diplomat, Moffett kept political relations in good order. Publicly committed to the ZR concept, cognizant that good relations with oversight committees defined appropriations, he paid close attention to Congress. Seen against the 1925–26 cries for abolition, his strategic skills are clear. Waging a brilliant campaign, Moffett steered his case through a tangle of hearings and entanglements to the five-year naval aircraft building program of 1926. This was the navy's so-called Five-Year Aircraft Program, which among other things authorized the construction of *ZRS-4* and *ZRS-5*. The naval aeronautical policy ultimately adopted laid down this mandate:

> To complete the rigid airships now authorized with a view to determining by operation, primarily with the Fleet, their utility for military operations. And,
> To continue the development of non-rigid airships, primarily for coastal patrol.[86]

In the pulling and hauling, Congress appropriated funds for an experimental craft with which to prove a "metalclad" theory and design. Its designer, Ralph Upson, a pioneer-engineer with Goodyear, had flown the *A-1* then taught navy pilots the art of flying blimps. In 1929, his vision was to fly as the ZMC-2 (see next chapter).

Amid the clamor, Goodyear-Zeppelin submitted trainer designs. The *YE16* closely resembled *ZR-3*; another, the 1.61-million-cubic-foot *YE15*, would have had to have been erected at Lakehurst or at Cape May (New Jersey) had its design been authorized.[87] But these overtures foundered. In Wilbur's assessment (to Moffett) of the prevailing mood, Congress was so unlikely to fund such a ship he considered it politic not even to ask for it. *Los Angeles* was yet young, viable, operational. BuAer, for its part, never tired of its goal to procure a rigid trainer, but neither did Congress abandon its general disdain for large airships. Authorization to procure was to remain just beyond grasp. Following a long, dark wait, a trainer was finally authorized—in 1938. It languished on the books, debated and unbuilt, into World War II.

In 1925–26, its project stumbling and underappreciated, the naval airship fraternity found itself in a box. With but one ZR in service, pressures were mounting for operations with the fleet—the platform's raison d'être. Yet, by agreement (in 1921, with allied powers), military deployment of *ZR-3* was barred. Any further construction would demand additional trained personnel, not to mention research and development, so as to effect improvements and advance the art. With no other dirigible to hand, *Los Angeles* became, in Bolster's phrase, "not only a fine training ship, but was also used as an important element of our forward-looking experimental development program."

In sum, *Los Angeles* assumed an outsized role, shouldering the project until the next generation of big ships. As for promotional duties, she conferred credibility, lifting airship stock with the press, public, and politicos.[88] The Navy Department was fortunate to have a second ZR with which to carry on. Without *L.A.* (and until another was constructed), the gains realized might well have dissipated.

However, her operators faced a series of fiscal, political, and logistical hurdles that, too soon, would undercut them.

CHAPTER 3

Balloons and Billets: Lighter-than-Air Training

We will build a machine that will fly.
—Joseph Michael Montgolfier (1740–1810)

In the young decades of powered flight, the word "aircraft" embodied airplanes and seaplanes, balloons and dirigibles—the two great principles of flight.

Having foreseen a market, Goodyear Tire and Rubber entered the fledgling aviation industry. As direct U.S. involvement in the European war neared, rush orders were placed for balloons as well as airships. A contract also was negotiated to train naval personnel at a flying field and manufacturing site southeast of Akron, Ohio—site of the firm's headquarters. By war's outbreak, Goodyear was an established brand in aeronautics.

At Fritsche's Lake (later Wingfoot Lake), construction commenced in March 1917. Under Goodyear supervision, contractors worked to clear and level a field from endless mud. A hangar was erected, a generating plant (hydrogen) built adjacent to it. Nearby construction embraced barracks for students, quarters for officers, mess halls, shops.

School convened. Warner L. Hamlen, Naval Aviator (Airship) No. 101, was one of twelve selectees for the original class of student aviators. "I learned that the Navy would provide the trainees and Goodyear would do the rest," he remembered. "This included furnishing the equipment and supplies, providing instructors [Goodyear balloonists] for flight training and for some of the ground school subjects such as elementary physics and meteorology, and responsibility for the day-to-day operation of the field." The syllabus included hands-on familiarization, including balloon and airship erection. The navy staff, for its part, instructed the students in seamanship, navigation, signaling, communications, drill, military courtesy, customs of the service, and the like.[1]

This was the navy's first airship-pilot program. Lt. Joseph P. Norfleet, Naval Aviator (Airship) No. 555, had entered lighter-than-air (LTA) during the war, serving under Lewis Maxfield at Paimboeuf and flying aboard British rigids. "Swifty" came also to Akron, as staff. "We got an awful lot of *very* capable young men trained as aviators in the First World War," he recalled.[2]

Heavier-than-air training was divided into ground school and elementary flight instruction, followed, in turn, by advanced ground school and flight training. Dirigible pilots steered much the same course. Ground school and preliminary training were originally conducted at the Massachusetts Institute of Technology, where the curriculum held a special LTA course. (Two other training sites were established in 1918.) Next came orders to Ohio.

For student aviators, classroom work commenced with

A kite balloon (ZK), 1929. Airships are exquisitely sensitive to atmospheric conditions; pilots must anticipate these effects. And if power is lost, an airship reverts to a balloon. Thus, the science of aerostatics was learned and practiced in balloons. (Lieut. Cdr. Leonard E. Schellberg)

theory of flight, meteorology, signaling and radio, and engines—much of which was practical work. With preliminary ground school well along, Lieutenant Maxfield, commanding the naval air detachment at Akron, issued orders involving "actual flying" of naval aircraft, starting with tethered balloons. Three ascents were stipulated—"primarily to accustom us to the sensation of being in the air but also to give us some experience in reading instruments," according to Hamlen.

Next on the curriculum: free balloons—the essence of aerostatics.

For much of the human career in the air, ballooning had been the sole means of transport. The first *un*tethered manned ascent was recorded in 1783. The science advanced steadily; soon, inflated shapes were being deployed for sport and for exhibition, in the service of science (exploration, data collection) and warfare (observation).

With the advent of dirigibles (*powered* balloons), balloons were assigned a further role—training for duty in airships.[3] Dirigibles are no more than streamlined balloons rigged with control surfaces plus motors for propulsion. Though easy-seeming, flight control is demanding; under way, static condition is subject to constant changes, all of which must be anticipated and corrected for. Upon loss of power or control, moreover, an airship reverts to a balloon. Hence, the science of aerostatics—the operation of displacement aircraft—was learned and practiced in them.

When she was torn from her mooring in 1924, *Shenandoah* confirmed the point. Drifting stern first before a full gale, *ZR-1* was gingerly ballooned—until motors could be started, a measure of control restored. When she was destroyed in 1925, the derelict was free-ballooned down—an impromptu action that spared lives. The small ships also were ballooned for varying intervals without power, sometimes saving the blimp.

A measure of skill acquired, student pilots progressed to nonrigids. The earliest trainer, Hamlen observed, "was simply a modified Curtiss *Jenny* fuselage complete with OX-5 engine, slung under the bag. Skids with small pneumatic bumper bags underneath were used instead of wheels. Three-place affairs, the forward seat was for the mechanic, the after seat for the aide and the center seat for the pilot." As experience lengthened, flights became longer and more frequent, including night sorties. In all, the course required eighteen flights for final qualification.

In September 1917, Maxfield reported the qualification of his first candidates. The secretary concurred with his recommendations: the first eight men plus Maxfield himself were designated Naval Aviators (Dirigible). (The remaining students qualified soon thereafter.) As ensigns, seven—including Hamlen, Naval Aviator (Airship) No. 101—were assigned overseas, directly to France. The rest were held stateside, ordered to stations along the Atlantic coast at now-forgotten air stations: Cape May (New Jer-

Inside the Free and Kite Balloon Hangar. The officer is examining a free-balloon valve; of laminated wood, this was fitted at the apex of the envelope. The man at right holds a load ring. Note the toggles, for attaching basket ropes and net. (U.S. Navy Photograph courtesy Navy Lakehurst Historical Society)

sey), at Rockaway and Montauk Point, in New York. One stayed on in Ohio, as an instructor, until transferred to Naval Air Station (NAS) Key West.[4]

To help prosecute the war, more than six hundred U.S. Army and Navy officers and men would be trained to fly and maintain the B and C-type nonrigids. Akron graduates designated as dirigible pilots reported to Pensacola for advance instruction, kite-balloon pilots to Rockaway, New York.

As wartime stations were closed, Pensacola headquartered the Airship School until 1920, when it was transferred to Hampton Roads. NAS Lakehurst was commissioned in mid-1921. Late the following year, in December, volunteers were requested for rigid-airship training. Forty-one responded, of whom nine were selected—Klein, Pierce, and Rosendahl among these. Accommodating a watchful army, Moffett agreed to let eight Air Service officers take the course. Among those reporting: Capt. (later general) William E. Kepner. A believer in airships, he held an exceptional reputation within the service and came to be well regarded by navy people.

School opened in March 1923—ground training until ZR-1 flew. "Naturally, there is no one qualified to instruct and I suppose they will devote most of their time into digging into reports and papers on file and instructing one another."[5]

Student naval aviators detailed for flying reported first to the flight surgeon, to determine fitness for aviation duty. If found qualified and "temperamentally adapted for duty involving actual control of aircraft," the peculiarities of aerostatics were usually introduced via balloons. Ground School lectures followed. The gunnery officer directed ground training, arranging the outlines of the (evolving) course, the schedule, and the list of instructors. The latter were experienced specialists, officers and chief petty officers.

In the thirties, preliminary plus advanced Ground School comprised about five months indoctrination—theoretical and practical work in aerostatics, aerodynamics, lifting gases, airship construction, free ballooning, blimps, aerology, and engineering. A written examination in each followed plus a final exam. ("I am getting the least bit tired of this constant studying," Cdr. Fred T. Berry [Class VI] was to write of the 1929–30 course, "and particularly so when the weather is hot.") Students maintained a school notebook plus flight notes; these

Personnel man-hauling winter stores at Lakehurst, about 1922. The Aerological Building is at right, the helium gasometer at left. Practically every organization operating aircraft was concerned with meteorological studies as related to air navigation. The New Jersey station played a unique role in fostering awareness; naval aviation and the Navy's weather service would grow together. (Elizabeth Tobin)

Fully rigged, crew embarked, a free balloon is made ready for the "Hands Off" order. As a ground crew of Marines and sailors wait, ACR "Bull" Tobin helps secure a pigeon box in ship's rigging. Note the sand-bag ballast. (Lieut. Cdr. Leonard E. Schellberg)

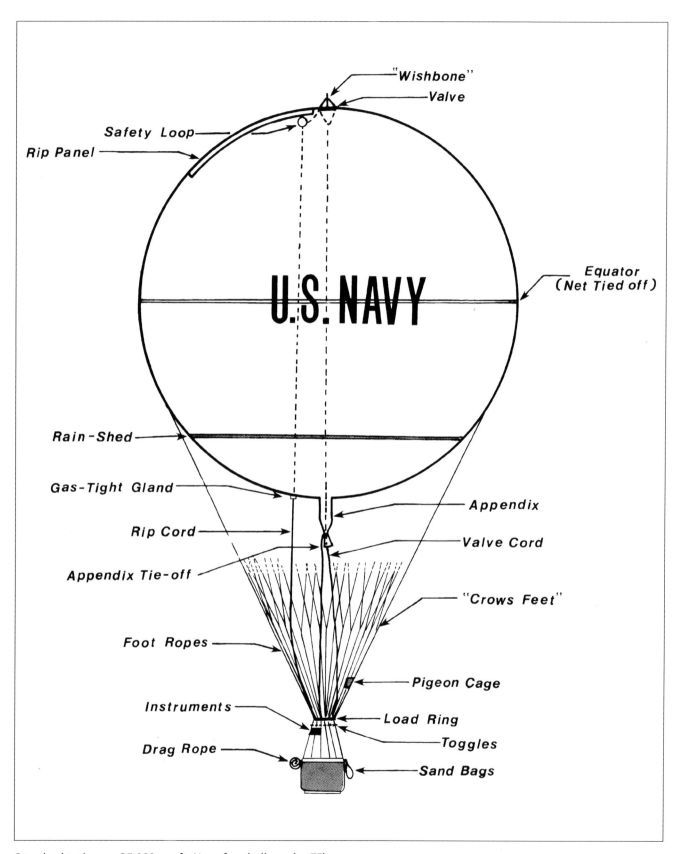

Standard trainer: a 35,000-cu.-ft. Navy free balloon (or ZF).

Table 3-1 Course of training for Student Naval Aviators (Airship), Naval Air Station, Lakehurst, New Jersey (1931–1932). Upon completion of preliminary Ground School, trainees reported for flight training. (Adapted from George H. Mills Collection, Smithsonian Institution)

I. Preliminary Ground School (one month)
   a. Lectures four days per week.
   b. Flight training during time not required by lectures.

II. Flight Training. (three months)
   a. Complete flight training except of final checks and solos in nonrigids and training in excess of 150 hours in rigid ships.
   b. Complete Gunnery, Practical Aerial Navigation, and Outside Reading Course of Advanced Ground School.
   c. Receive practical instruction in Mooring and Handling.

III. Advanced Ground School (four months)
   a. Lectures four days per week.
   b. Flight training during available time not required by ground school.

IV. Advanced Flight Training (four months)
   a. Complete requirements prescribed for flight training.
   b. Complete Advanced Ground School.
   c. Receive special instruction in the following duties:
      1. Mooring Officer
      2. Watch Officer (rigid)
      3. Engineer Officer (rigid)
      4. Assembly and Repair Officer
      5. Communications
      6. Nonrigid operation (extended flights)
      7. Aerological Officer

might be called for at any time and were submitted for grading as part of final qualification.

Classroom time was supplemented by practical works and tours at the hangars. Trainees took part in actual assembly, rigging, inflation, ground handling, and maintenance activities. Whenever ZR handling was scheduled, a zero hour was set—and all students were required to be ready for duty. Permission to leave the air station was seldom granted while any ZR was flying or when flight time in a nonrigid was possible.[6]

Lectures tended to occupy the winter months and other nonflying weather—a plight hardly rare at Lakehurst. "There isn't a darn thing going on around here," Berry jotted in June 1930. "The *L.A.* got us up at 4:30 AM and then the J-ships went out. The wind came up very strong and we brought them in. So here I sit trying to study."[7] Lecture instruction was cleared away early, though important topics might consume as many as twenty-five classroom hours.

All in all, Ground School held few difficulties for the conscientious. Duration of the training for officers: approximately one year. Class IV, for instance, reported aboard in April of 1927; the following spring, its students were signing *Los Angeles*'s log. With Commanders Alger Dresel and Berry on its roster, Class VI commenced its LTA training in mid-1929. Thirteen months thereafter, "Freddy" Berry secured his designation.

Follow for a moment Berry's course of instruction. Detached from duty at the Naval Academy on 15 June 1929, Berry reported to the commanding officer, NAS Lakehurst "for temporary duty under instruction in lighter-than-air craft." His inaugural flight came nine days thereafter—two hours and twenty minutes in the basket of a free balloon (ZF, in navy parlance), its pilot Lt. (j.g.) Wilford Bushnell. Four hops were promptly logged with the nonrigid (or ZN) J-3 (Tyler as pilot) along with a second ZF hop—at night, under Settle's master tutelage. Berry's aggregate time as July opened: twenty hours, fifteen minutes. Sorties

Trainers for the next generation of ZRs: (top) secured in berth; (right) on Lakehurst's East Field and (bottom) looking west from the hangar roof. (Navy Lakehurst Historical Society; Capt. Gordon S. Wiley; CPO Moody Erwin)

with *Los Angeles* did not come until mid-August, after which his hours jumped severalfold. Berry's balloon solo took place on 12 November. By month's close, this officer candidate had logged 186 hours, thirty-six minutes aloft.[8]

Kite or captive-balloon time does not appear until February–April 1930, during which eleven ascents were recorded in his aviator's flight log book. The longest was two hours and ten minutes, the shortest flight a mere fifteen minutes.

As the training advanced, further ZF plus powered hours were appended: on *J-3* and *J-4*, with *ZMC-2*, on board *ZR-3*. He soloed on 12 June with *J-3*—almost a year to the day since reporting for temporary duty under instruction. On 7 July, the qualification board (Pierce, Rosendahl, Clarke) had candidate Berry before it. The board found the commander "fully qualified in all respects." Total flight time at qualification: 306 hours, forty-five minutes.

At Lakehurst's LTA Training School, enlisted graduates were designated as "Qualified for Rigid Airship Duty." Indoctrination included lectures in elementary aerostatics and aerodynamics, aerology, airship nomenclature and layout, lifting gases, and other pertinent topics. Pupils went on to specialize the seaman (rigger), metalsmith, or machinist's mate ratings. The course required about six months. On 20 December 1930, for instance, a class of forty-four were graduated. Tragically, a half-dozen of its members would die with *Akron*.

To develop and refine LTA tradecraft, ballooning was essential. In 1921, three free balloons were approved for Lakehurst. (Pensacola had received its first five years earlier; in November 1918, sixty-three nonrigid and fifteen balloon pilots had qualified.) During the salad days of lighter-than-air, a dozen or more were in inventory. The standard training type held thirty-five thousand cubic feet of hydrogen, its basket three to four students along with a pilot-instructor.

The kite type was deemed mandatory as well. Kite balloons were elongated in shape and fitted with lobes for aerodynamic stability. Deployed throughout the war, the type had been used principally for artillery spotting and, at sea, for observation work (searching for mines and submarines) under tow by capital ships.[9] Temperamental beasts, these "elephants" had a tendency to swoop; more than one had dove in, sometimes killing the hapless observer. When airplanes gained reliability the type was deleted from fleet units and the leftovers transferred to the Army Air Service, the marines, and Lakehurst. Prospective officer-pilots suffered ten ascents and ten hours in their

Table 3-2 Ground School requirements, Student Naval Aviators (Airship), Lakehurst Class VIII (April 1932). (George H. Mills Collection, Smithsonian Institution)

Subject	Weight
COMMUNICATIONS	1
BALLOONS	1
INSTRUMENTS	1
PHOTOGRAPHY	1
GASES	2
AEROSTATICS	4
AEROLOGY	6
ENGINEERING	2
AERODYNAMICS	3
ORDNANCE	1
M. C. & D.*	4
NON RIGIDS	3
MOORING	3
RIGIDS	6
NAV. & FLIGHT INDOC.	4
GENERAL EXAMINATION	8
total	50

*Materials, Construction, and Design

dangling baskets, enlisted men at least five hours. Maximum height: 1,500 feet. The syllabus also called for practical kite work—inflation, deflation, ground and winch handling, assembly, and rigging.[10]

The first kite arrived on station in 1921. An R-type kite (37,500 cubic feet), the balloon had lain in storage at Rockaway Beach, New York—a (closing) seaplane and LTA patrol station. It arrived complete with all accessories, including an army-type balloon winch mounted on a truck trailer.

It was thoroughly unpopular yet mandatory; students were sent up to get little more than the feel of the thing. "Still bad weather with no flying," one student reported, "except that they did send the kite up this morning and I finished my time in that contraption and for good measure gave them fifteen minutes extra. Barring something unforeseen that was my last ride in that."[11] Upstairs, men got a workout. Swaying to its cable, the basket was anything but pleasant on all but the calmest days and could sicken stomachs. Pilots had scant control. But in calm, the ride could be enjoyable. That master balloonist, Tex Settle, even did paperwork while aloft, as he did on a September afternoon when, close aboard, *Shenandoah* cast off on her fatal flight to the Midwest.

An anachronism, the kite faded to footnote status, yet it afforded ancillary services. The type was exploited for the testing of parachutes.[12] Near the field, airmen used it to gauge wind velocity and direction. (One only was ever up.) On occasion, weather instruments were lofted, to record vertical gradients, especially temperature.[13] In fog, a kite was supposed to mark the (hidden) base.

Still another, unsanctioned service was available. Flying multiplies hazard. Hence the institution of flight pay—conditional upon four hours or ten hops per month. At the Lakehurst base, personnel holding flight orders exploited the type to log needed hours. A flight was defined as a thousand feet up and down. Loaded with the needful, therefore, its basket was allowed to rise dutifully on its tether ten times.[14] Settle and others would advocate the abolition of flight pay, convinced that it and the shore duty attracted deadwood into the program. In the meantime, flight-status personnel enjoyed the perquisite.

For officers, the training syllabus called for seven free-balloon sorties. At least one had to be a night flight, another a solo of one hour's duration—for which the student took charge of inflation and rigging. For enlisted men, three hops were stipulated.

Basket time was dictated by weather; most sorties were logged when the weather map was "flat"—that is, a slight gradient out of the east or southeast. In summer, the sea breeze could push inland as far as the Delaware River, fifty miles to the westward of the reservation—ideal for ballooning. The operations office was notified of promising forecasts, after which instructor and students assembled for briefing. At the hangar, ship's components were brought together and the envelope spread out, unrolled, and arranged for inflation.

Originally, a large hydrogen generator building had been provided. In 1922, however, the decision was made for helium, so a purification plant for the heavier gas was installed in the older building. To keep down losses of the gas, balloons continued with hydrogen. Accordingly, a small hydrogen plant was erected near the free and kite balloon hangar. With the danger from stray gas ever present, precautions were necessarily extensive. The plant itself—cheaply made (wood) and partially enclosed—was fitted with gastight wiring and switches enclosed against sparks. Open flames, matches or cigarettes were *verboten* where hydrogen was generated, used or stored. Though these rules were violated, accidents seem to have been few. It was a wise sailor who retained a healthy respect for "Mr. Hydrogen."[15]

The needed gas was liberated from water indirectly, by reacting ferrosilicon with a strong solution of hot caustic soda. These ingredients were readily available and, with due precaution, safely handled.[16] Production was slow, however; inflation times (two or three hours) were hostage to the rate of generation. Yet the method met the need, yielding a nearly pure product, and was comparatively inexpensive—a distinct advantage with funding a chronic consideration.

Free-balloon hops, Calvin Bolster remarked, "were always an adventure with no certainty as to where or how each would end." At the mercy of atmospheric whim, chance held sway: no two flights were exactly alike. According to Capt. M. M. "Mike" Bradley (Class IV), "free ballooning was one of the more exciting parts of the training, of course."

Assembled and fully rigged, the balloon was walked out, its passengers already embarked. At the point of takeoff, duties were allotted: sandbag ballast, ship's log, navigation (a portfolio of road maps) an assistant pilot to man the valve cord. Bags were removed until approximate equilibrium was attained. Meantime, the pilot-instructor calculated the ballast needed to clear any obstacles to windward. During a lull, orders to "Stand by to weigh off—Hands off—Hold" were barked. Weigh-off complete, the craft light enough to rise, its appendix was opened—to serve as an automatic valve during the ascent. When in all respects ready, the pilot sang out, "Let go!"—after which the handlers released their holds.

The balloon eased silently, eerily away.

At a safe altitude, the ship's log was opened. The crew settled into an informal routine. Devoid of propulsion, the aircraft promptly assumed zero relative airspeed. As students acclimatized, the pilot explained the fine points and issued instructions or suggestions as he rotated the conn among them. Students tried maneuvers or watched as the pilot handled the balloon. Mission? To find and exploit favorable winds and altitudes, loiter in level flight, prolong time loft. Below, most of the south Jersey backcountry was a flat mosaic crisscrossed by roads, rails, and fire trails—ideal terrain. For local daylight hops, a chase truck drawn from the station garage followed its wandering charge.

Balloons—hydrogen, helium, or hot air—are delicate to control and exquisitely sensitive to changes in weight, temperature, humidity, pressure. A slight change realizes a large effect. A handful of sand over the side (for instance) had an immediate effect on navy altimeters. To descend, the valve cord was pulled, the seconds counted off. Novices found that the craft was seldom in an absolutely stable condition. Future commanders had to recognize true aerostatic condition at all times, anticipate changes, and apply the proper control at the proper time, checking undesirable movements at the start, not waiting for the ship to "run away"—that is, gather undue up or downward momentum. Inattentive pilots found that their courses decayed to a series of ascensions and descents, the corrections for which demanded wasteful outlays of ballast and gas. Turbulent air raced the heart, certainly, complicating control. Above broken terrain, convection currents are quite normal. Over a plowed field, balloons tend to rise; conversely, above a wooded area, a downward impulse is imparted.

Navy ballooning was more than practical aerostatics, much more. A companion to the wind, the basket offered a classroom in applied meteorology. "Certainly," Reichelderfer reflected, "there's no better way to learn practical meteorology then to be a balloonist or an airship operator. You not only are exposed to it in a more lurid and lucid way—more tangible way—but, as a matter of operating performance and safety of the ship, and of life, you had to know what was going on in the atmosphere."[17] Under way, a *constant* appraisal of clouds, winds, temperature, terrain (and their tendency to change) was advisable. Trainees learned that the atmosphere is no homogeneous medium but instead harbors a highly variable (often hazardous) structure, vertical currents and gusts particularly.[18] The thoughtful student learned at last to integrate its inventories of cues and intuit a working knowledge of weather.

Ballooning is a scenic, unpredictable, highly social experience. Night air is comparatively stable, demanding minimal use of valve and ballast. Granted a bit of luck, a skilled pilot can log astounding distances. Settle and Tyler

Preparing for launch at the west end of the big hangar, 1940–1941. (Rudy Arnold Photograph courtesy Lieut. Comdr. Leonard E. Schellberg)

Lieut. T.G.W. Settle, July 1932. Naval Aviator (Airship) No. 3350, a gifted balloonist, and highly esteemed by all hands, "Tex" would command J-3, instruct at Lakehurst's ground school, and serve aboard ZR-1 and with Los Angeles. *(U.S. Navy Photograph courtesy Navy Lakehurst Historical Society)*

were supremely accomplished balloonists. ("Tex," one admirer said, always "went quite a distance.") In February 1927, Settle reached Lisbon Falls, Maine, in 21.5 hours—a sortie of 478 miles. In April 1936, free balloon number 8924 was let go with Tyler and two assistant pilots aboard. Seventeen hours later, the trio executed an emergency landing at St. Christine, in the province of Quebec—forty miles beyond the international border. Returned to duty, Tyler was ordered to explain. "Due to strong winds and mountainous country," the lieutenant reported, "unable to land prior daylight."[19]

Gliding in utter stillness was unforgettable. Flights later in the training particularly, and night hops, were characterized by unstructured, easygoing air. En route to nowhere, the basket's chaos of ballast, sandwiches, and thermoses squared away, all hands settled in. As the hours added, many a sea story was exchanged. Adm. Harold B. "Min"

Miller, an airplane jockey with the hook-on units of *Akron* and *Macon*, elected to take the Lakehurst course. "It was the fatigue of all-night flights combined with exhaustion of the [corn or gin] bottles that tended to bring most of us down the following morning!"[20]

Not surprisingly, there were hazards. Telephone and power lines, thunderstorms, and hard landings introduced adventure as well as accident. One crew reached the Adirondack wilderness and had to walk out—in winter. When another made an intermediate landing close by a Jersey hog farm, the animals stampeded. Enraged, farmhands began shooting at the intruder. Anxious to escape, the pilot ordered one man to drop a bag. Instead, *each* student did so—with instant result. Relieved of 120 pounds, the balloon shot to about ten thousand feet, the force of its rise virtually immobilizing the crew. Shaken and short on ballast, the crew landed the basket shortly after.[21]

Notwithstanding, the diverting aspects of navy ballooning were fondly recalled; airmen later prized these hours, about which all hands seemed to have stories: "dropping in" on startled civilians, "bouncing" on clouds,[22] irate farmers, the serenity and camaraderie of the overnights.

No balloon can float indefinitely. For the landing, wind, ground speed, and direction were carefully assessed, along with rate of descent. Once down, the envelope was flattened, its components disassembled and packed, arrangements effected for the return. As the valve was disconnected, though, static sparks could ignite residual hydrogen. "This would cause a minor explosion," Bolster recounts, "and I have a good friend

Table 3-3 Weights and lifts for the 9,000-cubic-foot and 35,000-cubic-foot Navy free balloons.

	9,000 cu. ft.	*35,000 cu. ft.*
Envelope	184 lbs.	225 lbs.
Net	49	101
Basket	66	97
Load Ring	8	20
Valve	5	15
Valve Cord	4	4
Drag Rope	30	30
Instruments	7	7
Pigeon Box (four birds)	9	9
Weight Empty	362 lbs.	508 lbs.
Gross Lift*	676 lbs.	2629 lbs.
Useful Lift	314 lbs.	2121 lbs.

*Hydrogen at 100% purity, standard conditions

BALLOONS AND BILLETS

The balloon's essentials packed inside, a basket is loaded into a chase truck for the return. A pigeon has been shoved off with a report of the landing. (Rudy Arnold Photograph courtesy Lieut. Cdr. Leonard E. Schellberg)

who temporarily lost a very handsome mustache which was literally burned off when this happened."[23] Seemingly empty fields could swarm abruptly with the curious, some with lighted cigarettes.

Flight training advanced to powered craft. The BuNav syllabus required a minimum of twenty sorties and twenty hours in nonrigids. Prior to his first instruction hop, the student was obliged to inspect his craft and familiarize himself with its various parts as well as the rules of the ship. "He shall practice moving the controls and handling valve ballast lines," a syllabus notes, "so their feel may be coordinated." Under way, trainees spent one hour as radio operator and one as mechanic. Having logged ten hours' instruction, students were checked as to steering courses, keeping constant (envelope) pressure, and flying a constant altitude. Five satisfactory landings under varying conditions were stipulated. Before qualifying, two flights had to be logged without an instructor or qualified pilot on board.

However, statements of policy stressed the rigid type; until 1934, the general naval policy was to operate only such nonrigids—two to four—as were needed for training. The department had practically ceased small-ship operations.

The B-type had come after *DN-1*, followed by the larger C ships. Designed for antisubmarine work, the latter carried depth charges and boasted twin power plants—to lessen the chance of breakdown and to improve endurance for convoy escort. By Goodyear's account, aircrews held the Cs in high regard. With a bit of luck, indeed, *C-5* (Scotty Peck, copilot) might have been first aircraft to cross the Atlantic. But pilots had lodged criticisms. These centered on the control car—crowded, cramped, awash in noise and propeller blast. Hence the D-type; its envelope was enlarged and the car wholly redesigned, most notably by removing the fuel tanks. The D-type was fitted with two Union engines, and these were mounted close together on outriggers at the stern. Fuel was carried in tanks hung from the equator of the bag, a change that bettered in-flight comfort but griped maintenance crews. Envelope volume: 190,000 cubic feet. Authorized in July 1918, four of the six built were turned over to the army.

The last design airborne prior to ZR operations: the J class.[24] In the early months of 1919, the CNO had requested the preparation of plans for a twin-engine coastal ship with a boat-type car suitable for water landings. The design was worked up jointly by BuAer and Goodyear, the contract being let in 1921. *J-1* took to the air in 1922. At Lakehurst, the first liftoff was logged on 16 May 1924.[25]

ZR-1 was expected to absorb much of the program's energies, so no further nonrigid procurements were projected. Progress in design and operational technique was ceded to the army, and to Goodyear-Zeppelin. (In 1933, naval policy as it pertained to the type still read, "To build only such nonrigid airships as may be necessary for training purposes.") Until its deflation, *J-1* was the only blimp in service until *J-3* and *J-4* reached Lakehurst in 1926–27. The Js were wonderful for the purpose—stable, easy to fly (though *J-1* was temperamental), and forgiving of inexperience. Further, their open cockpits were

J-4, a nonrigid airship or ZN. The J-type logged yeoman's service for training and experimentation. Excepting ZMC-2, J-3 and J-4 were the only blimps in inventory during 1926–1931. Prior to flight tests using Los Angeles, *in April–May 1926, the only airship pressure-distribution investigations in the U.S. had been made on C-7. (Mrs. Charles M. Ruth)*

ZMC-2 berthed in the big room, Hangar No. 1. NACA tests on specimens cut from her hull after five years' service, in 1934, showed that corrosion had not appreciably lowered the strength of her Alclad sheet. Note the fabric envelope of J-3, under inflation. Thanks to her metal skin, ZMC-2 gained or lost superheat rapidly, making the "Tin Ship" a notorious trainer. (CPO Moody Erwin)

exhilarating, and reminiscent of wartime patrols.[26] Conceded to have had excellent flying characteristics, the J-ships were exploited also for limited experimental work.

A prototype for "metal airships," *ZMC-2* first flew in August of 1929, its pilot "Bill" Kepner. Following trials, she reached Lakehurst that September. The brainchild of Ralph Upson, *ZMC-2*'s hull was composed of an aluminum alloy sheet cladded with a thin skin of pure aluminum on each side. "Alclad" offered high corrosion resistance along with high tensile strength. Hull internals—five major structural rings—could hold shape without pressure; during handling or when under way, however, *ZMC-2* had to be pressured so as to withstand the stresses imposed.[27] Rather than a four-fin configuration, the design had eight fins spaced equally about the stern.

The "Tin Ship" could bully the deck watch. During daylight, *ZMC-2* could pick up more superheat than the fabric ships, going to pressure height *in berth*; at night, the pressure might drop too low, causing the skin to "oil-can" (buckle)—something that could be heard throughout the hangar. The watch would be obliged to start the auxiliary blower, to return ship's internal pressure to normal.

ZMC-2 was intended to help prove the feasibility of the all-metal concept, but limited funds and commitment to the fabric type was to defer development. Pressed into training service, *ZMC-2* was a notorious teacher.[28] With a short, roundish hull, the unique control surfaces and metal skin, she proved tender to the controls—becoming close to uncontrollably light after but a few hours. "Many a pilot," one airman recalled, "made many a pass at the ground crew to get it down."[29]

Mere weeks after delivery, the stock market crashed, ushering in the Great Depression. At Lakehurst, the inventory would not change until 1931; *K-1* was the navy's first nonrigid with its control car flush with the envelope or bay.

The Navy Department was not the sole force for progress in this era, however limited its small-ship deployments. The army had a program of its own, and (as we have seen), the Air Service was nursing grander aspi-

BALLOONS AND BILLETS

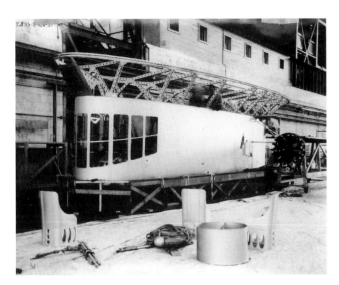

Control car, K-1. Delivered in 1931, K-1 was the Navy's first nonrigid with its car suspended flush with the envelope. (National Air and Space Museum, Smithsonian Institution (SI Neg. No. 2000-9496))

A training airship (or ZN), L-1 was the first nonrigid of her class. To help meet the demands of the 1941–1945 war at sea, a total of twenty-two L-ships would be manufactured. (Author)

rations.[30] Goodyear was also deploying a squadron. Its fleet was barnstorming the country, keeping airships before the public as it logged experience, refined concepts, and trained airmen for the commercial rigids in prospect.

Eventual qualification in the big ships was the express aim of the Lakehurst course. For student-officers, a minimum of fifteen ZR flights and two hundred hours were required. Further, satisfactory performance of duties at various stations—during ground handling as well as under way—was needed for qualification. For this service alone, *Los Angeles* proved invaluable. Operational doctrine begun with *ZR-1* was exercised with *ZR-3*; between 1925 and 1931, moreover, *L.A.* provided much-needed refinement and continuity to the program.

Hence, typically, *L.A.* operated with trainees aboard. When a zero hour was set, those on the flight list reported to the executive officer for duty assignments during handling and flight. (All other students reported to the mooring officer.) "It didn't take 'em long to get us into the *Los Angeles*," Mike Bradley said. "Matter of fact, you hardly knew when you were going to go. You'd just 'stand by' down there, and if they had enough lift for a couple of students, at the last minute they'd name a couple and you had to jump aboard. Sometimes there were more than a couple." For Fred Berry's inaugural sortie, the flight list held fourteen names. "Today has been a flying day and I have had my first hop on the *L.A.* It was certainly a marvelous experience and I would not have missed it for anything[.] We left here about 11 AM and landed at 6 PM. Went to Philadelphia, Atlantic City and then over New York City flying down the Hudson. I wish you could have been along. You would love it."[31]

At the hangar, students attended all musters, respectfully assisted in morning weigh-offs, and helped prepare the aircraft for flight—a process requiring meticulous attention. As well, each was obliged to familiarize himself with ship's condition and its orders, the names of the crew, its organization. In class, written examinations awaited. "Finally I have the old books out once more and am burning the midnight oil," Berry wrote. "I expect to get my final examinations the last of this week. . . . Will sure be glad to get them behind me."[32]

G-1 on the west landing field, 1937. This ship, the former Goodyear "Defender," was purchased in September 1935. Following modifications, G-1 was delivered to the Navy at Lakehurst. (Lieut. Cdr. Leonard E. Schellberg)

To finish up, a qualification board passed judgment, making recommendations as to whether wings should be awarded each candidate. With the "Naval Aviator (Airship)" designations conferred and new orders in hand, graduates reported for duty aboard a commissioned ZR, one of the nonrigids, or to the station command. For enlisted personnel, duty was assigned according to the available flight billets and to those billets assigned to the station.

For their denizens, life within the great hulls centered upon the keel corridor or "catwalk." Arrayed along this tunnel-throughway were ship's vitals—fuel and ballast tanks, water and fuel distribution lines, control cables, oil, storage, living quarters, bunks, head.[33] Along this passage (and the "ratwalks" to the wing cars), all was raw function—a spare, enclosed world of routine. Above and around, the bulging cells pressed against the wire bracing and netting. Beneath, the cover suggested safety. This was misleading: save for a few diagonal wires, all that interposed was doped cotton cloth. More than one sailor misstepped and punched through. On one sortie, a large hole was reported beside the keelway. Man overboard, aft? "We were relieved," the skipper related, "to find the civilian who had punched the hole still aboard. He had gone exploring aft, and put a foot through the envelope into thin air. It hurt neither the ship nor him, but it did give him the thrill of his life."[34]

At specific locations along the keel length, water ballast was distributed port and starboard. *Los Angeles* carried twenty-two bags: fourteen service bags of three-ply rubberized fabric holding 2,200 pounds, plus "emergencies" near bow and stern of about 550 pounds each. A rigger sounded the levels then compared his readings with a scale on each bag. The results, in turn, were either phoned in or reported verbally to "control," where they were transposed onto a chalkboard record of the water remaining and its distribution. Near to hand, rows of toggles regulated releases. The amount dropped was estimated by pulling, then counting seconds—or, for helium, by how many calculated pounds the lifting force of one of the cells had been decreased.

Patrolling riggers saw to it that the cells were kept clear of ship's controls, netting, and structure. In the middle of each bay (thirteen on *L.A.*), a graduated tape was attached to the inner circumference of the outer cover. When a cell was 100 percent full, one could "just barely" see the bottom of its tape. As volume decreased, the excess fabric—normally blossomed out—would fold in over the keel, thus exposing more of the tape. This was read once an hour and logged (along with ambient conditions and altitude, from instruments at frame 100).

The fuel system resembled that for water—a main fuel line running fore and aft, fitted with a hose connection at

The keel corridor of Los Angeles. *The view is aft, just forward of Frame 55. Here Cell No. 4 fills the bay overhead. Note the water-ballast line (right); used for filling ballast bags, this ran along the top-port keel girder from Frame 185 (an outside host-connection) aft to Frame 25. The planks were for emergency damage-control repairs. (San Diego Aerospace Museum)*

The flushing-water (portside) and drinking-water tanks at Frame 145, above the main car. The climbing-shaft ladder is at left and, behind, the gas manifold. Of canvas, the manifold connected thirteen of the fourteen gas cells lifting ZR-3. Opposite, following the keelway, is ship's water-ballast line. (Charles E. Rosendahl Collection, University of Texas)

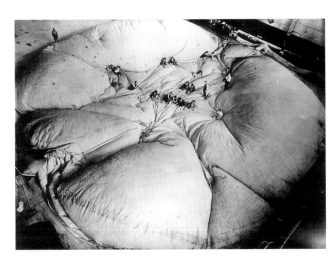

A gas cell from Los Angeles *under repair. Approximate cost for one of these fragile chambers: about $18,000. (National Air and Space Museum, Smithsonian Institution (SI Neg. No. 2000-11191))*

the bow. The main connected groups of sheet-aluminum tanks suspended upright by wires on opposite sides of the keelway. Some were spaced well forward and aft, so as to provide good trimming moment when fuel was shifted. On *Los Angeles*, each tank held about 650 pounds of gasoline. Fuel was usually grouped in threes, with the center unit a "slip tank"; in emergencies, it could be cut free (to drop through the cover). The first lieutenant directed

Fuel-tank group at Frame 135. The center unit is a slip tank—quickly dropable in emergencies. Running fore and aft above the tanks are ship's telephone and other electrical wires and, keelward, the four control cables to the movable tail surfaces. (Charles E. Rosendahl Collection, University of Texas)

which tanks were to be used and, as necessary, requested shifting of fuel to maintain trim. Reserve was checked by sounding each tank through its top, with the results logged by the engineer's keel watch. (Concurrently, the main line was checked for leaks.) Near the power cars, a gravity tank above the keel fed each Maybach. These were always kept full; it was the duty of the keelman (a chief petty officer) to keep these replenished, directing and redistributing fuel using retractable windmill pumps cranked out into the slipstream.

Moment is the product of a load and the distance from an axis. Unequal distribution of lift and load caused the introduction of shearing forces and bending moments. The load of the ship was, of course, transmitted to the frames by the keel corridor structure. In operating, calculations—to find the maximum shear and bending moments—were made frequently to determine the best loading plan. Both fuel consumption and water distribution were checked once every four hours—each watch—so as to keep the combined moment of fuel consumed and water made equal to the original fuel moment. The German routine had been comparable.[35]

As must now be apparent, underway duty was an demanding regimen—a grinding, relentlessly repeated, and unforgiving routine. No idlers were aboard. Thanks to training, discipline, and the chain of command, the big ships were a structured, bound-together world of interdependence.

For climb-outs, all hands manned landing stations, with an officer (customarily the first lieutenant) standing by the ballast controls. *L.A.* had no loudspeaker system, so "the word" was passed via telephone to four keel stations.[36] From there, orders were passed along by the keel watch, verbally, man to man. In the main car, the junior officer of the watch passed the word to the officer spaces. During ascents a careful inspection was essential, especially in the keel. ZRs were complex machines. Breakdown or damage might occur to any of a myriad of systems: gas cells, valves or ballast, steering controls, telegraph gear, engines, fuel. Hence, keelmen were at their stations, ready for orders. During the first hour or so of each flight, crewmen carefully inspected the entire ship. At altitude, the radio aerial, with its "fish," was reeled out, the after engine cut in, and, if needed, the water-recovery condensers.

The control car was nerve center, or like the bridge of a surface ship. The rudder wheel was mounted at its forward end. Here also—port side, facing outboard—was the elevator wheel. ZRs were low-altitude craft. The ele-

Aviation Machinists Mate 2nd Class (AMMC2c) Frank T. Karpinski on the ladder to No. 1 power car, at Frame 55. (ADC John A. Iannaccone)

Lieut. Cdr. Bertram J. "Bert" Rogers, Naval Aviator (Airship) No. 3413, on the bridge of ZR-3 (starboard side). Note the engine telegraphs; the OOD signaled speeds to the power cars, and machinist's mates on watch executed the signals. The telephone is to the keel corridor. A graduate of Class IV, Rodgers joined ship's company in 1929, was engineering officer aboard Akron *and, in 1934, executive officer of* Macon *(ZRS-5). (National Archives)*

vator, since it controlled up and down movement, was of paramount importance and its operator a respected personage, as the *Rigid Airship Manual* reminds: "It is urgently necessary for the commander and the elevatorman to be on the lookout all the time, and to carefully watch what the ship is doing dynamically, and to inquire with the most meticulous care into every change in the dynamic performance of the ship."[37] The skipper, for his part, kept a wary eye on the inclinometer and on the operation of the elevators as indicators of static equipoise.

Bumpy air demanded an expert elevatorman. When the going got rough, a relief was called. (If conditions were serious, the crew was sent to landing stations.) In this regard, Reichelderfer had a vivid memory. "Whenever we'd get into rough air—turbulence, cumulus, thunderstorms—we'd always have a good elevatorman. That was the most important job, because it kept the ship on an even keel. And there were some *excellent* chief quartermasters and enlisted elevatormen. But whenever the ship would begin to go up and down more than usual, the skipper would jump out of his stateroom, go to the control room, size up the situation. 'Send for [Chief Warrant Officer] Buckley!'" The man was large and powerfully built, and his skill and touch were "tops."[38]

Bridge routine resembled that afloat. The officer of the deck (OOD) conned the aircraft by commands to the steersmen and by the engine telegraphs. He notified the keel watch of projected changes in altitude of five hundred feet or more (or if the ship was near pressure height). The OOD watched ship's changing condition, the steersmen, the weather; he also kept the log. When routine changes in course or altitude were made, he applied to the navigator on watch. Two officers navigated and kept the watch officer informed of the ship's condition in all respects. Thanks to weather charts and radio data, and a weather eye, the captain knew what changes to expect. As for the navigator, it was his duty to note every clue as to changes in winds. "He must at all times [the *Manual* continues] be able to give information on wind force and direction, and must clearly record the time the last wind correction [to the ship's course] was made. Strict orders must be given for regular marking of the ship's position on the chart, according to bearings or by dead reckoning." The executive officer stood watches as navigator and supervisory pilot. (If the "exec" was himself an engineering officer, a navigating officer was detailed to stand watch with him.) The assistant engineer officer handled most of the practical in-flight problems, such as weight and trim, under the supervision of the engineering officer.[39]

An airship, most assume, descends by valving lift. In fact, however, "the normal way of landing *Los Angeles* [or any airship] after a trip is to fly right in. Heading into the wind, the motors are slowed until the ship is nearly stationary and then ropes are dropped to the ground crew. They haul us down by main force against the buoyancy of the ship. Occasionally, some of the ground crew get a shower bath when the ship, coming down too fast, checks its fall by releasing a little of its water ballast."[40]

Big-ship piloting was an intricate operation, requiring that trim and lift be constantly observed. It demanded focus and rigor, unceasing alertness, and swift, adept decision making—controlling, trimming, ballasting. Reacting to patterns of experience, submerged in their task, the captain and his officers became perfectly at home in the arts of both static and dynamic ballooning. As with all vessels, final authority and responsibility lay exclusively with the captain. A rigid airship was not flown by any one man; it was commanded.

In contrast to airplanes of the day, lengthy missions were standard duty. Even "local" runs consumed from six to eight hours, with full days hardly unusual. *Los Angeles*'s longest operation (to the Fort Worth mast, in 1928) had the aircrew aloft for slightly more than forty-eight hours. In 1933–35, *Macon* ranged nearly six thousand nautical miles on strategic searches. Her longest sortie, in July 1934, lasted eighty-three-hours—more than three days in the air.

Ship's organization was tailored accordingly. The enlisted force comprised three sections: two flying, one at the hangar. Under way, flight sections alternated watches—four hours on, four off. Each section was so divided that it could man all stations and make minor repairs—that is, it had sufficient men for the rudder, elevator, and keel watches. Similarly, each engineer's section had a man for

An aviation chief rigger (ACR) at *Los Angeles's elevator, portside. Just above the wheel are two inclinometers. An American altimeter is mounted at center, a German altimeter alongside. At top-right, beneath ship's outline, black toggles operate emergency bags fore and aft, white the large 2,200-lb. bags spaced along ship's keelway. The blackboard held a running record of water-ballast at various stations, fuel consumed, and water recovered ("made"). Note the flashlight. (National Air and Space Museum, Smithsonian Institution (SI Neg. No. 78-8050))*

Sailors share an idle moment. With 54 officers and men assigned, a normal flight crew comprised two watch sections—about 40 men. Rigid airships were not flown by any one man; they were commanded. (Cdr. Claude Roof courtesy Navy Lakehurst Historical Society)

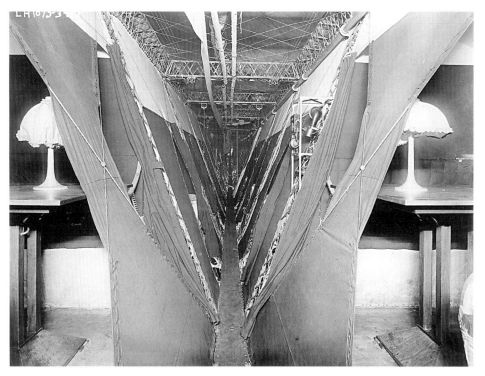

Looking aft along the tube-like keelway, near the crew spaces. This was the main avenue of communications between the bridge and all stations, including the power cars. (National Air and Space Museum, Smithsonian Institution (SI Neg. No. 2000-9083))

each car, with another up in the keel checking fuel and related matters.[41] When not flying, the deck force was divided into special details: gas cells, structure, outer cover. In berth, *L.A.* was weighed-off daily by the OOD and the hangar watch, after which a meticulous record of the distribution of weight on board was submitted to the first lieutenant.

The various demands of flight duty were considerable. Take berthing. *Los Angeles* had berthing space for half the crew. Bunks were fitted off the keel in small tentlike enclosures ranged along both sides of the keel—officers forward, just aft of the main car, and aft between frames 100 and 85. When the watch was relieved, that section could turn in, using sleeping bags in bunks just vacated ("hot bunks") until, turn and turn about, it was time to go back on watch. (Each man had his own bag liner.) Visualize, say, a change of the watch after a half-day out. Forsaking a warm bag to his opposite number, a groggy "mech" dons bulky flight gear. Bundled up, he navigates the close, dim corridor to his hatch and ladder. The noise rises as he climbs quite easily—no parachute—into a frigid slipstream and down into the hot din of his duty station beside the hammering valves of his Maybach.

The cabins were notches up from the bunks. Unlike the communal keel, the passenger car had semiprivate (hence privileged) spaces. It was four to a compartment, with the captain and his executive officer having their own bunks. "When bedtime comes," one guest wrote, "the super-pullman beds are made, fitted with springy pneumatic mattresses for a peaceful night's sleep while the big ship sways slowly in the breeze." Squabbles did occur. Rating the sleep-space accommodations, one officer later remembered, "Occasionally there would be some argument, particularly when we started getting more senior officers in training." The skirmish might prompt a query, "What's your signal number [relative seniority in rank]?"[42]

Lt. Howard N. Coulter, Naval Aviator (Airship) No. 3910, reported aboard in late December 1929. His inaugural hours in *L.A.* were logged on 20 January—the dead of northeastern winter. Hence heavy flying gear. "This was," Coulter told the writer, "a period of rigid economy for the Navy and there was little money for necessities and strictly none for refinements. There was no cold-weather flight clothing available[,] . . . so on this flight I wore a raccoon coat to keep from freezing. I wondered at the time what some of the crew members might be thinking—what was an apparent collegiate playboy doing on board a navy airship? It was, of course, strictly four hours on, four hours off—regardless of season. On winter nights, it could get pretty cold."[43]

Privations aside, ZRs (save pioneer *Shenandoah*) were no hardship gastronomically. The chow—regular, well planned, well prepared—was the best in the navy. Under way, energy depletion coupled (often) with low temperatures produced huge appetites. On board *Los Angeles*, the galley was on the starboard side abaft the compartments. It was aluminum-lined on bulkheads, deck, and overhead, and all cooking was by electricity. Later, a four-burner commercial gas range was adapted, fueled by liquefied gas. The entire outfit, including portable oven for baking, weighed in at about a hundred pounds. Small but serviceable, the galley space was fitted with a sink and running water plus drawers and shelves for provisions, mess gear, utensils. Here the ship's cook worked tirelessly; until the *Akron*, he was the navy's only flying chef. "Fortunately," Coulter recalled, "we had a splendid cook, Ship's Cook [First Class Richard S.] Peak—a name no crew member could forget."

Once heated, meals were issued in aluminum Dutch-oven type containers. These were carried to the crew spaces aft—two rooms amidships off the keel (one for riggers mess, one the "mechs"). There were two sittings in each. Each had a table and seats. "By the time you got [the food] back there, it was stone cold," one rigger recalled of winter chill. Dishware were aluminum and paper (to save weight and reduce washing). Rubberized fabric bags received waste and were removed upon landing. Amid banter and bravado, the section coming on watch ate first, then relieved their counterparts so *they* could eat. Mess was egalitarian—the officers ate the same food, from the same galley, but did so in the main car. Wardroom mess: a small table and seats set up in one of the passenger compartments.[44] Each officer was his own steward—that is, they served themselves.

With attention to weight and spoilage, menus were drawn up under an officer's supervision. The food and commissary supplies had then to be assembled and stowed. Under way, with three meals daily to prepare for fifty-odd sailors, plus coffee and sandwiches in between, seldom was the cook inactive. He had landing and postflight duties as well, such as extinguishing galley fires when "landing stations" sounded, disposing of garbage, and dumping the car's waste-water and septic tanks.

On board *ZR-1*, the fare was soup,

The passenger cabins of the control car. The view is aft. The ladder is to the keel corridor, as is the door. Built as a commercial ship, officer's aboard Los Angeles flew comfortably, in contrast to Shenandoah's spartan spaces. (National Air and Space Museum, Smithsonian Institution (SI Neg. No. 2000-9084))

L.A.'s small but highly serviceable galley. The view is forward, towards the cupboard spaces and basins. Note the aluminum-lined bulkheads and the four-burner electric range. The tank is for water. Regular, well-planned meals were essential to the well-being of the crew. (National Archives)

beans, sandwiches, sardines, coffee, chocolate. *Los Angeles* traveled more elaborately. During one passage in 1926 to *Patoka*, wondrous meals were spread. "When evening comes," a guest-passenger reported of the quality and ambition of ship's mess, "the radio tunes in a distant broadcast station and dinner is served to the accompaniment of an orchestra in a faraway hotel. It starts with piping-hot soup, followed by fried chicken, mashed potatoes, pickles, tomato salad, bread and butter, cake, fresh-made pie and coffee, all prepared on the electric stove." On 6 September 1928, during a twenty-nine-hour run up the Hudson Valley, Peak conjured the remarkable bills of fare:

Breakfast

Stewed Prunes	Fried Bacon and Eggs
Cocoa or Coffee	Bread and Butter

Lunch

Fried Pork Chops		Buttered Rice
Apple Sauce		Baked Corn
Cocoa or Coffee	Oranges	Bread and Butter

Dinner

Cold Sliced Ox Tongue		Baked Beans
Tomato and Cucumber Salad		Pickled Beets
Cocoa or Coffee	Pears	Bread, Butter and Jam

For night rations, the watch savored hot coffee and sandwiches; chocolate bars were issued between meals. Ice was kept to hand not simply to preserve food but for lemonade, limeade, and ice water and tea.[45]

Regarding Peak's galley talents, Coulter was unabashed in his praise:

> He was indefatigable in his efforts to stay our constant hunger; huge sandwiches of cold roast beef or ham, and gallons of coffee kept all hands going during the nights. For breakfast the standard menu was steak and scrambled eggs and toast and more coffee. The steaks were small, about one-half inch thick and maybe two and a half by four inches in size—tender enough and always delicious.[46]

Day to day, flight implied a measure of risk, for Peak as well as his shipmates. In the afternoon watch of 11 October 1930, over central Jersey, the ship took an extreme diving angle, spilling a pot from the stove. Hot water burned his foot.

One factor more biased morale. With smoking banned under way, the longer missions proved especially trying for those so addicted. Consolations: a dry smoke, chewing tobacco, or gum. Along the keelway, men could be seen carrying paper cups into which, as needed, they spat their plugs. When smokes were stolen (in defiance of all prohibitions), it was most often in the engine cars. The pods hung outside, away from patrolling eyes. On the inboard (ladder) side of each wing car the access-hatch sliding door could be closed.[47] In 1931 a space was set aside aboard *ZR-3* for smokers, and smoking rooms were designed into *Akron* and *Macon*.

Throughout the winter of 1925–26, helium was accumulated. That March, *Los Angeles*'s cells were reinstalled and inflated partially; by April, *ZR-3* was fully reinflated—the first weigh-off in months was logged on the 8th. The midwatch of 13 April found *L.A.* docked on cradles, ready for flight. At 0735, assembly was sounded for a ground crew.

Maybachs warmed and men on the handling lines, the ship was weighed-off at berth. Eight minutes thereafter, *L.A.* was walked through the northwest door. Thirteen minutes more were consumed in hustling her to the tower. Riding to the mast wires, she rose statically at 0942. The main wire pennant carried away, so the ship was hauled down on the yaw guys. A new pennant installed, a mooring was completed at 1105.

Steele made ready for letting go. Throughout the afternoon watch, as the crew filed aboard, water was valved in three-hundred-pound increments—three thousand pounds altogether—nearly all at frame 100. Following weigh-off, he cast off for a local postrepair hop. It was 1701. Noiselessly, the machine drifted downwind; after two minutes, bells clanged, her VL-1s barked to life, then came to cruising speed. Of two hours and thirty-five minutes duration, this trial was the first sortie of 1926. It was also a poignant one: save for those transferred from the station, all officers and enlisted men on board were survivors of *Shenandoah*.

Minutes out, a hole was discovered in the cover, near number-two car. Inspection disclosed three punctures in number-six cell, adjacent. Repairs were effected. About 2 percent of its volume had been lost—roughly 5,300 cubic feet. (A postflight inspection would reveal a dent in number-two propeller where it had been hit "by some object.") Having pushed three miles offshore, Steele ordered a westerly course, inland, toward the station. At 1853, he secured again to the high-mast cup, after having flown forty-two miles.

Throughout its career, *ZR-3* would host innumerable experiments, tests, and analyses. Next day's trial saw preliminaries for a particularly salient series of tests. The aerodynamic forces acting upon airship hulls held keen interest for designers. In these years, the best way of obtaining pressure-distribution data was by full-scale measurements—that is, in actual flight. Hitherto, investigations had been limited to tests on *R-32*, *R-33*, and *R-38 (ZR-2)* in Britain, plus an unpublished German study on *LZ-126*. In the United States, work was confined to information gained on the nonrigid *C-7*.[48] With this in mind, and in conjunction with NACA, BuAer instituted an elaborate series of tests on *Los Angeles*.

A bit of background. Aeronautical design consists of certain fundamental elements. One is knowledge of the forces that will act on a structure—the loads it will bear in service. These must be known as accurately as possible. Next, a factor of safety must be set as low as practicable, or the structure will be too heavy. These considerations are the foundation of all aeronautical design. "And if one is lacking in any respect, you're in trouble."[49] Design data pertaining to airships was largely empirical; as yet, knowledge was far from complete. Trials with *ZR-1* had proved that aerodynamic bending moments in straight flight in gusty weather exceeded any produced by maneuvers in still air. Prior to modern aeronautics, information with respect to forces imposed by various maneuvers and turbulent air (especially gusts) had held scant interest. As BuAer's C. P. Burgess put it, "Existing data on the structure of gusts is very meager." Now, though, the conditions that most interested the designer (and airship navigator) were storm winds, squalls, and sudden gusts. Reichelderfer explains: "The airship designers always wanted the meteorologists to define a gust in terms of the qualities involved and the mathematics involved. They were never happy with the meteorologists because, to put it quite practically, there *is* no standard gust, nor is there a limit to a gust—which is what the airship designers wanted to find, so they would know the wind stresses they would have to design for. Now they were really in a dilemma because an engineer has to design for certain stresses. If you don't know what the stresses are—[it's] pretty hard to design."[50]

The startling breakup of *Shenandoah*, occurring at a time when proposed airships of unusually large cross-sections were under study, had underscored the need for much greater hull strength. Plainly, existing design theories were wanting. "It is obvious," NACA advised, "that as much additional data as is possible should be obtained from each new design." In this context, *Los Angeles* proffered a full-scale testbed—a flying laboratory for experimental research. (At Lakehurst's aerological office, special studies using new types of instruments and recorders also were in progress.) *L.A.* proved a fountainhead of free-flight pressure-distribution data, to be used in rigid-airship structure design.

Preliminary strain-gauge measurements had been taken on *L.A.* during May–June 1925, to determine stresses and angular accelerations in flight. But the tests had been abbreviated, owing to the projected Minneapolis flight. More careful planning marked the much more extensive work scheduled for 1926, including the use of improved recording-type gauges. The specific purpose of this test program: to measure actual aerodynamic forces and their (simultaneous) distribution, the resulting stresses and bending moments in certain structural members, and ship's motion when these loads and stresses were applied.[51] Pressure was measured by orifices installed at desired points and connected by tubes attached to multiple manometers. (The investigation on *C-7* had been analogous, the equipment cruder.) Near the close of the extended layup and overhaul, the devices were affixed to points on the passenger car, on the forward portion of the hull, and on the tail surfaces. Aloft, the program comprised measurements in both smooth and rough air, turning trials, deceleration runs, and stress measurements in representative girders in both rough air and still-air maneuvers. Active throughout the work was Burgess. A remarkable talent, C.P. was masterful in the fields of airship stress analysis and project design.

For the 14 April run, Burgess joined the ship's company along with three officer-passengers: a naval lieutenant and two marines. (No army, this hop.) In a nod to training, five students accompanied, among them Lieutenant Commanders Zeno Wicks (Naval Aviator [Airship] No. 3477), Vince Clarke (3399), and Myron J. Walker (3378)—the latter pair from Class III. The next run, on 27 April, had Lt. Cdr. Frank C. McCord aboard, also from Class III. (Clarke would assume command of *L.A.* in 1930; and in 1933, McCord of *Akron*.) The ride proved brief: Steele was back after slightly more than two hours, training crew and ground handlers. The four local and short-duration sorties that followed emphasized the NACA program and training. C. P. was on board for each, assisted by five or six bureau experts. The controls set at various angles, severe still-air maneuvers were flown, as well as straight-flight tests in rough air.[52]

The measurements were worked up and tabulated,

and curves of pressure distribution plotted. The final results showed "quite conclusively" that loads in rough air "greatly exceed" those resulting from maneuvering the airship. Indeed, the resultant forces confirmed that tail-surface loads exerted by gusts were "considerably larger than any obtained during maneuvers in smooth air, which indicates the importance of gust information for airship design." The report continued, ominously, "In this investigation, the tail surface loadings caused by gusts closely approached the designed loads of the tail structure."[53]

These were Steele's final hours. He commanded four flights more that April and his last on 7 May—strain-gauge tests in conjunction with passes over a camera obscura (to help determine turning characteristics).[54] The need for new leadership was plain. Steele had proven overcautious. Partly for reasons of health, he had ceded shiphandling to his more assertive second in command.[55] Further, Steele had lost interest; his wife was encouraging him to leave lighter-than-air.[56] Earlier, he had requested sea duty, to further his chances for promotion.[57]

A time of transition was at hand. Moffett's options: either to promote Klein into the position he had been filling or bring in another officer. The latter cut two ways. Though a more senior skipper might "draw more water" (have more influence outside the command), a new man would block the advancement of more seasoned junior officers. Wiley, for one—a member of ship's company since the commissioning, he was demonstrably qualified. So was Maurice Pierce; ship's navigator and a senior lieutenant commander, the man's skills were plain.

Klein, for his part, was much disliked; although energetic, his aggressiveness and ambition had abraded his associates. Moffett, for one, had been rankled by his demand to accompany the *ZR-3* on its transatlantic jump. So the admiral hesitated. As Fulton had it, Moffett "did not feel competent reliefs of suitable rank were available. He went through a lot of soul-searching before recommending Rosendahl to the *Los Angeles*."[58]

On 15 March 1926, Lt. Cdr. Charles E. Rosendahl, Naval Aviator (Airship) No. 3174, was ordered to *ZR-3* as executive officer, with the expectation that he would succeed Steele when the latter was ordered to sea. Klein (whom Rosie replaced) was punished for his temerity; he received orders to command of the *Chewink*, a "Bird"-class minesweeper. Also reporting aboard was Lt. Roland G. Mayer (CC), a *ZR-1* shipmate close to Rosendahl. Naval Aviator (Airship) No. 3244 and one of the best structural and motor men in the service, Mayer had assisted with free-ballooning *Shenandoah*'s ravaged bow back to the ground. Capt. Edward S. Jackson, an earnest officer with no LTA experience, received orders to command NAS Lakehurst, with Pierce his second in command. Certain long-serving naval aviators carried on with *Los Angeles*: Arnold (No. 3184), Buckley (3223), Settle (3350), Thurman (3351), Tyler (3376), Wiley (3183). As for enlisted personnel, such stalwarts as Baker, Boswell, Bishop, Copeland, Jandick, Miller, Malak, Patzig, Russell, Schellberg, and Swidersky—and scores more—continued to serve ably.[59]

Moffett's judgment was to be vindicated when, in May 1926, a new discipline—and dynamism—began to move *Los Angeles*.

CHAPTER 4

Rosendahl's Reign

One of the very intriguing things about airship duty was [its] creative aspect.... It was an experimental and research organization almost as much, perhaps more, than it was an operating organization.

—Cdr. F. W. Reichelderfer, USN (Ret.)

On the tenth day of May 1926, Charles E. Rosendahl succeeded Steele as Commanding Officer, USS *Los Angeles*. With crew at quarters, and before the duly assembled dignitaries, families, friends, and shipmates, Steele read his orders detaching him from command. Then his relief read his orders, dated 20 March 1926, to assume command.

An admirable if problematic figure, Rosendahl had served in destroyers during the war. A naval-aviator student in Class I, he had reported to Lakehurst in 1923 "for duty involving actual flying." (For this officer's inaugural hop, in a kite balloon, Weyerbacher had been pilot.) Tall, straight, serious, Rosendahl possessed a trying personality—demanding, competitive, combative.

An arch-believer, the lieutenant commander pursued the project with a devotion bordering on obsession; his faith in lighter-than-air became a consuming cause. The man's courage was undeniable, his advocacy tireless, his pace inspired. Propelled into prominence by the *ZR-1* wreck, a bachelor and self-dramatizer, Rosendahl, thirty-three, would retain the spotlight, facing down opponents, exercising a steely, exacting command style, and eclipsing his fellows; no operational LTA officer had anything approaching his influence.[1] As it happened, the

Ship's officers and men pose for photographers, May 1926. Commanding officer, USS Los Angeles: Lieut. Cdr. Charles E. Rosendahl, Naval Aviator (Airship) 3174. A hard-charging advocate of big airships, "Rosie" had succeeded Steele as skipper. Two of the three sections of ship's complement flew together. (Kevin Pace courtesy Navy Lakehurst Historical Society)

A group of Los Angeles' officers circa 1926. l. to r. front: Lieut. Cdr. Charles E. Rosendahl; Cdr. Maurice R. Pierce; Lieut. Cdr. Joseph C. Arnold; unidentified. Back: Lieut. Raymond F. Tyler; Lieut. Herbert V. Wiley; Lieut. Charles E. Bauch. From Lakehurst's Class I, Pierce, Arnold, and Tyler were contemporaries of Hunsaker, Lansdowne, Rosendahl, and Wiley; each appears in the earliest pages of ship's log. Pierce would twice command the Lakehurst station. Arnold, by mid-1933, was executive officer aboard Macon. Tyler had served at Paimboeuf (France), at Akron, with the Atlantic Fleet Kite Balloon Division, and at Pensacola before receiving orders to Lakehurst thence to ZR-1, *and, in 1924,* Los Angeles. *(Louise Tyler Rixey courtesy Navy Lakehurst Historical Society)*

years 1926–29 were a critical span: the man's particular talents fit that particular time. As his authority swelled, Rosendahl was anointed the navy's number-one ZR commander. Deliberately, this all-business personality strode across the stage, guiding, scripting, directing, and promoting lighter-than-air—and to its next two decades this officer's destiny was to be tied.

In the meantime, with ZR-3 as its instrument, Team Rosendahl turned to the task with due deliberation—heavy doses of initiative, innovation, and derring-do, to say nothing of keeping ZR-3 (the sole commissioned aircraft) flying. Between his assumption of command and the close of 1926, nearly three hundred hours were recorded in the ship's log—more than doubling its time in the air. Seven moors were flown to Patoka that year, adding sixty-three hours with the tender. The high mast was exploited as well: altogether, 156 hours, twenty-nine minutes at its cup. A trial moor to Henry Ford's tower at Dearborn realized ten hours, four minutes more. The man's energy and pace were frantic, his zeal obsessive.

The star of Herbert V. Wiley, the ship's navigator since April, also was in the ascendant. Wiley—serious, studious looking—had reported to Lakehurst within days of Rosendahl. There he had helped fit out Shenandoah and had served aboard. Designated on 5 February 1925, at midyear Wiley was ordered to Los Angeles. This pair would log hundreds of shared hours. Further, Wiley assumed temporary command of ZR-3 when Rosie joined Graf Zeppelin for its first Atlantic crossing. When Wiley succeeded to the command, Rosendahl temporarily relieved *him,* so Wiley could make a passage to Friedrichshafen on the peripatetic zeppelin. Though junior in rank to Rosendahl, Wiley was the sole officer to serve in each of the navy's four commissioned rigids. Meantime, the lieutenant commander was shortly to receive orders to become executive officer.

But this spring, already, Los Angeles's infirm cells were bringing trouble. On 21 April, exploiting inclement weather, the crew partially deflated numbers two, three, and eight—due to low purity (hence loss of lift)—then re-inflated them the next day. Between 3 and 6 May, the ship's entire volume of gas was withdrawn, repurified, directed back.[2] This was running hard to stay in place. ZR-3 lost twenty-six thousand cubic feet daily thanks to aging cells. Less than half that rate—twelve thousand cubic feet, on average—arrived from Fort Worth. To help address the problem, much work was under way to replace skin-lined fabric, principally to reduce costs and to improve impermeability. No satisfactory substitute was yet in use; by 1927, as discussed earlier, two fabrics showed promise: cloth lined with cellophane and cloth coated with viscose latex.[3]

Rosendahl's first liftoff as skipper took place on 13 May—an unhurried, ten-hour cruise taking L.A. east then north up the coast to New York Harbor. The muster and walk-out were customary midwatch fare—ground-crew call at 0310, weigh-off at 0345, walkout to the mast ten minutes thereafter. The ship rose vertically on the mast wires, and ballast streamed out (including all emergency bags) as the bridge adjusted to a temperature inversion at seventy-five feet. Secured to the cup at 0505, Los Angeles began taking on water. At 0645, section two stood relieved; at 0750, the flight crew commenced filing on board.

Cast-off was logged at 0940. Total personnel: forty-nine, plus a dozen passengers. Along as "officers for instruction" were Lieutenant Commanders Clarke and Wicks, Lt. J. B. Carter (Class II), and two lieutenants more. New enlisted men got a bit of training as well, on

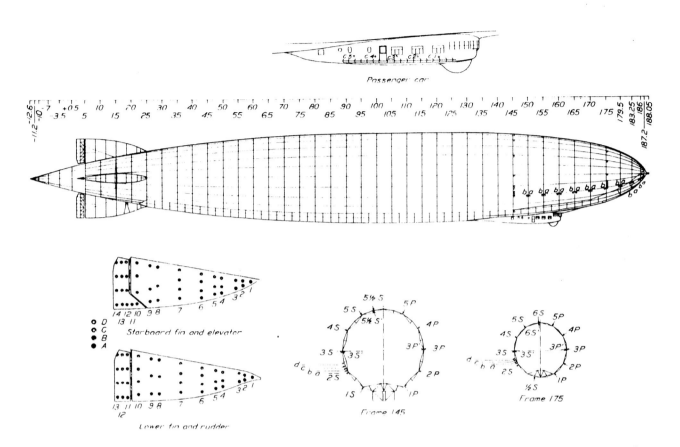

Location of orifices installed for the April–May 1926 in-flight program. Objective: to measure actual aerodynamic forces and their distribution; the resulting stresses and bending moments in certain structural members; and ship's motion when those loads and stresses were applied. (NACA Report No. 324)

the rudders and elevators. Pierce and Weyerbacher, along with five civilians from the NACA (led by Burgess), also were flying this day. Those five had work to do. But with conditions unstable over the shore, Rosie turned to sea; at 1044, *L.A.* passed over Seaside Heights. Offshore, five tests were run, including steady turns to starboard and port. Afterward, steadied on course 325 degrees for New York City, the dirigible stood up the harbor at half speed, all engines. The Statue of Liberty was put astern, then Hell Gate Bridge and the Brooklyn Navy Yard before a heading was taken for Sandy Hook. With the men at landing stations and a weigh-off effected, further tests commenced—this time of various combinations of engines at various speeds. On the return leg, one mile off the beach, near Long Branch, a full-power run on five engines realized an airspeed of 59.5 knots. Back over base, a camera obscura test was conducted, followed by deceleration tests. Then yet another trial: slip tank number seven was let go, to test emergency release gear. Despite its 640 pounds of fuel, the tank did not explode upon impact.

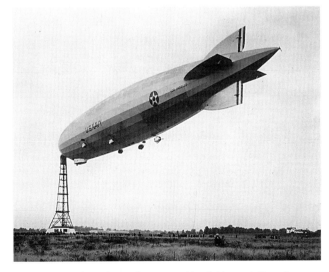

ZR-3 moored to the tall mast at Ford Airport, Dearborn, Michigan, 15 October 1926. Though not without merit, the high-mast concept required that ships be "flown" continuously while at the masthead. (James R. Shock Collection)

Following a three-hour hold in the open ("Standing by in hands of ground crew to Westward of hangar, waiting for favorable opportunity to dock"),[4] the flight crew ended its day at 2315, with *Los Angeles* secured in the big room. Not until midsummer did they again fly her. Next day, in berth, tests of a new single-wheel taxi car were conducted.

Team Rosendahl had a large problem. The airship had to be laid up awaiting new cells, and until an adequate supply of helium could be banked on station. Her cells were losing about 1 percent lift per day. Accordingly, two days later the hull was shored up and the overhead suspensions attached topside, preparatory to deflation and repurification of ship's supply (save for 2 or 3 percent left in some cells, to sustain their weight). Hostage to her medium of lift, *Los Angeles* assumed the role of hangar queen. Her inoperative status was to last more than ten weeks.

The postrepair flight was logged on 26 July, during which *ZR-3* calibrated the radio-compass stations at both Manasquan and Sandy Hook. Of three officers along for instruction, two were from Class III: Lieutenant Commanders M. J. Walker and Frank C. McCord.

In the meanwhile, the CNO had approved a schedule of operations: radio-compass calibrations (nine more such flights that August–September) and rendezvous with *Patoka*'s stern mast. On 4 August, a floating moor was logged off Newport—the first of seven such couplings that half-year. While in New England, radio-compass stations in Rhode Island and in Massachusetts were calibrated, after which *Los Angeles* returned to her mobile base. She pounded homeward on the 6th. Just before sighting the Jersey coastline, the skipper shed altitude, dropping to twelve hundred feet to take up moisture from fog—his ship being light.

Publicity—always a desideratum—was indulged. Having cast off from *Patoka* at Plantation Flats on 8 September, *L.A.* was photographed from the army's *TC-9* (up from Langley), which then cruised on her starboard beam as Rosendahl prepared to calibrate the Polyner's Hill radio-compass station in North Carolina. Two days later, a jaunt was flown to Philadelphia with a Fox Reels man and 105 pounds of cameras and gear along. (To direct a landing, Lt. J. M. Thornton, Wiley, and an enlisted ground-crew party had been sent ahead.) While the ship swung to the wind after leaving the hangar, however, the four struts on number-one car broke. Temporarily repaired, *ZR-3* was circling the Pennsylvania city at 1330. Landing at Model Farms field (the site of the National Air Races), the crew connected a hose (from a fire engine!) and took on ballast. In the forty-seven minutes all that took, the ship's after bumper bag (number-one car) sustained damage. It was removed and placed on board. Rosendahl retrieved his ground-crew party, weighed off, then lifted away.

Mail went over the side from number-four car not long after. Its unauthorized sender, mechanic R. E. Ruefus, implored, "The finder of this note will do me a great favor by delivering same to my Mother." Found in Delaware Park, the missive was posted (one presumes) to its Easton, Pennsylvania, addressee. On the bridge, Rosie elected to cruise upriver northward of the Delaware Water Gap before reversing course downriver. The airship lay abeam the east door at 2015.

Disembarked, the crew was mustered—no absentees—a hangar watch was set, and repair work was set about.[5]

Now came an excursion to the Midwest. Courtesy of Henry Ford, a tall mast had been erected near Detroit; the city was anxious to have its workings tried out. And so a flight "to the mooring mast at the Ford Airport, Dearborn, Michigan," Rosendahl's orders read, "is authorized to be made beginning 11 October 1926, or as soon thereafter as the flight is in all respects practicable. The choice of routes, and decisions as to start, continuation, or termination of flight, rest with the Commanding Officer." Early on 14 October, a casualty interposed. As the ship rode to the Lakehurst tower in stiff, gusty air, idling number-one engine, the heat from a bend in the exhaust line ignited the top of that engine's car. This was extinguished with

Full house at Lakehurst, including the U.S. Army's TC-5, October 1926. Throughout the twenties, the Air Service would push hard to acquire a rigid airship. (Airship International Press)

"very slight damage." Fueling began, and before long a trio of civilian workmen came on board to repair the cover. Gassing commenced. Three hundred pounds of provisions shunted up the elevator. At 1025, the workmen departed, and Moffett clambered through the bow hatch, whereupon his flag was broken.[6] Nigh on the field, the army's *TC-5* lifted off. Inside *Los Angeles*, landing stations sounded and the tail drag let go. At 1107, *L.A.* unmoored with forty-four persons: a flight crew of forty plus the admiral, Lt. Cdr. Vince Clarke, Lt. Anderson, and a Pathé News representative.

"Very unstable" air plus a squall line were penetrated en route. At Dearborn, Rosendahl completed a mooring at 0538—having first fouled his main wire in trees on the approach.[7] Fueling commenced. Erected at the field's southeast end, Henry Ford's tower was rather more complex than that at Lakehurst: upon mooring, aircraft could be shunted to the ground, to take on passengers and supplies. While riding to its masthead, E. J. Riley, AMM1c, was transferred to hospital with pneumonia. The dirigible cast off in the afternoon watch, letting go 1,200 pounds of water on the escape. The flight track eastbound touched Cleveland, Buffalo, Rochester, Syracuse, Albany, and (following the Hudson) Kingston, Poughkeepsie, West Point, New York City. At 0550, 16 October, one mile west of Lakewood, the watch had the naval reservation in sight.

Before the year was out, seven sorties more—all brief, all for indoctrination—were affixed to the log. However, in light of the chronic helium crisis and the late-autumn weather, BuAer felt obliged to recommend another layup. The year's final hop fell on 3 December—a tri-state tour west to Philadelphia thence into northeastern Jersey and over New York City before the return leg, course southwest.[8] The layup commenced on the 17th. That New Year's Day, *Los Angeles*'s log totaled sixty-nine flights and 591 hours, eighteen minutes airtime.

L.A. lay shed-bound till April. Nonetheless, her commander managed 162 flight-hours in the first half of 1927.[9] Instigated by BuAer, additional speed and deceleration tests—both with and without the water-recovery apparatus in place—were flown come September. Ashore, for more than two months the entire facilities of the wind tunnel at the Washington Navy Yard were devoted to research work on three rigid-airship models fitted with various control surfaces.[10]

Along with two moors to *Patoka*, the first tentative steps toward marrying ZRs to the airplane were taken (see chapter 5). Late in March, shed time had been devoted to experimental devices, including docking *L.A.* in a spring-type contrivance to continuously indicate in-berth buoyancy. During 1928, a large heating unit was experimented with—to create excess superheat in-berth, so as to permit the ship to get away quickly with more than normal load.

This was a trial-and-error era, with the bureau and Lakehurst spawning grounds for new ideas. One initiative: better ground handling. For *ZR-1*, this had been largely manpower—an iffy business that left the ship vulnerable. "We had to have cross winds of six miles an hour or less to take the ship out of the hangar during those early flights," Bolster said of this bugbear, "and accurately predicting them was a difficult task." Gust watches were therefore instituted. Station records disclosed the period between 1000 and 1400 as the worst for a docking/undocking, between 2200 and 0200 the best. Accordingly, zero hours usually were set for nocturnal lulls, when relative stability prevailed.[11] More hours could have been flown had it been possible to tolerate even cross-hangar blows of velocities that, really, were quite moderate.

In short, *terminal* rather than *flying* weather was confounding operations. As operators, Rosie and his officer team were preoccupied with the shortage of handlers. The need for mechanical aids was plain. In March 1927, for instance, a train backed into the ZR dock and then pulled out again with 187 Lakehurst marines, reassigned to duty in China. One outcome: the ship was confined to berth for two weeks, after which *ZR-3* operated tentatively for months, thanks to a reduced—and raw—ground force. There had to be a better way.

The Army's TC-13 *at March Field, California, June 1936. The Air Service would cede lighter-than-air operations to the Navy in mid-1937. (Collection of Mr. and Mrs. Joseph T. N. Suarez)*

Further, the "high" handling technique was also wanting. Although a 167-foot tower conferred some independence, operations from there called for precise airmanship, and access was restricted. But the fundamental concern was trim: the ship had to be "flown" at the masthead—kept within a few degrees of inclination or horizontal regardless of fronts, thunderstorms, squalls, and sea breezes. The tail drag proffered a measure of pitch control, but it was only an expedient—and a nuisance in itself.

On 25 August 1927, these tyrannies were tellingly displayed. On the mast since 0535, *Los Angeles* vaned into light and variable winds—causing her to swing uncertainly about the tower. Save for scattered stratocumulus, the sky was clear. As daylight progressed, a great deal of superheat accumulated within the ship. (One indicator: the tail drag lifted clear.) On the bridge, the rudder was needed at times to keep bow-on to the wind, and the elevators and ballast adjustments to maintain trim—normal routine. At 1315, 125 pounds of food were received aboard. An early afternoon cast-off was expected in almost ideal weather. The station forecast for "today": "Partly cloudy, high clouds, variable winds, mostly northeasterly, less than 12 mph."

From 1100 to 1329, wind direction proved unusually variable: observations taken on smoke (from the Power House), the kite balloon, from instruments at Aerology, and at the BOQ and on board *L.A.* herself at times showed no agreement.[12] During the noon hour, a fairly definite wedge of sea air—a mild cold front—began to work its way inland. This "sea breeze" approached more rapidly aloft than at the surface. On a training run, *J-3* passed through this boundary about five or six miles east of the station. Upon its return, *J-3* again penetrated this shift; on approach, its pilots faced a south*east* wind, with *L.A.* seeing a north-north*westerly* flow. *J-3* had met a very steep temperature gradient. Weighed off at five hundred feet, she was found light; yet at the height of the mast *(and L.A.)*, the blimp commenced to "fall through," its bumper hitting hard.

Time: 1329. The cold, denser air had arrived.

Los Angeles had a heading of about 330 degrees (westward) at the masthead. The air mass enveloped her stern, abruptly granting lift. A tail-rise began. Further, the wind freshened from the southeast—that is, from astern. This added an upward impulse, shifting more of the hull into cold(er) air.[13] The gust had thrown the rudders hard left, apparently jamming them. In the control room, the acting first lieutenant, Cal Bolster, and the engineering officer, Tex Settle, were discussing conditions "which seemed (Bolster told me) a little unstable since we had little or no rudder or elevator control." As OOD, Settle phone-ordered men aboard and also began the process for loading water. Shifting slowly now to port, the dirigible continued its tail-up travel. (To check the rise, the elevators were pushed hard up.) Ripples of alarm spread through the crew; the rise "was so rapid and so steady that there was no doubt we were going to go up fairly high."

The unusual angle canceled any water. Settle's call to the masthead got seven men on board, to shift aft. Along the catwalk-tunnel, their world sloping crazily, men scrambled uphill or sought handholds. Pressing aft, Bolster himself got as far as amidships before being obliged simply to hang on. The tail-rise had passed beyond remedy. (Usual oscillation: eight to ten degrees.) "The inclination of the keel rapidly became steeper and steeper [Bolster continued] and soon those of us in the keel were hanging on to the transverse frames which were now horizontal. . . . The fuel tanks began to lose fuel out of the vent lines at the top, and all manner of loose objects such as spare engine parts and tools came falling down the keel as they would in an elevator shaft." On the mast and field, confusion reigned. Someone was yelling to "trip" the ship, to cast it free—"which was the last thing we wanted, since the engines could not operate and there was a great fire hazard with all the spilled fuel."[14]

Among the engineering watch, David Patzig, AMM1c, had been cleaning spark plugs in the forward end of number-two car. As the angle enlarged, the "mech"—absorbed in his work—kept shifting himself forward, toward the radiator. When the angle became unusual, he thought to check his outboard window. *L.A.*, he realized, was at a "terrific angle"—and still rising!

Overwhelming every system, a minor event had yielded a major result. The combined effect—wind and dense air—had upended the hull. At the same time, she was changing azimuth slowly—rotating west toward north. Everywhere inside, crewmen clung desperately. On the bridge, the watch had not lost its nerve; just now, though, there was nothing more to be done. Behind (now above) them, mess gear and provisions were going adrift in the galley and crashing into the bulkhead. Along the keel, various items were adrift as well, adding to the clatter. Aft, gasoline and ballast were spilling; soon, water began dripping into the control cabin. Braced alongside the rudder wheel, Settle and Aviation Rigger First Class (AR1c) R. H. "Ducky" Ward were staring *down*—through their knees—at the ground some two hundred feet beneath the forward windows.

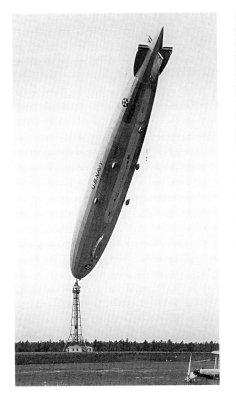
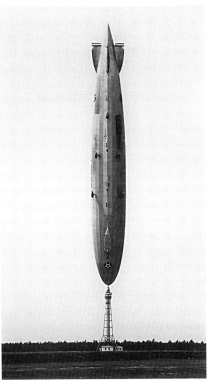
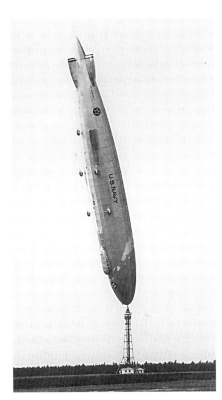

1329–1336 hours, 25 August 1927. The view is west. A strong, relatively cool sea-breeze has replaced a light westerly gradient. The combined effect of wind and dense air has increased lift and added an upward impulse, shifting Los Angeles toward the vertical (1st photo) before she could answer the wind. Still upended but rotating slowly to port (second and third), the stern begins to settle at about 1335 (fourth and fifth). Damage proved slight. (National Air and Space Museum, Smithsonian Institution (SI Neg. Nos. 2000-9507 to 9511))

L.A. stood on its nose. Lift and wind now opposed each other—the former tending to hold the hull aloft, the breeze trying to blow it over. *Los Angeles* poised like a ballerina. In his car, Patzig was *atop* his radiator as, nearby, the clatter of loose gear was distinctly audible. His chief concern was that the ship might fail structurally and impale itself upon the mast. Outside the hull, not far from his vantage, the tail drag had swung in close aboard. No panic was evident, just amazement. Aviation Chief Machinist's Mate (ACMM) Martin Miller retained remarkable composure. At the ladder to Patzig's car he shouted to the mech, pretending to accuse him of responsibility for their predicament. The mech did not answer.

Just back from lunch with Wiley, Rosendahl was heading mastward, intending an early cast-off. His photographer, ACMM Walter Johnson, happened to be taking pictures of nearby parachute jumps, using *L.A.* as background. Sensing the moment, Johnson snapped a sequence. From their own various vantages, employees and visitors watched, stunned. Aviation Chief Rigger (ACR) "Bull" Tobin thought *ZR-3* would certainly be destroyed. Faye Settle had just left the mast with a Texaco representative who had had fuel and related business to discuss with her husband. Von Meister, the American agent for Maybach Motors, watched from just outside the officers' club. ("Willy" visited the station every few weeks, to lend engineering advice concerning the engines.) Melvin Cranmer, a Public Works electrician, was changing red obstacle lights on the station gasometer; from the "hill," he had a superb view.

Time seemed on hold. At about 1335, still rotating to port, the stern began to settle, and to accelerate. The ship's helium had cooled considerably, shedding lift. Now, as crewmen pressed forward within, Settle dropped eight hundred pounds of emergency water and other ballast at frames 40, 55, 70, and 100 to check her—2,500 pounds altogether. Downward motion ceased. *Los Angeles* leveled off, resuming normal riding position parallel to the wind, heading southeast. Total elapsed time: seven or eight minutes.

Water from the mast was immediately added, to maintain trim. The discomfiture eased. "Except for noting the spilled fuel and oil and observing great caution to avoid any danger of fire," Bolster concludes, "those on board returned to a normal routine. We did not leave the ship." Though flushed with adrenaline, all hands were much relieved. However, only an inspection could ease the anxiety and confirm their luck. At 1400, a "general quarters" sounded, a ground crew mustered out. Not until 1545 was *ZR-3* positioned and the hangar watch set.[15]

Damage proved slight. A number of wires had snapped, tearing number-ten cell in two places. Four gasoline tanks had shifted from their guides, and four girders had sustained damage from speeding tools and whatnot. Almost unbelievably, *L.A.* was still flightworthy—a tribute indeed to the builders. Zeppelin's superb reputation seemed justified. Rosendahl (in his relief) admitted to Moffett that the twenty-two-thousand-pound tail drag "plus intelligent ballasting" should have coped with any situation under such calm conditions.

This escape was remarkable.[16] The airmen's confidence in their prize soared. ("It was a great thing to know she had that strength," Rosendahl exulted, reminiscing.) *Los Angeles* had "proved herself unquestionably strong," his report averred. "I think it is one of the most remarkable demonstrations that an airship has ever given. Of course it was very uncomfortable for the crew, but it is valuable to know what punishment the ship can stand and remain practically intact." Modern airships evidently were very strong. Also, plainly, investigations into new mooring-out methods were fully warranted.[17]

Publicity was another matter. Johnson's photos—of an unchosen test at an unwelcome time—were dispatched to Washington, for sequestration. This self-censorship was damage control. As Bolster put it, "such pictures would not enhance the rigid airship's popularity." Funding for the next generation of big airships—what would become *Akron* and *Macon*—was navigating a myriad of shoals. Theatrics might endanger the appropriation. The core of the matter was the military, political, and budgetary environment in which the airship had to operate. Lighter-than-air was never to enjoy the broad consensus accorded heavier-than-air. Nor did the naval service as a whole share the airship men's enthusiasm for their technology. Starved for funds and keenly sensitive to derision in as well as outside the department, the program retained a defensive obsession about the forces arrayed against it.

Los Angeles's gymnastics pointed up the merits of a *short*-mast system, the development of which now accelerated. Seven high moors more were made in 1927, three the following year, the last in 1929. By then, Rosendahl (in Fulton's phrase) was the navy's "number-one operator." The official correspondence is a reflection of the man's views and convictions. In this instance, no one was eager to repeat the experience. "In my opinion," Rosendahl observed, "this [incident] proves the necessity of going ahead more seriously than ever with the stub mast project." Initiated that spring, the trial demonstration was mere weeks away. As experiment officer and assistant to the airship's first lieutenant (Roland Mayer), Bolster bore the brunt of the development work. He had served at Lakehurst in June 1923, helping to erect *ZR-1*, after which orders had taken him elsewhere. Bolster had returned early in 1927 as part of *Los Angeles*'s complement. Though not as an official crew member, he would fly with *Akron* and *Macon* as well.

Los Angeles moored to a prototype "stub mast" within five weeks. Erected southwest of the big hangar, the device was nothing more than a sixty-foot radio mast supported by guys and anchors about the base and fitted with a mooring cup at the top. Also installed were water, fuel, and telephone connections. To provide a rolling connection to the ground and allow *ZR-3* to answer the wind, a single-wheeled contrivance was attached to number-one car, aft. In this way, the stern could describe a circle about the nose-hold. The after car had not been designed to handle the thrust loads it would receive fitted beneath with a "taxi car"; indeed, the limiting condition while riding to short masts was stress at the after car. So aluminum handling frames and struts had been fabricated to resist side-loads. Installed on both sides and detachable before flight, these members ran from the car up to the ship's structure.

The morning watch, 5 October, recorded the inaugural trial:

First field moor to a "stub" or low mast: 5 October 1927. Trials proved that, riding out, the airship answered easily to sudden gusts and wind shifts. The "A" frame struts counteracted side loads. (Charles E. Rosendahl Collection, University of Texas)

0500 Sounded assembly for ground crew and opened hangar door #2. 0510 Warmed all engines. 0540 Ship's handling lines manned by ground crew. 0555 Attached single-wheel taxi car aft. 0605 Embarked flight crew. 0608 Started walking ship out of hangar, east door, wind NW 8 mph. 0610 Ship out of hangar. 0615 En-

The after car riding a newly-devised taxi car, 1927, preparatory to mooring tests with an "experimental stub mast." Mounted forward is a handling frame and struts. Note the sand-bag ballast. (San Diego Aerospace Museum)

Close-up of the taxi car. Wind-structure studies by Lakehurst's aerological office were underway to assess (for example) the comparative merits of low and high masts. In 1930, a special station was opened at Akron Airport designed especially to collect weather data bearing on aircraft design and operation. (National Air and Space Museum, Smithsonian Institution (SI Neg. No. 2000-9502))

gine #4 ahead half speed for 30 seconds. Ship walked to experimental stub mast on west landing field. 0630 Arrived at stub mast, water ballast line and mast wire attached. 0637 Added water ballast at frames 100 and 40. 0645 Water ballast line detached. 0649 Ship left ground on handling lines and trail ropes. 0651 Ship attached to stub mast, 500 lbs water ballast valved from frame 145, wind WNW 8 mph. 0705 500 lbs water ballast added at frame 145. Wheel tail drags attached at frames 70 and 25.[18]

Though conditions resembled those experienced during the nose-stand, the airship responded well. Engines warmed, her automatic-valve covers removed, ship fueled, a flight section began embarkation at 0945. After twenty-five minutes, handlers manned their lines and ZR-3 was weighed off—valving ballast as necessary—preparatory to cast off. At 1020, the "A" frame handling struts and taxi wheel were disconnected. (Now the handling party was controlling the stern.) Rosie unmoored at 1025. Walked clear in shifting surface winds, landing stations sounded. At 1037, the aircraft was released. Fifty-one people were flying, including New Jersey senator E. I. Edwards.

This and subsequent trials proved that L.A. answered easily to sudden gusts and wind shifts.[19] Though crude and in need of refinements, the short scheme looked to be effective. Charles Roland, a Class VI graduate, assisted the mast project. "It was destined to become," he said, "the cornerstone for the development of a mechanized system for handling airships on the ground. . . . Solving that problem entailed considerable engineering and experimental work at Lakehurst in the late 20s and early 30s."[20] Next advance: a circular track for the wheel. In the predawn hours of 2 July 1928, the "taxiing car" device secured to the centerline car, *Los Angeles* was undocked onto the East Field. Well clear, she was walked around the shed and secured to the stub mast. At 0850, Rosendahl unmoored. Fifteen days later, on the 17th, prior to casting off from the circle, L.A. was swung by hand through a full circuit, to test the track. No serious defects were noted.

Using this new scheme, ZR-3 could fuel and reprovision (adjusting ballast accordingly) *on* the field—and, when so desired, be off again, or rehoused. On 15 August 1928, for instance, L.A. cleared her berth at midmorning and began a moor-out. That evening, though, lightning was observed bearing southwest. All engines were warmed, a ground crew mustered, landing stations piped—and the dirigible strained back to berth.

A purely mechanical system—a *mobile* mast—was a logical evolution. But the prototype suffered teething problems and was not tested until March 1929. On that occasion, returning from a local run testing radio gear, ZR-3 eased into the hands of the landing party, then walked to an awkward-looking, makeshift "crawler mast." Designed by the Bureau of Yards and Docks, its mast tower was adjustable in height (sixty-nine to eighty feet) and mounted in a triangular steel frame, supported under each corner by a caterpillar-type track. Weight: a hundred tons. At 0722, the main wire was connected, and after four minutes ship's mooring cone locked in. In this novel arrangement, the flight crew disembarked at 0740.

The first mobile unit was tractor towed. Once tests had confirmed practicability, a self-propelled version was initiated. On 22 June 1929, under a roof, *L.A.* was towed by the vehicle (tractors assisting). Detached from spring balances (these measured buoyancy in berth), the aircraft was moored and mechanical handling gear rigged aft. Under watchful eyes, ZR-3 was shunted from mooring position, then back. ("1019 Commenced moving ship thru west hangar door while moored to the mobile stub mast with the aid of tractors. Stopped as stub mast cleared concrete apron. 1035 Reversed direction and returned ship and mast into position in hangar.") Late in September, *Los Angeles* cast off the device for a local flight. Two weeks later, she was landed, walked to the mast, secured, and then docked with the contraption.[21] On its June 1930 visit, *Graf Zeppelin* was handled in and out by the "Iron Horse" and side trolleys.

Manual ground handling had never sufficed. Now the

A purely mechanical system: L.A. moored to Lakehurst's original tractor-towed "crawler mast," about mid-1929. The mast tower was adjustable in height. Note the airship's bow hatch. (Capt. Gordon S. Wiley)

To provide a greater measure of control, BuAer adopted a mooring scheme mounted on rails—self-propelled masts, circles, and ancillary gear. Here the after-car rides to a mooring-out circle, May 1930. (Charles E. Rosendahl Collection, University of Texas)

View forward beneath L.A., while riding to the mast at the West Field rail-circle. Note the ladder from No. 1 car to the keel. (National Air and Space Museum, Smithsonian Institution (SI Neg. No. 2000-9503))

matter was urgent. With *ZRS-4* in prospect, BuAer wanted experimental work and the portable system completed. The new scheme, including the first motorized unit, seemed satisfactory with a craft of *Los Angeles*'s displacement.[22] But the Goodyear-Zeppelin ships would have more than twice her volume. As Bolster observed, "The problem of holding and moving these huge, relatively fragile airships against fairly high side forces, which were estimated to reach a total of fifty thousand pounds in a strong cross wind, clearly indicated the need to devise a system which would travel on rails." Early in 1930, in order to provide a greater measure of control, BuAer decided (over Rosendahl's protests) to adopt a system mounted on rails—"a natural and logical development," in Bolster's view. "Thus the problem of holding the bow portion of the ship for any mechanical handling system was neatly provided [for]."[23] It was this scheme—with its masts, circles and ancillary equipment—that was finally installed. A self-propelled railway mast, equipped with a telescoping mast top, was used by *Akron* for her operations, by her sister, and by *Hindenburg*.[24] Even *ZR-3* exploited a docking track-plus-trolley system and, during 1934–37, used the light rail mast.

One seldom-cited aspect of the ZR program is here exemplified: the endlessly experimental nature of the technology. BuAer had not hesitated to authorize trial installations when, it appeared, valuable lessons might be learned. As this pertained to *Los Angeles*, a liberal policy was adopted concerning modifications. One outcome: homegrown departures from the original interior arrangements, equipment, and so on.[25] The working atmosphere, this openness to innovation, lent a certain sweetness. Reichelderfer perhaps expressed it best. "One of the very intriguing things about airship duty," he remarked, "was the creative aspect of that whole operation. It was an experimental and research organization almost as much, perhaps more, than it was an operating organization. So we were all busy trying to think of new ways to improve airships and new ways to define and measure and quantify the elements that came up in airship operation. . . . The research input was very strong, and this made it a *delightful* kind of duty."[26]

Lighter-than-air hummed with innovation. The hectic evolution recited above was but one of innumerable schemes, each designed to advance—and thus serve—the cause. Initiated by Rosendahl and by idea men such as Bolster and Wiley, the mast innovations are illustrative of the man's zeal, energy, determination—and his growing influence. Thrust into prominence, this handsome figure sought (and received) an unusual degree of attention.[27] As commanding officer, he was instrumental in keeping *ZR-3* before press and public as well as the naval community at large. Prior to 1928 especially, a strong body of opinion held that restrictions on her military operations did not justify keeping *L.A.* in commission. Yet missions were found. "This was really a matter of survival," Reichelderfer later mused. "The constant pressure for funds, the competition between heavier-than-air and lighter-than-air showed itself in the fact that the [HTA] proponents would emphasize the limitations" inherent in the airship—large ground handling crews, low speed, and so on—"while we in the airship field would seek *always* to find some new use, particularly a use that an airplane couldn't do."[28]

Los Angeles's 106th flight and attendant publicity were vintage Rosendahl. The fleet carriers USS *Saratoga* (CV 2) and *Lexington* (CV 3) had just been commissioned. The lieutenant commander proposed to demonstrate their utility as "emergency fueling stations" by landing on one of the forty-thousand-ton vessels. On the closing day of 1927, Rosendahl visited *Saratoga* (at the Philadelphia Navy Yard) and conferred with her commanding officer and other officers, who proved willing to attempt the project—provided it did not interfere with ship's itinerary. Early in January, the two skippers directed a request to the CNO's operations staff, seeking approval. Final arrangements between the carrier and Lakehurst were effected via radio.

The experiment was initiated on 27 January 1928, thirty miles off Rhode Island. This was the year's first sortie; exceptionally poor weather had held *L.A.* indoors.

After departure at 0354 (Fulton on board as observer), course was set for the coastline. Rosie had Newport below by 1200. Here he maneuvered with bare steerageway in the vicinity of *Saratoga*, at anchor in Narragansett Bay. At 1223, both stood out the channel.

At 1415, well to sea, landing stations were piped and preparations made for touchdown. With the carrier steaming into the wind, *ZR-3* executed a "rehearsal landing approach" one hundred yards on the carrier's port beam. At 1440, helium was valved for one and one-half minutes; the approach commenced at 1453. Altitude was cut to 225 feet as Rosendahl maneuvered at various speeds to overtake *Saratoga*, which, thanks to a light but gusty wind and rough seas, lay pitching and rolling. (For the trial, a steady, stronger wind would have been better.) At 1524, he had the flight deck beneath, but a gust carried *L.A.* abruptly up to four hundred feet. Another valving, this time for forty-five seconds, as again Rosendahl maneuvered in. At 1526, trail ropes were dropped to waiting sailors, and the dirigible struggled down. During this hauling-in, a gust drove it downward and to starboard, a movement checked by dumping a thousand pounds of emergency ballast—thoroughly drenching some deck crewmen with water and denatured alcohol in fifteen-degree air. A landing of sorts was realized at 1535.

Success was short-lived. The carrier was maintaining bare steerageway, but the wind continued very gusty, breeding unwanted accelerations. Because of the flight deck's pitch and roll, holding on to the control car proved hugely difficult. "The vertical movement of *Saratoga*'s

USS Saratoga (CV-3) *as seen from her sister,* USS Lexington *(CV-2). To hold any military future, the naval rigid would have to integrate with aviation as it was evolving around the carrier. Despite suspicion, even hostility, the promise of the large airship was compelling enough that the Navy could hardly ignore it. (ADC John A. Iannaccone)*

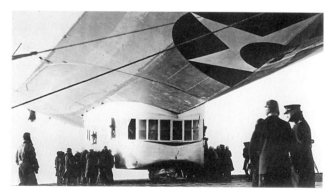

Thirty miles off Rhode Island, in rough seas and gusty wind, L.A. is held to the flight deck of USS Saratoga, *27 January 1928. Time in contact: seven minutes. In this rendezvous Moffett saw movement toward the fulfillment of his idea of big airships steaming with the Fleet. (National Air and Space Museum, Smithsonian Institution (SI Neg. No. 2000-11189))*

flight deck as the ship pitched," Bolster said, "was in direct and violent opposition to the airship's desire to stay fixed in space. Fortunately, the [air]ship had a resilient bumper bag under the control car, but it soon became evident that the ship would be severely damaged if we lingered there."[29]

At 1542, after seven minutes in contact, a gust lifted the forward part of the car out of the hands of the party, the sudden strain parting the port trail rope. Rosendahl had had enough. "It was deemed expedient to cast off entirely," the log dryly records. The abrupt departure stranded Wiley, who had disembarked to assist with the landing party.

The operation had consumed about an hour—a fine piece of shiphandling. Maybachs at "Ahead Half," course was set for home. The New Jersey coastline was crossed at 1831; by 1847 ZR-3 was circling the field, cutting out water recovery on all engines. Back in berth (2000), the airship was ballasted with sand and water, and flight personnel disembarked. Ground crew was dismissed, the west doors shut, the hangar watch set. Inspecting all maneuvering valves, heads wires, cells and other parts, the watch found all to be satisfactory. "Entirely superficial" damage had been sustained—several bamboo members of the bumper-bag structure broken and about six duralumin members of the control-car bottom showing deflection.

Not until 11 February did ship and crew again push aloft.

Rosendahl's benign account of the experiment reveals the undoubting, generous, upbeat faith in the platform that was his signature: "In spite of the fact that this was the first attempt with a large airship and although hasty arrangements and only crude equipment were available on this initial attempt, this operation is considered to have proved the feasibility of landing an airship on the deck of a large ship."[30]

In this rendezvous Moffett saw a movement toward the fulfillment of his idea of ZRs steaming with the fleet; he assessed that it "greatly increased the possibilities of refueling dirigibles at sea, and thereby means a tremendous increase in the radius of action of such cruisers of the air." When Fulton returned, he found the admiral radiating satisfaction. Previously impatient with Lakehurst, he greeted his aide with the remark, "At last they're really beginning to do things there."[31]

The experiment was not repeated. LTA exploitation of aircraft carriers would be revived late in World War II, but only to replenish blimps.

Late in January 1928, CNO directed that *Los Angeles* sortie to *Patoka* in the southern Chesapeake, off Old Plan-

Rear Admiral William A. Moffett (center) congratulates Lieut. C. C. Champion, USN, for his climb to 33,455 feet—a new record for C class seaplanes, Hampton Roads, April 1927. An ardent supporter of naval aviation, Moffett's defense of the airship was a holding action, intended to buy time to evaluate its promise for the Fleet. (National Archives, 460041)

tation Flats, Virginia, on about 10 February. Later, when the tender had steamed to Cuba, *L.A.* was to moor to her there. Rosie had a more audacious plan—a leap clear to Panama, using *Patoka* as an intermediate fueling stop. If successful, it would be the first nonstop flight by airship to the Canal Zone—pioneering a route, perhaps, for future commercial rigids. Moffett was enthusiastic; after due consideration, the secretary approved, to Rosendahl's agreeable surprise.

Coco Solo, an old station commissioned for seaplane patrol, was too small. Hence France Field—an army base at the canal's Atlantic terminus—was made available. By way of preparation, the Lakehurst command hurriedly dispatched Scott E. Peck to the Caribbean. Arriving by steamer, the lieutenant had with him the wooden stub mast and a swivel pneumatic wheel for the ship's after car.

Peck held a designation as Naval Aviator (Airship) No. 976. A crew member (machinist) of *DN-1*, the navy's first airship, Scotty had later been relief pilot for the blimp *C-5* during its abortive transatlantic try. Reporting for refresher training in the spring of 1927 (Class IV), he would be redesignated in 1929 (No. 3476).[32] Among his Lakehurst classmates: Lt. Rodney R. Dennett, Lt. Cdr. Bertram J. "Bert" Rodgers (No. 3413), and Lt. (j.g.) George F. Watson. A year's training ensued; by May 1928, the trio were signing log entries. That July, Watson reported on board *ZR-3* for duty.

Table 4-1 Flight organization chart, USS Los Angeles (1928). (Rigid Airship Manual)

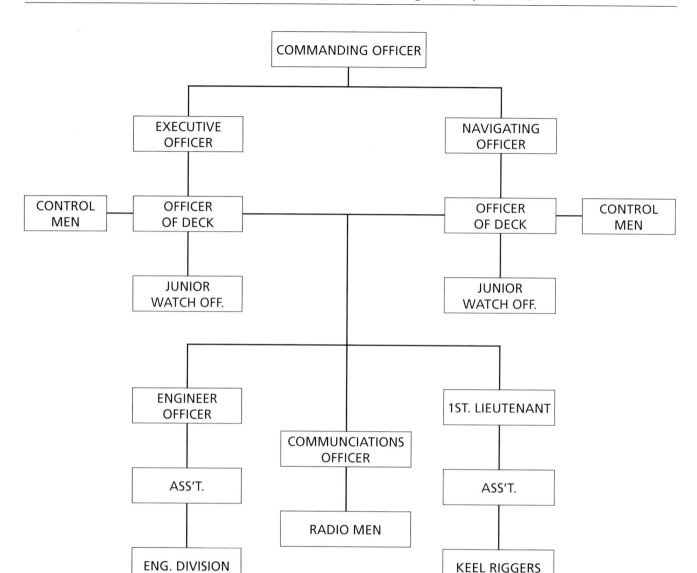

Peck did not join ship's company until January 1930. Meantime, he was quick to set about his work. Erected within days, Peck's Panamanian mast consisted of a sixty-foot guyed spar about eighteen inches in diameter. Atop, its mast cup was mounted on the nose spindle salvaged from *ZR-1*. No winches were installed; *L.A.* would have to be walked to the mast by the landing party. Accordingly, three hundred soldiers received instruction in handling *ZR-3*—and also in keeping back the curious.

Cast-off for Panama was taken at 0656, 26 February. The flight list held forty-one: nine officers and a crew of thirty plus two officer-students. The objective lay 2,200 miles distant—the airship's longest run since delivery and all of it, save for the Bahamas, Cuba and Jamaica, over water. "For this reason," one paper mused, "the flight is regarded as one of prime importance and a great test for the craft."

Once the ship had risen slowly to 1,500 feet,[33] all engines were telegraphed to "Half," and course was set for New York. From there, at 0800, departure was taken south. Air temperature: six below. The next hours brought snow, sleet, rain, and fog. Off Cape Fear was turbulent air—courtesy of following trade winds over the Gulf Stream. *L.A.* was plotted 240 miles offshore by 1600, between Savannah and Jacksonville; at 2000, she hung three hundred miles east of Daytona. In the midwatch of the

27th, floating in fog, visibility zero, position about fifty miles south of Great Abaco Island, Rosendahl was seeking radio bearings from Nassau, British Bahamas. Its lights were sighted just before 0105, nearly nineteen hours out. Pace: better than fifty miles per hour. Having circled and dropped greetings, the dirigible pushed on. The Cuban island lay beneath the keel by 0515. *Patoka* was sighted at 0632, at anchor near Media Luna Cay.

Rosendahl so advised the Navy Department and, with the northeast trades assisting, set course direct for Colón seven hundred miles farther on. In Washington, the navy communications office disclosed that *ZR-3* was "a little bit south of Cuba and proceeding to the Canal Zone." Early in the first watch, Colón was in view two points on the port bow, distance twenty miles. At 2130, *Los Angeles* was circling preparatory to landing. Trail ropes dropped, she eased into waiting hands at 2200, whereupon the ship was walked to the mast—her Panama "terminal."³⁴ The nose cup was locked in at 2247. Fueling commenced. In alternate sections, the flight crew went ashore, as the skipper had it, "to rush the shower-baths and also to send home the inevitable postcards." Meanwhile, the taxiing wheel and tail drag were attached.

Nonstop flight time New York to Panama: nine minutes shy of forty hours.

L.A. loitered but briefly. At 0600 28 February, Peck reported on board "for temporary

ZR-3 secured to an expeditionary stub mast at France Field, Canal Zone, 27–28 February 1928. The nonstop Lakehurst-to-Panama sortie (nine minutes shy of forty hours) was her longest since the transatlantic delivery. (Capt. Gordon S. Wiley)

Los Angeles lifts away from her Panama "terminal," 28 February. Time at the mast: 11 hours, 25 minutes. (National Air and Space Museum, Smithsonian Institution (SI Neg. No. 2000-9089))

L.A. *passing over the Gaillard Cut, Canal Zone, 28 February, en route to homeport via Cuba and the mast-ship* Patoka. *(H. J. Applegate Collection)*

duty and transportation to the United States." Four hours thereafter, flight crew embarked, the airship was weighed off in equilibrium; at 1012, it lifted away. With water recovery cut in on all VL-1s, she rose to 1,300 feet. Landing stations were piped down. Low cumulus and the rough terrain made for bumpy air as, on various southeasterly courses ZR-3 jaunted over the isthmus and its great canal. Above Culebra Cut, vertical currents tossed her up and down at high accelerations. Within ten minutes, at 1110, ZR-3 had the canal's Pacific end below. Having circled Panama City, the rudderman steered northerly courses, transecting the isthmus once again. France Field lay beneath at 1213—and a Panama departure was taken.

With the trades bow-on, a nonstop return run was unwise. Engines signaled to "Half," Rosendahl plodded through the tropical afternoon and evening. Objective: Cuba, and a waiting *Patoka*.

At 0615 on the 29th, the altimeter showed 1,200 feet, and Jamaica's South Negril Point lay eight miles west. The aircraft pushed on, five Maybachs pushing at half speed. Twenty-two minutes more had land in sight ahead, and at 1058 the tender was too. As Rosie maneuvered on various courses, speeds, and altitudes, lift was valved for over three minutes, water recovery cut out, an approach begun. Minutes after lowering a grapnel the device engaged; in eight minutes more, the main wire had been coupled. The yaw guys were joined and at 1303 a mooring effected. Average speed this leg: twenty-nine miles per hour.

Naval commanders have considerable autonomy. Intending to overnight only, Rosendahl postponed takeoff when advised of unfavorable conditions at homeport. Swinging lazily to her floating mast, ZR-3 was refueled, oil and provisions were put aboard, and the maneuvering valves were inspected. Sundry pleasures were savored: the airmen enjoyed shower baths, a movie, and, on *Patoka*'s deck, the balmy tropical air.

Midmorning, 1 March, found preparations being made for getting under way. All engines warmed, *L.A.* was weighed off and 778 pounds of fuel valved; cast-off was

Los Angeles *moored to* Patoka. *(Capt. Gordon S. Wiley)*

logged at 1040—with twenty-three degrees of superheat. (The 1,300-mile return, her skipper had radioed, would take about twenty-five hours.) Course was set directly north. Over the island, though, the air proved choppy and turbulent. Despite the jettisoning of eight hundred pounds of ballast, ZR-3 became three *tons* heavy from loss of superheat. The captain ordered his rudderman to steer west, to return to the blue-green skirt of sea: they would cross after the heat of midday. So *L.A.* cruised the southern shoreline, the sea contrasting with brick-red mountains and the sharp coast. At 1450, number-four engine was stopped to repair a broken hose clamp. The beach again was crossed at 1710, at Port Batabano. An exotic terrain scrolled before navy eyes. "We headed for Havana," Rosendahl marveled, "across fine looking agricultural country checkered with tobacco fields, farms, modern railways, and numerous cleanly villages, but otherwise tropical in all respects."[35] Belying a clear sky with scattered cumulus, unstable conditions persisted.

Just before 1800, Havana lay beneath. At sunset, the aircraft again turned to seaward, steering toward Tampa—355 degrees.

Bow pointed up the coast, four engines telegraphed to half speed, the great hull punched through an unblemished sky. Seventy-five minutes into a new day, the lights of Sarasota shone below. Heavy surface fog set in, forcing altitude changes up to 2,400 feet. With dawn, repeated up-down height changes were ordered, seeking helping winds. Near 0900, *L.A.* lay 125 miles southeast of Brunswick, Georgia; at 1232, Charleston spread away beneath. Conditions again proved unstable: severe gusts and vertical accelerations. But a strong southwest gradient now sped the airship along. Norfolk filled the starboard-bow windows around 1900; during the first watch, the Chesapeake was overflown, Maryland and Delaware crossed, and at 2250, south Jersey gained. The lights of the naval reservation were sighted at 2300.

The Panama venture was not yet done. On leaving Cuba, the department had ordered *L.A.* to assist in the search for a missing amphibian plane with T. G. Ellyson (Naval Aviator No. 1) aboard. Rosendahl reached the search area after dark, so he elected to fly on, refuel at Lakehurst, then hunt by day. A cold front was approaching; dispatch was essential. With luck, *L.A.* would be away before it came. But the front was near.

At the mast, conditions proved exceedingly turbulent, with gusts to fifty miles per hour. The mood was tense and uneasy. Throbbing engines could be heard through a cutting wind while "at the end of the field, the brilliance of the lights on the mooring tower held out a promise of a landing." For night moorings, the position of the mast's cable end—on the ground, downwind—was marked by a man swinging a light. This told the captain where to drop his cable, maneuvering against the blow. Not until 0014 was ship's main wire dropped. The port and starboard yaw wires followed; these, in turn, were attached to cables led from winches. Hauled from side to side by the yaws, dipping with each pull, the floating mass "crabbed" instead of "watched" (varying) into the wind. The rudder jammed, preventing correction for a gust. Then the starboard yaw parted; this was knotted back together. As the main cable was taken up, the wind, in seeming rolls and waves, attacked the cranky hull—oscillating, pitching, yawing, resisting. With *L.A.*'s bow still fifty feet from and slightly above the masthead, her main cable (nine-sixteenths-inch wire) snapped in protest. Time: 0120. His mooring gone awry, Rosendahl sang out: "Cut the yaw lines—cast off." *Los Angeles* shot to a thousand feet as, wearily, men on the ground plodded back to the ZR hangar. The aircraft hove to and recovered her dangling cables. On the ground, as radio messages passed between ship and station, "impatience gave way to disgust, and awaiting scores below scanned the heavens with anything but favor in their looks. . . . Within the hangar, the customary gloom, vast and cheerless in the cold space, was broken only by the coming and going of officers, men and visitors—for visitors there were, even at that early hour." After an interval, a ground crew scuffled back outside—and waited, in edgy impatience, stamping feet and swinging arms.[36]

At 0335, landing stations were piped anew along the keelway. *L.A.* settled into tired hands, the main car scraping sand and dry grass. Time: 0358. A trio of officers disembarked—to assist matters on the ground—"but some sort of intuition," her commander recalled, "kept me in the control car, contrary to my usual practice." *L.A.* was being walked to the south trolley rail amid the bark of orders when an abrupt snow squall ambushed from the northwest, hitting the hull almost broadside. ZR-3 lay just short of the shed. Before she could pivot to the thrust she wallowed sideways, toward the trees, dragging the clinging heels of three hundred attendees. The colder air had generated lift—several thousand pounds' worth. "A blinding curtain of snow let down at the same moment." In driving white the voice of Lt. Cdr. Vince Clarke rang on the wind: "Hang on—hang on!" Then a bark from Rosendahl: "Let go everything!" Safety, again, lay aloft.

In seconds, the order was repeated, signal bells clanged,

motors throbbed in reply—four set to "Ahead Half." As *ZR-3* pushed away, the door to the bridge snapped open and an officer rushed forward, breathless. "My God, Captain, we've carried a bunch of men up on the handrails. We've got to get them in at once, sir!" Six handlers had not released their grip: four had been lofted on the lee side of the control car, a pair on the after car. One of the two aft dropped at fifteen to twenty feet, spraining a toe; the other, BM2c Bruce McClintock (a member of ship's company) had hauled himself inside. The imperiled four were told to hold on as, at five hundred feet, the engines were stopped. The word "help forward" passed, and the off watch produced ropes, improvised rescue parties, and punched out windows. Using hands and lines (and verbal encouragement), human chains began pulling the danglers up and in. Crawling out and along the main rail, Seaman Second Class Donald L. Lipke (holding on by one hand) reached down and helped two men gain purchase. Lipke (later an *Akron* casualty) was the last man back on board. "Always on the job, Richard Peak, our efficient air-going cook, soon had hot cocoa and soup for all hands."

Noses were counted. On the ground, tension vanished when the number was radioed down.[37] By 0500, *Los Angeles* hove to—ground speed zero—awaiting daylight.

With break of dawn, the dirigible concluded its emergency two-hour, eighteen-minute getaway and alighted for another go. The field lay snow carpeted. Walked to the east trolley rails, *L.A.* was shackled to the taxiing cars. But more than half her length still protruded when, ever capricious, the wind shifted to north. Gusts swirled snow about ship's stern. Two emergency bags drained empty—1,100 pounds—as *ZR-3* was hustled inside.

Weary crew retreated to bunks, Rosendahl to his office as reporters sought morsels. "It was great," Peck volunteered. "We had a wonderful time and didn't have the slightest sign of trouble until we reached here. And, of course, the difficulty here was caused by faulty weather conditions. . . . Cruising over Cuba in the warm sunshine was like a fairyland." In his relief, Rosendahl praised the Weather Bureau's cooperation and called for new devices to land dirigibles. "I believe some mechanical device can be developed that will bring the ship down and run her in safely and without any such ground crews as are necessary now. You can see what a big saving that would be in overhead and men." As for *L.A.*, she would never be used for transatlantic runs, he told the reporters. "It is too small," he said. "She's just about the right size for this kind of trip. Four thousand miles is about enough for her. When it comes to regular transatlantic service we'll have to have bigger ships."[38]

Arriving naval messages saluted top-flight airmanship. In the big room, a watch turned to policing the dirigible, inspecting maneuvering valves, installing covers over the automatics, putting the walkway cover in place. The septic system was drained, tanks flushed, miscellaneous upkeep attended to. At 1010, refueling started—and two customs officials insisted on going through all returning suitcases.

Now came an extended pause. The northeastern winter kept *L.A.* in dock for the balance of March. Complaining to the CNO about Lakehurst's location, Rosendahl pushed for a more southerly, less stormy haven. Permission was obtained to land at the only ready alternate, Langley Field, in Virginia. In 1929, at Parris Island, South Carolina, an expeditionary mast (portable stub mast) would be erected—equipped with a path for taxi-wheel travel and anchor blocks for handling yaw lines. Meantime, Rosie fidgeted. During April, his command was laid up for a general overhaul and for installation of new cells. *Los Angeles* did not again leave the ground until 2 May—an eight-hour, thirty-five-minute sortie along the Atlantic coastline (during which tests for output of the cooking-generator propeller were run). A four-week hiatus ensued; this ended on the 30th, when seventeen hours were logged into Pennsylvania airspace as far Pittsburgh.

Midday, 30 May, found crew and captain circling Bettis Field, Pittsburgh, where the National Balloon Races were under way (two navy entries). Next came a Memorial Day flight—the essence of public relations. Rosie unmoored from the stub mast at 1017 with nine passengers, including members of Congress. Again he pushed west, this time to West Chester, Pennsylvania. At eighteen hundred feet "Stop" was rung up as *Los Angeles* hove to over the grave of the late Hon. Thomas S. Butler. Floral tributes were dropped. Butler, a "big navy" Republican, had chaired the House Committee on Naval Affairs, helping to push through the legislation that had authorized construction of *ZRS-4* and *ZRS-5*.

June saw less than fourteen hours' flying—a run to *Patoka* off Newport in company with units of the Scouting Fleet, and the return.

The month of July held a half-dozen flights—and a change in command. Eckener had invited one officer to Friedrichshafen, where *Graf Zeppelin* was nearing completion. Rosendahl was chosen, and Wiley assumed temporary command. Detached from his duties, Rosendahl

The British R-100 *at the St. Hubert Airport mast, south of Montreal, 1–13 August 1930. Despite the apparent success of this intercontinental run, development of imperial links via airship evaporated to nothing. (Elizabeth Tobin)*

returns, *L.A.* secured to the high-mast cup. Next day, 25 July, unmoored at 0206, Wiley again steered for New York Harbor.[40] Over Governors Island, the rudderman spun his wheel to port, to follow the Hudson. The river path was quit at West Point, when the ship's bow was pointed northwest, toward Oneida. As the topography climbed in elevation, altitude was increased toward pressure height. (The alarm functioned at 5,225 feet.) At 0955, *L.A.* reached a maximum altitude of six thousand feet; at this point, the dirigible had lost 1 percent of lift through the automatics. Wiley circled at Oneida, after which he ordered an easterly course; the Hudson was regained near Albany. At 0035, 26 July, *Los Angeles* stood secure in the hands of a ground crew, which walked her inside.

Undocked at 1834, 30 July, the airship steamed south, to follow the Philadelphia-to-Baltimore airmail beacons. Late-nighters in Delaware and Maryland could see its lights. The homecoming brought a lesson in aerostatics. Landing stations were piped (at 0345, 31 July) and the craft weighed off—approximately 250 pounds light at four hundred feet. Five minutes thereafter the approach commenced—but was quickly aborted. Wiley maneuvered back in, dropping the trail ropes over the landing party. To combat a fifteen-degree inversion, seventeen ground crewmen embarked via Jacob's ladder. Forty-one minutes after his initial approach, Wiley settled his command into waiting hands.

arrived in Germany at midmonth. Flight trials were still two months off, so the officer proceeded to England, to confer with members of the *R-100* and (rival) *R-101* organizations—each eager to establish imperial links via airship.[39] Meantime, Lakehurst received orders to make itself ready to receive this newest zeppelin. Wiley's first liftoff as skipper took place on the 16th. Following the coastline southward, he took her as far as Atlantic City. There course was reversed and the shore followed to the northward; at midnight, *ZR-3* hung over New York City. By 0500, 17 July, Wiley had secured to the stub mast, only to cast off at 0957—again for New York. On the 24th, a third appearance was logged over the port city. During the revisit, *Los Angeles* exercised in formation with *J-3* and *J-4*. On both

Lieut. Cdr. Herbert V. Wiley, Navial Aviator (Airship) No. 3183, at a portside window of the control car with son David. Wiley and Rosendahl were to log hundreds of shared hours until Wiley succeeded to the command of ZR-3, in 1929. "Doc" Wiley was the only officer to serve aboard each of the Navy's four commissioned rigids. (Capt. Gordon S. Wiley)

Herbert Wiley retained command until late October, when Rosendahl returned with *Graf Zeppelin*. Meantime, the lieutenant commander pursued a strong schedule of operations. When he could. On 7 August, ground handlers were cut by a draft of fifty men; until replacements arrived, the force remaining was insufficient. Four sorties were recorded in August,[41] and for September again four.[42] After landing on the 10th, a hurricane in the south washed out further flying. By the 27th it was decided that conditions would not improve, so the overhaul scheduled for 1 October commenced. When it concluded, Wiley skippered *Los Angeles* on its longest flight since the transatlantic leap.

The desire of Fort Worth's citizenry to see a big airship—the *"Graf"* or ZR-3—had caused BuAer to cable the Bureau of Mines helium plant (on 10 August) to put its mast in shape to receive the latter "as promptly as practicable." An American Legion convention slated for early October in San Antonio caused the bureau to issue orders for a cross-country. At Lakehurst, *L.A.* was pronounced "ready for flight" during the afternoon watch, 6 October. Within minutes, the vessel was undocking for Texas.

General assembly had been sounded at 1730—the call for ground crew. At 1800 the ship was weighed off in equilibrium, one section of flight crew aboard. Walk-out was through the west doors into kindly weather—a west-northwest breeze and clear sky. The forward car and trolleys released, and the second section scrambled aboard. At 1825, *Los Angeles* lifted off with eleven officers and thirty men. (To superintend arrangements for receiving her, including daily tabulations of weather conditions, Peck had been dispatched ahead.) With four hundred pounds of emergency water dropped fore and aft, "Cruising" was rung up on the annunciators in cars two, three, four, and five; number one cut in at 1830. Ten minutes later, water recovery was cut in on all units, and a course was set for Richmond.

L.A. would be away until the late hours of the 10th, daunting duty only ship's crew could truly appreciate.

Richmond lay beneath *ZR-3* just after midnight, Atlanta nine hours later, and at 1250 Montgomery, Alabama. Mobile was traversed that evening en route to New Orleans. In darkness, the Mississippi River was crossed at the coast. At 1932, New Orleans spread before the ship. Here Wiley circled. About 2115, a falling meteor was observed to southward, then, at 2132, severe lightning from the same quarter. As the log records, "Observed lightning to the southward, evidently over the gulf but moving inland in a northeast direction; changed course to maintain safe distance from the storm center." The flashes disappeared astern. That midnight, over Lake Charles, Louisiana (and closing on the Texas line), the big ship hung at 2,200 feet, engines one, three, and four at "Half." The ship was approximately three hundred pounds heavy, with the following disposable weights stood logged: 13,776 pounds of fuel, 9,545 pounds of water ballast, 1,200 pounds of oil, and sixty-eight pounds of kerosene. Forty minutes into the new day, 8 October, *L.A.* crossed the Sabine River; within ninety minutes, the sighting of Beaumont, Texas, was logged. Course was set for Houston. The city's lights shone at 0340, whereupon a new course was steered—for San Antonio.

Los Angeles arrived in mid-morning, just as the parade was forming. Steering various courses and speeds, she circled for eighty-one minutes and dropped greetings at Kelly and Brooks Fields. Departure was taken at 1056. Austin went

Los Angeles *in her prime—here steaming northward near San Antonio, Texas, 8 October 1928, Wiley in command. (San Diego Aerospace Museum)*

into the log at 1220 then, almost four hours thereafter, Dallas. Wiley circled briefly before a new course was set, this time for Fort Worth and his objective: the Naval Helium Production Plant and its mast.

The nation's sole facility of its kind, the plant had commenced operation in 1921. Thereafter, depending on the volume of natural gas available for processing, varying amounts of helium had been extracted monthly. In three years, costs had been trimmed to sixty-five dollars per million cubic feet at 94 percent purity. This cost reduction was insufficient. "It was believed that the lowest cost had been reached for the process as originally installed and various ways for improving the process were studied." Changes in the carbon dioxide–removal system realized especially high efficiencies, enabling a major modification to the liquification system that, by 1925, increased the recovery by nearly 40 percent.[43]

When Production Plant One was closed, most personnel were transferred to the Bureau of Mines' new Amarillo plant. The first separation unit was placed in regular production in April 1929. Within days, one full tank car of 98 percent purity was shipped—the highest-grade helium yet produced.[44]

By 1715, Wiley had the Texas high mast below. A semiportable version of the one at Lakehurst, it had been erected in 1924 on land adjacent to the helium plant. Essentially a sixteen-inch shaft of steel tubing 160 feet in height, the mooring mast was supported in a concrete foundation and braced by three sets of guys. At its top were the upper and lower operating platforms and the mooring mechanism. A climbing trunk as well as gassing main, water, and fuel lines were built in, along with a machinery house (three winches, lighting panels, pumps and other gear).

On board ship, helium was valved for twenty seconds; at 1741, landing stations were ordered. An approach was under way within fourteen minutes. Again all maneuvering valves opened, for fifty-five seconds—but *L.A.* was still too light. Wiley stood away. The sun set at 1806. For his second try, forty seconds more were timed on the valve cords until, at 1819, cables were lowered. With twenty-five thousand people marveling, a difficult moor was completed at 1837. Refueling got under way.

Time in the air: forty-eight hours, twelve minutes.

During the layover, local aviators were advised to stay at least two thousand feet away.[45] As for autos, all were kept beyond a thousand-foot radius of the mast. Army cavalrymen and dismounted guards maintained order, assisted by police on special duty. For ship's officers there waited a suite of rooms at the Fort Worth Club. As the midwatch log recorded, the earliest hours of the new day, Tuesday, 9 October, passed more or less routinely on the bridge: "Moored to mooring mast, Fort Worth, Texas, with the following disposable weights on board: fuel 16377 pounds, water ballast 10900 pounds, oil 1957 pounds. Wind SSE, 10 knots steady. 0105 Completed gassing having received 130000 cubic feet. Adjusted water ballast to compensate for changes of lift."[46]

Wiley lingered just shy of fourteen and a half hours before lifting off—at 0858, on the 9th. Instead of setting course straight for home, he exploited the circulation then prevailing aloft by steering northerly courses, en route Chicago, before he steamed east. Oklahoma, Kansas, Missouri, then Iowa were crossed in turn. At 0039, on the 10th, the Mississippi River at Iowa Junction lay beneath the keel. Over the Windy City, Wiley ordered course eastward, toward the barn, via Toledo and Youngstown to Pittsburgh—a familiar track.

In her peregrinations, *Los Angeles* often followed airmail routes and established airways. Planes presented a constant hazard, so a paralleling off-freeway course usually was steered. In October 1929, for instance, the first watch had the dirigible "cruising on various courses following about 2 miles to W of airway beacons Newark, New Jersey, to Philadelphia, Pa." En route back from Parris Island in January of 1930, the 1800-to-2000 watch recorded *ZR-3* "over Washington, D.C., following Washington–New York air mail route, to northward." When it was necessary to cross an air lane, traverses were made quickly, then the flight track resumed.

The New York–Philadelphia airway had seen a near miss. At 2154, 31 August 1928, while *Los Angeles* was crossing the Delaware River at Yardley, a head-on collision was narrowly avoided when a mail plane approached close aboard, heading directly for the airship. Swerving, the airplane passed within a hundred feet. In a letter to Moffett, the Pitcairn Aviation Company reported *ZR-3* as not carrying lights—when in fact all navigation lights were on. In his report, Wiley stated that the pilot was on the wrong side of the airway.[47]

En route from Texas via Chicago, the bridge had Philadelphia's lights in sight at 2010, 10 October. Within forty-eight minutes, homeport lay below. Settled to a ground crew at 2133, *L.A.* was walked toward the hangar's east end, where, under the glare of carbon-arc lights, she was positioned. The forward car and docking trolleys were secured, whereupon she was struggled across the sill into berth, the space ablaze with light.[48] The ship's log had her docked on cradles forward and aft by 2232. The walkway cover was put in place. Forty-three men disembarked, more than glad to be back. Time from Fort Worth: thirty-six hours thirty-seven minutes flying—slightly more than one and a half days.

ZR-3 now boasted 134 journeys in her log, aggregating about sixty thousand air miles. This bespoke genuine progress and intimated the potential of lighter-than-air travel. As (some) reporters acknowledged, astonishing progress had been gained. The midcontinent sortie alone stood as a remarkable performance. The airship and airplane *both*, it appeared, should "prove indispensable to mankind and become commonplace. . . . The future of the commercial airship holds brilliant promise. In less than five years transoceanic passenger traffic will be on an established basis."[49]

Five days following the Texas return, as if to underscore that augury, *Graf Zeppelin* reached Lakehurst—the first of its four arrivals. Concluding a 111-hour crossing—to the immense relief of base personnel—the new zeppelin pounded in at 1738—dusk, Monday, 15 October.

The ship had been slated to arrive Sunday morning, and by Monday the field lay awash with visitors—as many as fifty thousand. Automobiles had begun collecting on Saturday and throughout that night, filling designated parking. Special one-day excursion trains also had descended. Augmenting the regular watch, base personnel on special detail had charge of the parking and of keeping clear a road to the dock. Sunday, however, brought neither zeppelin nor word of its position. The invasion became occupation. To divert the milling mobs, the air station filed visitors into the berthing space, for a look-see at *Los Angeles*. By noon, supplies of food had been exhausted; not a morsel more could be had. For about fifteen miles out, all roads leading to and from the base lay blocked.

Dr. Eckener made landfall—the Virginia Capes—early on the 15th. An airman of flamboyance and style, he steered north, parading his ship before Washington, Baltimore, Wilmington, Philadelphia, New Brunswick, Trenton, and Newark before arriving over New York City at lunch hour.[50] Here Eckener circled. Then came the last miles, hanging low and close—at about one thousand feet, eyewitnesses could see the rent fabric of the port horizontal fin. (The damaged cover had been corrected by well-advertised mid-air repairs.) When Lakehurst sighted the "*Graf*," heightened emotions soared. Held back by tired ratings, thousands gaped as broadcasters described the scene from perches about the hangar. Lines were seized and run athwartships, "spiders" connected. Further groups grabbed on, helping to spring the aircraft down-

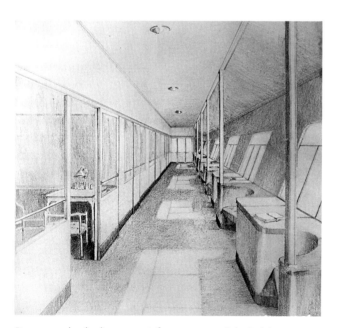

Promenade deck concept for commercial airships, 1928—"from which the ground can be watched through outward-tilting windows." The promenade spaces designed for Hindenburg *resemble this portrayal. (San Diego Aerospace Museum)*

A passenger stateroom as conceived by Goodyear-Zeppelin, 1929. (San Diego Aerospace Museum)

ward; the handrails dropped near and were seized. *LZ-127* hung just clear of the sand, almost opposite the doors and facing a slight breeze—twelve miles per hour, according to the portable anemometer.

Visitors surged in. When Eckener emerged, he was shouldered by a pair of burly policemen, then carried to the dock and a waiting press room. Inside, he saw *Los Angeles*. Pointing proudly, "There she is," he called out. "There's my sweetheart."[51]

Not until 0200 was the German zeppelin hustled into the south berth, through a shed-door clearance reduced by half (owing to the presence of *ZR-3*). Prior to official entry into the United States, the air passengers met a gauntlet of requirements: public health officials, then immigration, then customs. As for baggage and such, the navy men turned to—impressing Eckener and his party. Their duties done, German and selected U.S. Navy guests—including Bauch, Callaway, Carter, Jackson, Moffett, Peck, Pierce, Richardson, Rosendahl, Settle, Thornton, Tyler, Watson, Wiley, Wicks—were shuttled to Manhattan, to be feted by the glitterati of New York City.

Eckener was back.

Reprovisioning plus repairs consumed a dozen days. In the meanwhile, an average of twenty thousand people trekked to the reservation daily. On the 21st, a Sunday, an estimated 150,000 packed in to marvel. *Graf Zeppelin*—776 feet long, its hull embracing 3,700,000 cubic feet—seemed to fill the great room. Alongside sprawled *ZR-3*. Overhead, the naked fin—stripped of its fabric, which was being replaced. Because of this work, the erection of a mobile mast plus *J-3* and *J-4* snuggled beneath, *ZR-3* (wedged into one corner) lay effectively immobilized.

The station played host to delegations of every stripe. Open houses were huge entertainments. For weekends as well as holidays and special occasions, a platform usually was

Dr. Hugo Eckener, airship operator par excellence, at Lakehurst, spring 1927. At that date LZ-127 was under construction; it would take to the air in October 1928 as the Graf Zeppelin. Standing, l. to r.: Lieut. Cdr. Herbert V. Wiley, Lieut. Roland G. Mayer (CC), and Cdr. Garland Fulton (CC). Seated: Capt. Edward S. Jackson, commanding officer of the station, Eckener, and Lieut. Cdr. Charles E. Rosendahl, commanding officer USS Los Angeles. As well as its military mission, the Navy had accepted responsibility for furthering airship development for commercial applications. The liaison between American and German airship men was very close. (Charles E. Rosendahl Collection)

rigged alongside *Los Angeles*'s control car. This enabled the public to step up for a look-see inside. The keel was off limits, although special credentials could procure entrée. "And of course, many officers had personal guests," Reichelderfer reminisced. "I know many a time, good friends of mine or people that I would meet, who expressed an interest in airships, came to see me, and I either got permission to go aboard from the officer on watch—the duty officer—or, if I happened to be the duty officer, then it was simple. I could just take them aboard." The base had become popular. "The airship," he continued, "the whole Lakehurst installation, the drive through the pine forests . . . was an enjoyable outing for people from New York, Philadelphia, Washington, Baltimore. And there were rather large crowds."[52]

Graf Zeppelin represented a great step forward, halving transoceanic travel times. Investors understood. On this layover, Eckener met with interested bankers and industrialists (Walter Chrysler, for one). His assistant and translator: Willy von Meister. In Washington, Edward P. Warner, assistant secretary of the navy in charge of aero-

Admiral Moffett and Dr. Hugo Eckener, captain of Graf Zeppelin, *amid lovely company at the Lakehurst hangar. A superb airmen and publicist, Eckener helped to demonstrate the feasibility of intercontinental, nonstop air transport. (U.S. Navy Photograph courtesy Navy Lakehurst Historical Society)*

nautics, espoused the intercontinental airship: "There have been thirty attempts to cross the Atlantic by airplane in ten years," he remarked, "and nearly two-thirds have failed. Every airship, on the other hand, that has tried the same thing has met with success. On the side of practicability there is no comparison."[53]

In the first hours of 29 October, the visitor was undocked for its return crossing. Embarked were twenty-five passengers and a trio of observers: Pierce, Settle, Bauch—and von Meister. Bilingual, knowledgeable of the Americans and Germans, Willy had made himself indispensable, hence the invitation. Anointed as Eckener's special representative in the United States, he would look after technical, commercial, and related matters during forthcoming transoceanic flights—fuel, spare parts, hydrogen, propane, whatever.

That same day, Rosendahl resumed command of *ZR-3*.

Reports had reached the press that *Los Angeles* would begin hook-on experiments. "Within the next few weeks," one paper had it, "the Navy will attempt to have a plane launched from and reattached to the gigantic dirigible *Los Angeles,* it was learned today. . . . It will be the first such attempt and, if successful, will mark one of the most remarkable and useful developments in aviation, Navy experts say."[54] But the final quarter-year saw little flying, thanks to poor weather and the transfer of about fifty ground crew. Three sorties only were flown. On 30–31 October, a ten-hour mission[55] tested a new three-bladed propeller on number-three engine. (Number-four engine went out of commission due to a split radiator.) In the first watch of the 31st, near Englishtown, various courses and altitudes were flown testing six-meter transmissions from the Radio Corporation of America laboratory at Rocky Point, Long Island. November saw no flying. In berth, an experimental steam superheat radiator was installed in the keel and winter flying gear was put at the ready. Varnishing of structure also was effected.

Most urgent, tests crucial to the Goodyear-Zeppelin designs were pressed. The new ships would carry airplanes. That same month, a trapeze built by the firm's engineers was attached to *Los Angeles*'s bottom girders, at frame 100, for an experimental program of hook-ons. A mockup of the water-recovery gear intended for *ZRS-4* and *ZRS-5* was installed portside, just forward of frame 115, outboard of the first longitudinal. Though the program of tests originated at Goodyear, the work was conducted under Cal Bolster's supervision, as the ship's experiment officer. With a manometer installed in the lat-

En route to the East Coast, Graf Zeppelin *meets the American Southwest on the final leg of a world circumnavigation, 27 August 1929. A fantastic demonstration, the flight altered air-transport thinking of the day, by serving to focus attention on the potential of long-distance air navigation by large airship. (San Diego Aerospace Museum)*

eral walkway at frame 115, *ZR-3* lifted off on 4 December for a series of speed runs—drag tests of the dummy recovery panel.[56] Her own water-recovery apparatus, NACA/BuAer tests disclosed, upped resistance about 20 percent and reduced propulsive efficiency about 7 percent.

Forty-one minutes into the 5th, *L.A.* secured to the stub mast and stayed out, engines warmed intermittently. The sky became overcast, and the wind increased from the southeast. Within hours, rain plus gusty conditions were challenging the on-board section. At 1245, the handling rails aft carried away, allowing ship's stern to kite to sixty feet. The crew went to landing stations as the elevators brought down the stern—gradually. The standby flatcar was hurried beneath, handling lines at frame 25 taken in hand, crewmen *and* ground personnel called onto the field. Despite improvised measures aft, it was decided to take on fuel and retreat to the air. At 1652, *Los Angeles* cast off. As she hung over the base, the struts and suspensions for number-one car were inspected and its engine tested—all OK. Rosendahl steered offshore, where he loitered before shoving off to the northeast, following the Long Island shoreline. From the New London area (0800), he elected to return roundabout via Bridgeport, New York City (1130), Newark, Princeton, and Trenton. At Camden, a sound experiment was tried. A huge "superdirectional" horn had been installed in Victor Talking Machine Company plant. As various courses, altitudes, and speeds were ordered, music and oral greetings were thrown skyward. At 1450, 6 December, course was set for home.

Amid the excitement, Lieutenant Reichelderfer reported for duty. Assigned (by Moffett) as Lakehurst's aerological officer, he would serve as in-flight meteorologist as well. Since the latter counted as sea duty, the stratagem kept Reich in aerology *and* in line for promotion.[57]

The final weeks of 1928 were used for completing installation of gear for attaching an after carriage. The hook-on equipment received shed tests, and an experi-

Via fabric sleeve and balloon, hydrogen is replenished for the German visitor from the station plant. Commercially, *Graf Zeppelin* represented a great step forward, a carrier which halved transoceanic travel times. (San Diego Aerospace Museum)

mental viscose latex cell—in the number-two position—was removed due to deterioration (after six months) and replaced with a new unit. Two days after Christmas, the press disclosed that *L.A.* had been authorized to conduct a training cruise to Florida. At year's end, her log boasted 130 flights since the commissioning (137 from the first builder's trial). Altogether, the dirigible had flown just under 1,534 hours.

The new year's first sortie was an extended operation with *Patoka*. The rendezvous proved to be the second-longest stint of Rosendahl's tenure: an excursion to northwest Florida, where the surface ship had anchored in the sheltered waters of St. Joseph Bay, west of Apalachicola. (The return from Cuba, in March 1928, was his longest single voyage.) Takeoff was at 2040, 8 January 1929, with forty-seven aboard: ten officers, a crew of thirty-two, plus five student officers. The journey proved largely uneventful, save for a (temporary) shutdown of engine number two over Maryland, due to a burned piston and scored cylinder. Cape Fear was crossed at 1200, on the 9th; Parris Island, South Carolina, lay below at 1610; Brunswick, Georgia at 1905. Altitude was upped to fifteen hundred feet, and, minutes later, a radical course change was ordered. Rather than proceeding directly to Jacksonville and Miami, *L.A.* turned inland, on a heading for the mooring ship awaiting her. Head winds had increased. In the midwatch, over interior Florida, the sky overcast and the air turbulent and gusty, ground speed dropped to ten knots—with airspeed registering forty-five. *ZR-3* became overdue. Winds moderated, but *Patoka* was not sighted until the forenoon watch. The mooring was logged at 1246, 10 January. Time aloft: forty hours, six minutes.

Port routine entailed securing the "mooring gear" (tethered drums that could fill/empty as the tail rose/fell, thus dampening vertical oscillations), draining oil from three engines, adding ballast, and inspecting all maneuvering valves—plus receipt of fuel, lubricating oil, helium, and provisions. Bridge watches that afternoon and evening (of the 11th) proved trying—gusty with frequent rain squalls, which added tons to the cover. In such downpours it took prompt action to shed ballast and then to *replenish* it (anticipating reappearance of the tropical sun). "On several occasions, the rain fell in such heavy quantities that in spite of the dropping of water ballast and even fuel, as fast as we could, it forced the tail . . . into the water. Never before or since, did the *Los Angeles* get such a thorough soaking." With as much as ten feet submerged, the lower fin took on considerable water. Upon the return to Lakehurst, ice was found within it.[58]

In Florida airspace, two excursions to the southward were taken, the first commencing four hours after mooring. But unsettled conditions thwarted a push to Miami. On the 12th—a thirty-nine-hour, thirty-one-minute cruise—St. Petersburg-Tampa, Fort Meyers, Miami (at last), Fort Lauderdale, Pompano, Boca Raton, West Palm Beach, Orlando, and Jacksonville were granted inspections of the silver ship.

Pushed by south-southwest winds, the ship took less than fourteen hours for the return to New Jersey. By the 1200-to-1600 watch of the 15th, ship's force was engaged in postflight work: placing a walkway cover in the keel, securing all automatic-valve covers, flushing out ballast bags and water-recovery filter tanks, cleaning out the refuse tanks, and policing the galley, control car, and passenger spaces. All drinking-water tanks were drained, and fifteen five-gallon tins of alcohol were removed, plus 420 pounds of food. At 1330, "Bert" Rodgers reported for duty.

Back in Lakehurst, the next month and a half witnessed no flying. Two VL-1s were unshipped for overhaul,[59] and cell seven—found to be leaking and patched in Florida—was replaced. The layup proved unexpectedly long; corrosion was discovered in areas of the hull that, inopportunely, were inaccessible with gas cells inflated. So the overhaul was exploited to clean and varnish about 35 percent of the structure and also to clean the fuel system of the residue of dirty gas. Matters were compounded when a workman, sent down between cells three and four to repair a broken wire, tore a hole in the former, necessitating an emergency deflation, removal, patching, and reinstallation. The overhaul period was complete on 20 February. *L.A.* stayed shed-bound, though, hostage to vile weather. So a leaking number-three cell was changed.

A postrepair flight entered the log on 1–2 March. Two days later, President-elect Herbert Hoover was inaugurated. Along with an assortment of other aircraft (including army blimps from Langley), *ZR-3* had orders to Washington's quadrennial pageant. *L.A.* rode to the wind overnight; in the morning watch of the 4th, at 0714, the mast crew released her. Rosendahl steadied on course 245 degrees, altitude 2,500 feet, four engines at "Cruising," one at half-speed. (Along were two hundred pounds of reporters' and photographers' equipment.) The bridge saw Townsend, Delaware, below at 1042, then Maryland. En route the sky deteriorated; over Annapolis the airmen found light rain and poor visibility. Minutes later, at 1240, Lakehurst's *J-3* and *J-4* (in company with *TC-5* and *TC-10*) were sighted a few hundred feet below, crossing the bow. All engines slowed, *L.A.* loitered at various courses and altitudes in deplorable skies—still fog and rain—before taking up column formation with the pressure ships. "Quickly falling into column, we proceeded as planned, beginning our part at the Capitol at exactly the appointed moment." It was 1500. In a providential break in the clouds, the aerial parade appeared over the Capitol as scheduled and glided up Pennsylvania Avenue. (Tuned in, Rosie and Company could hear the broadcast description—nationwide hookup—of the murky scene.) At 1505, over the Naval Observatory, radioed orders set each skipper free to proceed at his own discretion; on *Los Angeles,* course was ordered for Baltimore.

Typically, contact courses (using ground landmarks) were flown. On this run, though, thick weather plagued navigation. Ghosting along, keeping below the clouds, the bridge watch glimpsed the deck but once between Washington and Philadelphia. Following the Mount Holly–Lakehurst road, *L.A.* groped the final miles home. Not by chance was the south Jersey topography (for mile on mile about the reservation) undisturbed by natural or manmade heights—a distinct advantage for a low-flying airship feeling its way. "We hoped it was always quiet" in the control car, Bradley told the writer. "And it generally was. If good weather was with us—and we hoped that was true—it was just a question of standing there and steering the course and keeping the ship in level flight or whatever altitude you wanted to go. And keeping track of where you are, and where you've been, and where you're going. But normally it was a calm atmosphere. If we got into fog, we generally had a lot of worried people up there who couldn't do anything about it much anyway—but just hope, and ease your way toward the ground to see if we could find it."

Without the customary circling, Rosendahl landed. "As we touched the ground, the clouds had lowered to the top of the hangar."[60] On the field, helping out, was an enlisted man fresh from Aviation Mechanical School: John Iannaccone. Before him was the first airship he had ever seen.[61]

The inaugural gloom proved emblematic of Herbert Hoover's presidency. That September–November, the stock market jumped a precipice: the *New York Times* averages for fifty leading stocks almost halved; the averages for twenty-five leading industrials fared worse. Billions of dollars' worth of investments evaporated.[62] By year's end, the industrial indices had sunk to a low level, and three million men had been thrown onto the streets. In May 1930, Hoover pronounced that "we have now passed the worst." That same month, the governor of the Federal Reserve Board admitted that the country was in "what ap-

Table 4-2 Weekly Progress report (aeronautical work) for the week ending 15 June 1929, Assembly and Repair Officer to Commanding Officer, NAS Lakehurst. (National Archives)

USS LOS ANGELES	—Install triplex glass control car.
" " "	—Install folding deck for navigator.
" " "	—Repair keel walkway.
" " "	—Streamline landing gear struts.
" " "	—Install thunderstorm indicator.
" " "	—Removed dummy condenser.
" " "	—Install airspeed meter securing device.
Airship J-4	—Repaired deck in car.
Airship J-4	—Repaired fuel tank.

pears to be a business depression." The Coolidge-Hoover prosperity was done.

At Lakehurst, training and testing held high rank. On 27 March, ceasing (due to darkness) search operations for a missing airplane, Rosendahl proceeded to a point east of Barnegat Light, there to begin a test run collecting data on a new condenser. Hanging to sea, the sortie approached one and a half days aloft. Then a month's hiatus. In the faded hours of 23 April, an excursion was flown past New York Harbor, four student officers observing. The city pronounced itself thrilled: "All her cabin lights gleaming, and the silver cloth of her bulk reflecting

Table 4-3 Distribution of useful load in USS *Los Angeles* in normal flight condition, December 1929. (Design Memorandum No. 92).

Frame	Fuel	Ballast Water	Drinking Flushing Water	Alcohol	Oil	Crew	Students	Spares etc.	Oil in Power cars	Water in Power cars	TOTAL
25		800				175		62			1,037
40	1296	725				175		352			2,548
55	1992			150	725	700		189	160	415	4,331
70	4288		80								4,368
85	2016		150	300		1050		100	390	830	4,836
100	5832				825	1225		579			8,461
115	3192		300			700		100	390	830	5,512
130	5832					175					6,007
145	1944		600			175		235			2,954
160		825				1925	1750	557			5,057
175		800				350		405			1,555
186								444			444
TOTAL	26392	3,150	830	750	1550	6650	1750	3023	940	2075	47,110

the light of the full moon, the big air cruiser played to a roof-top audience."⁶³ Landing minutes into a new day, the trainees and flight-crew sections were exchanged on the field. By 0040, *ZR-3* was again aloft—a short trip to sea. Back in berth, the newly appointed assistant secretary of the navy for aeronautics, David S. Ingalls, made an "unofficial visit." The navy's only fighter ace (five victories), Ingalls was a fervid proponent of airships. In 1931, at his instigation, permission would be granted for *Los Angeles* to participate in a fleet war game. Meantime, usually with Moffett or Secretary of the Navy Charles Francis Adams, Ingalls was a frequent passenger-guest.

Rosendahl's time as skipper was nearly through. Following April's pair of local excursions, his last sortie was logged on 8 May—again offshore, thirteen officers on board (one a student). The change-of-command ceremony was held in the afternoon watch, next day.

The lean years seemed through. A larger, modern ZR was due for completion in 1931 and, within fifteen months, a sister. With a myriad of research-and-development investigations under way or in prospect, the CNO in April had ordered Rosie to new duties, elevating him to the post of Commander, Rigid Airship Training and Experimental Squadron (ComRATES), with additional duty under the commanding officer, NAS Lakehurst, in charge of experimental projects.⁶⁴ By the spring of 1929, the thirty-five-year-old lieutenant commander was the dominant personality on the LTA operating scene.

At 1440, 9 May, with ship's crew mustered at quarters, Charles Rosendahl read his BuNav orders detaching him from command to duty as ComRATES. Herbert Wiley then read his orders, which directed him (the ship's log noted) to report to ComRATES "for duty in command of this vessel and assumed command." For his part, Vince Clarke was ordered to *Los Angeles*, to duty under instruction as prospective CO. Clarke would succeed to command upon Wiley's detachment—expected about 1 September.

Strong on operations, Wiley was a man in a hurry. Within minutes of the ceremony all engines were run up, food placed aboard, the aircraft unmoored from the scales and placed on the taxiing cars. At 1526, twelve officers, twenty-seven crew, and four student officers were under way—a roundabout up the Hudson, thence west, across New York State, to Niagara Falls (circled) and the lakeshore southwest of Buffalo before the rudderman took her eastward, across Pennsylvania.⁶⁵ Flight sections pushed aloft three times that May, four in June, once that July. An overhaul begun on 8 July was completed on 10 August.⁶⁶ That month proved active: a half-dozen excursions, for 108 hours.

This same August, *Graf Zeppelin* again graced eastern skies, to begin a round-the-world flight. At 0100, the 5th, the east-door leafs aside and the west ones "cracked," *LZ-127* was walked inside and secured. (Air station skipper at this time: "Maury" Pierce. Wearing two hats, Wicks was serving both as his exec and as assembly and repair officer. Beyond his regular duties, Wicks had gone through LTA flight training, receiving his designation that May.) Two days thereafter, during the forenoon watch, ship's force was engaged in sealing the outer cover and assisting Ger-

Los Angeles "on the sticks" for maintenance and repair, July 1929. Note the cradles for the main and wing cars, the shack for the hangar watch, and the panel of outer cover being applied to the hull. (Author)

man crewmen in doping the cover and gassing their *Luftschiff*. A few hours later, Rosendahl and Jack Richardson "reported on board the *Graf Zeppelin* as Naval Observers for the flight around the world."⁶⁷

A fantastic demonstration, the circumnavigation with a payload of passengers and mail (three intermediate stops) was an epochal achievement—"the outstanding aeronautical event of the year," *Aeronautical Engineering* announced. It altered air-transport thinking, serving to focus attention on the potential of transoceanic navigation by airship. The West Coast (for example) had been awakened to the fact that the Orient need not be three weeks away by steamer. The large airship, as a vehicle, represented immense possibilities for trans-Pacific commerce.

The airship's challenge lay in the transportation of large payloads over long distances. Commercial acceptance hinged on its ability to transport economically at reasonable speeds and with a degree of regularity and safety. But already obsolescence loomed—even as the platform was being developed. Their sister craft—airplanes—had multiplied by thousands, unlike airships. Reason: unit costs. With investment capital lacking, government orders constituted the sole market. Compared to heavier-than-air, airships were becoming a neglected engineering art. The data pertaining to airplanes filled libraries, whereas design information on airships was meager. In the onward flight of aeronautical progress, refinements in the aerodynamics of airplanes were fast eroding the airship's range and payload advantages. Their speed was about to soar.⁶⁸ But the airmen of 1919–39 were not clairvoyants; nor were the denizens of the engineering world. Ever attentive to the problems of flight, "with a view to their practical solution," NACA was stimulated in 1929 to declare: "If the United States is to take full advantage of the possibilities of air transportation, especially in the field of transoceanic air travel, the importance of the development of rigid airships can not be overemphasized, as in transportation by air over long distances, and especially over water, rigid airships have marked advantages over airplanes."⁶⁹

At the Langley Aeronautical Laboratory, in Virginia, many flight researchers had become outspoken advocates of airships.

> It was not clear at all to them or to anyone else at the time that the airplane would win out over the airship, let alone as totally as it soon did. Airplanes of the early 1920s were slow and small. . . . LTA advocates believed correctly that airships had enormous unproven capabilities: they were not much slower and could carry many more passengers in far greater comfort than airplanes, most of which still had open cockpits[,] . . . and with their extreme range and low operating cost, they could be used not just as military weapons but also for transportation of heavy commercial and industrial loads.⁷⁰

During October, *Los Angeles* logged a high-mast moor. Having taken off at 2024 on the 7th, Wiley loitered in the area before securing to the stub mast, at 0622, in lull conditions. After little more than two hours, he again let go. Near the station, at two thousand feet, Wiley ordered a turn into the wind for simulated hook-ons. A Vought VO-1 airplane then closed. Chief Boatswain's Mate E. E.

Graf Zeppelin *berthed in the ZR hangar at Lakehurst, 15–29 October 1928. Note the bow hatch to* Los Angeles' *keel-corridor. (U.S. Navy Photograph courtesy Navy Lakehurst Historical Society)*

Reber (attached to the Naval Aircraft Factory) commenced making approaches for landing on the airship's "plane landing device" (trapeze) using various speeds. Exercises completed, the dirigible cruised in south Jersey–Philadelphia airspace before returning home, arriving at 1610. Eased to eight hundred feet, main cable led out, Wiley commenced his approach to the tower at 1701. Water recovery was cut out, cable dropped, and at 1717 the line connected. At 1734, the cone was locked in the mast cup, and minutes later the tail drag was secured.

L.A. swung to her mooring but briefly. Two thousand pounds of fuel plus four hundred of drinking water were taken on, after which the flight crew embarked. Landing stations sounded within twenty-five minutes; at 1945, ten officers, eleven students, and a crew of twenty-nine pushed aloft. Altogether, *Los Angeles*'s log recorded 535.15 hours at Lakehurst's tallest structure.

It never again held an airship. An outworn hazard to operations, the tower was pulled down in 1935 for scrap.

For the officers of *ZR-3*, the year's closing months were marked by a pair of concerns: the deepening economic collapse and, close to home, the material condition of their vessel. The matter of service life is crucial. The only aircraft in a commissioned status, *Los Angeles* had flown the flag more than five years. Her log was nearing 2,500 hours—about a hundred thousand air miles. (Already, she had worn out her original cells and several sets

Charles Lindbergh at Lakehurst, 11 December 1929. Just disembarked from his Vega, the Lone Eagle signs the air station's log—here presented by CPO William "Bill" Cody. Accompanied by Wiley and the air-station skipper, Cdr. Pierce, a tour of Los Angeles followed. (ADC John A. Iannaccone)

of replacements.) In addition, *ZR-3* had operated hundreds of hours at various masts, the technology for which she had nurtured. By the fall of 1929, this record constituted a longer useful life than any other rigid, anywhere. Those responsible for her condition faced a unique case: a near-total lack of precedent to guide them in appraising the fitness and remaining useful life of their vessel. There was no body of reference, much less any broad experience, to draw upon. *L.A.* was breaking new ground. (Inherent loss of strength due to corrosion and bending of lattices, Burgess had calculated, however, did not exceed 10 percent.)

That October, Wiley wrote the bureau. Since 1926, structural samples—channel and lattice sections—had been cut from the hull and shipped to the BuStandards for examination, to assess changes in physical properties, particularly with respect to corrosion. To Wiley, the results were evidence of a worrisome trend: progressive deterioration. Though it was not unexpected, given her age, a more complete scientific investigation by a special board was suggested to ensure that *L.A.* was not unsafe. This body would not only inspect the aircraft but review the test results on file and consult with outside experts. By close study of all correspondence, reports, and available data, the rate of deterioration would become evident, Wiley believed, and a date upon which operations should cease could be forecast. In his view, this would be about July 1931—"but it may be considerably extended, or perhaps antedated."[71]

Rosendahl forwarded this report, signing off on the appointment of a board.[72] In his endorsement to the secretary, Moffett agreed to the basic recommendation as "sound." Although there were "no precedents to govern," it seemed logical to follow the applicable parts of navy procedure for surface vessels—that is, to have the inspection made by a Board of Inspection and Survey or a sub-board. This therefore was recommended. Further, BuAer was ready to suggest officers who could serve as members of it or as advisors—men prepared to assist at no cost to the government.[73] (Informal assurances had already been received from the builders, the Aluminum Company of America, and from other organizations.) Four days later, the CNO ordered the president of the board to conduct the inspection; he so informed the secretary on that date.

On 4 March 1930, the board met to "in order to determine the material condition of the airship, to investigate the rate of deterioration of her structure, and, if practicable, to recommend a date upon which operation of the aircraft should cease by reason of the reduction in strength due to such corrosion."⁷⁴ The final report was submitted that September. In general, with the operating log approaching three thousand hours, material condition was found to be good. Considering the varied and demanding conditions attending the ship's career and the almost total lack of precedents, this reflected credit on those responsible for *Los Angeles*'s material upkeep. The requirement for a quarterly hull board report—a statement of condition as found by a board of three officers attached to her—was judged to be sound; the practice had yielded useful results. Nonetheless, the board found it impracticable to predict with certainty the useful life remaining. If operated as prudently as before, and if carefully and continuously inspected by qualified personnel, the board saw "no reason why the airship should not continue in service for two to four years longer."

The report went on to make recommendations (such as limiting maximum airspeed to sixty knots) that, in its view, would not only increase useful life but promote the collection of data needed to continuously appraise that remaining life. "The price of a successful airship is careful inspection and continual attention to material upkeep," particularly as age rises. Hence the near-annual practice of sampling and testing was endorsed.⁷⁵ Continued cleaning and varnishing of the structure also was urged. In view of the difficulties in appraising the condition of equipment and materials being used in new ways and under new conditions, officers trained along engineering lines were recommended for inclusion in ship's complement.

Finally, by way of emphasis, the board looked forward to the annual resubmissions of data to it. The first such report was anticipated on 31 March 1931. This amounted to making the quarterly material report for the period ending each March unusually comprehensive, by including data for the prior twelve months instead of only three.

Los Angeles and USS *Lexington, 1930. In the play of politics and interwar competition for funds, the Navy proved unwilling to fully assess the ZR's promise—a doomed battle with entrenched interests. The platform would fail to win acceptance as an instrument of naval warfare. (National Air and Space Museum, Smithsonian Institution (SI Neg. No. 80-7738))*

In this way, the material records would be made more complete, condition more easily appraised, and thus a better forecast of probable life assured. At that point,

> assuming the airship will continue in service for several years, the Board considers that the most useful purpose that the airship can finally serve is to destroy her through tests designed to yield data on devices for mechanical handling, landing, and mooring airships. The board makes no suggestions at this time as to the nature of such tests, but believes a real service to the art of airship design, construction and operation can be served through eventually conducting destruction tests on the U.S.S. LOS ANGELES.⁷⁶

That October, eight missions entered the log, including experimental airplane hook-ons in the air near the Lakehurst station. Next month, there were a half-dozen more sorties, then two in December—both on the 9th, the latter with Moffett and five officers on the passenger list. Driven down through an inversion into the hands of the landing party, a midnight moor (0020) concluded the year's flying.

In berth, each four hours, the closing days of 1929

were dutifully logged by the watches. Leave was granted. But first, ship's force engaged in routine maintenance and postflight inspection, while from five to as many as twelve "civilian air mechanics" effected sundry repairs (e.g., the roller hatch to number-two car, the radiator shutter on number three, the ladder to number one, a galley window). At 1420, 10 December, fueling concluded—10,080 pounds, A hundred pounds of alcohol entered at frame 175. Gassing was completed at 1555 (about 88 percent full), the cells having received 151,450 cubic feet of helium.

L.A. did not fly again until 16 January—a thirty-hour excursion into Pennsylvania followed, on the 20th, by a nine-hour sortie to the nation's capital. These were the first big-ship hours for Lt. Howard N. Coulter, who had reported late in December.

During Rosendahl's tenure, 113 flights totaling more than 1,400 hours had been annexed to the logbook. *ZR-3* had secured to *Patoka* on eighteen moors. Sorties were flown for NACA, eleven calibrating radio-compass stations. Then there were the publicity sallies. This collectivity of operations conferred valuable experience, and, further, tendered a keen appreciation of ship's capabilities. These results, in turn, pointed the way toward improvements in subsequent designs.

Thanks in large measure to Rosendahl's hard-driving initiative plus that of fellow officers—Wiley and Settle, notably, but also lesser-known key players like Arnold, Bauch, Bolster, Buckley, Pierce, Thurman, Tyler, others—*Los Angeles*'s precarious status had eased. As had the supply of helium, somewhat. Most crucial for the future, contracts for the next ship had been signed; at Akron, *ZRS-4* was under construction. Since the necessary personnel had to be trained, the pace accelerated; training, indeed, would approach the manic. The air station and flight commands were active with research and development projects and in helping to train the influx of both officer and enlisted students. Already (spring 1929), the volunteers of Class VI had reported for duty, to go through the course. The fourteen officer-students constituting Class VII reported the next year, and in 1931 Class VIII (twenty-two senior officers).

Navy lighter-than-air tended to be a hotbed of personalities and factional turmoil even in its best years. In part, this was characteristic of any big collaborative project. (Such infighting, Scotty Peck observed, was a shame; they already were surrounded by "snipers.") A blustery, compelling figure admired as well as resented, Rosendahl exercised a ruthless, autocratic leadership when he gained hegemony, holding colleagues to his own strict standards.

(He was known to dress down officers publicly, an unusual affront.) As he saw it, everyone was united for one aim. Single-minded in commitment, zealous in advocacy, demanding of loyalty, this one officer resounded throughout the ZR project. Rosendahl's advocacy and force of will are evident in the hail of lengthy, seminar-like memoranda he lobbed into BuAer.

As everyone learned, Rosendahl did not shun notoriety. Quite the contrary: he projected himself center stage. ("A publicity seeker," one shipmate scoffed.) His penchant for self-promotion, his good looks, and his evangelical zeal made the future admiral a favorite among the public, toward which he developed a proprietary relationship. "This guy Rosendahl," Will Rogers remarked in his column, "has done a fine job of it." No leading figure delivers change single-handedly. Rosendahl, though, tended to dominate his profession; an irrepressible cheerleader, he eclipsed every contemporary, overshadowing associates no less gifted than he. But for Rosie, Settle (for one) would have held ZR command.[77] Loyal if resentful, most officers stayed on; others, frustrated with shop politics and mindful of careers, quit the program. Rosendahl enjoyed Moffett's full support; the admiral usually went along. Fulton proved rather more critical: maintaining a certain detachment, Froggy dampened some of man's stronger views, and Rosie didn't cross him.[78]

The record admits of no argument. A towering presence, this officer helped to shape the course of U.S. naval aeronautics. By sheer force of personality and command presence, by luck and by labor, Charles E. Rosendahl became the most famous airship officer in the United States, his name virtually synonymous with lighter-than-air.

However, he tended to discharge his passion (and celebrity) counterproductively, frustrating his own legacy. Intensely political, in the art of political war this officer was no artist: intolerant of opponents, brusque, inflammatory, challenging, Rosie proved fractious both up and down the chain of command, seeding personal and political enmities. His rhetoric (and published views) inspired further distaste; the man's popularity compounded the resentment. The ranks of anti-airship forces were thus enlarged, undercutting in certain corridors support for his beloved program. As per his detractors, the goad Rosendahl would be made to pay—personal grudge as policy.

It is an irony. Yet, seemingly, Rosendahl was indifferent to these consequences.

Advancing in fits and starts, these were introductory years, a time of transition and experimentation. Though barred from fleet work, *Los Angeles* had added much to the canon. At no time yet, however, had the type been a useful *naval* adjunct. Those hard operations were at hand. Thanks to ZR-1 and ZR-3, this further experience would be built upon at least a modicum of performance.

Elsewhere, globe-trotting countries and continents, *Graf Zeppelin* was establishing its own remarkable record, attracting a hullabaloo wherever it appeared. This spoke well of the type. Of all the great rigids, its career would be the most extravagantly chronicled. For its part, *L.A.* was logging hundreds of needful hours. Now two big ships more were in prospect. And Britain was building, Germany preparing to do so.

Its successes mounting, the large airship seemed destined to become a fixture in U.S. military and commercial plans. Yet the platform faced uncertainties and hazards—and not only the meteorological variety. The disdain of fiscal competitors was proving a powerful force; among the various stake-holding constituencies, carrier aviation in particular viewed airships as a drain on funds. "We naturally were greatly enthused over the future possibilities of the U.S. Navy's rigid airship program," Bolster remarked, reflecting on the long-ago. "I guess it was the golden period of an era which was unknowingly approaching an untimely end."[79]

Fading platforms of sea power: battleships of the U.S. Fleet, 1934. Properly, the naval rigid was an instrument of very-long-range reconnaissance over vast ocean reaches. ZR ships, proponents held, could perform vital scouting service for capital ships. (ADC John A. Iannaccone)

CHAPTER 5

Testbed for the New Ships

This place has been boiling with activity.

—Cdr. Fred T. Berry, USN (Lakehurst Class VI)

In the end, the large airship proved stronger on promise than on performance. But in 1928, the promise was yet robust. On 6 October, contracts with Goodyear-Zeppelin were signed for two fleet-type ZR scouts. Nine days later, *Graf Zeppelin* completed her transatlantic leap, arriving over the Lakehurst station to worldwide acclaim.

For a brief, hopeful span, the black clouds were dispelled. Many sensed a fresh wind, the brink of a new era.

Within thirteen months, at Akron, the "Airdock" erected by Goodyear-Zeppelin[1] was nearly complete. In November 1929, Admiral Moffett formally inaugurated construction of *ZRS-4*.[2] Workmen soon were clocking day and night shifts. As materials and parts passed through their various stages, all were inspected and reexamined by company representatives, then by the navy force under Tex Settle, INA, assisted by Lt. George V. Whittle.[3] Vice president in charge of engineering and chief designer: Dr. Karl Arnstein. Wicks was here as well; on a visit to advise on helium storage, he had received an offer.

Everything about *Akron* was superlative—even her gas cells, which weighed in at more than twenty-five thousand pounds, including valves. Cell VI amidships would hold 970,000 cubic feet of helium—nearly 40 percent the volume of *Los Angeles*. *ZRS-4* would also be the first airship in the world to be equipped with a hangar for carrying airplanes.

Backstage, a parallel drama—an effort to realize a national commitment to trans-Atlantic and trans-Pacific *commercial* service. Between 1919 and 1935, the naval rigid dominated lighter-than-air aeronautics in the United States. In 1920, the navy had assumed responsibility for *all* big-ship development. The commercial future of the type—assuming it was to have one—was likewise in Navy Department hands. This dual stewardship was to persist, remaining essentially unchanged for two decades, viz., "To build and operate naval airships as necessary to determine their usefulness for naval and other governmental purposes *and their commercial value*" (emphasis added).

This twofold charge is evident in the deployment of *Los Angeles*; indeed, her sorties to Bermuda and Puerto Rico had been simulations, to assess commercial value over water routes. The ultimate failure of all commercial planning (as Fulton remarked) was to have a "potent influence" on airship history in this country.

Rather than two aircraft, then, the department's specifications (and its contract to build in a non-navy plant) were intended to found an *industry*.[4] This vision was due

Unmoored from the mobile mast, Los Angeles rises statically off the East Field, 0911, 14 October 1929. Thirty-eight crewmen are at station with a dozen officer-students accompanying. Note the "airplane hook-on device" drawn up against the keel. Within minutes, Lakehurst's UO-1 landplane will be making approaches to the trapeze, continuing tests initiated that August. (Lieut. Comdr. Leonard E. Schellberg)

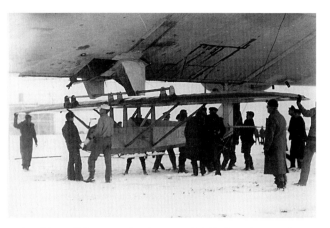

A Prüfling glider attached to special fittings beneath the keel of ZR-3 at Frame 120, 31 January 1930. Pilot-to-be: Capt. Ralph S. Barnaby, USN (CC), holder of Soaring Certificate No. 1. (National Air and Space Museum, Smithsonian Institution (SI Neg. No. 2000-9512))

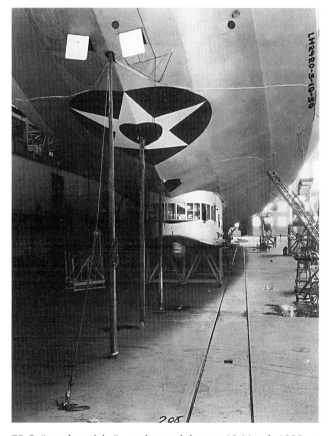

ZR-3 "on the sticks"—a six-week layup, 10 March 1930. Ship deflated, the keel shores sustained live loads in the corridor while work was underway in the hull. Hull weight was suspended from the roof-truss, the car weights relieved by cradles. Labor cost for a docking and deflation (three days): about $800. (National Archives)

largely to Goodyear's Paul W. Litchfield, who believed airships would enjoy a place in commerce as well as national defense. Interests coincided here: the navy was as intent on obtaining a builder as much as Goodyear was on being that supplier. Ensuing orders from private interests would sustain the art and, as well, secure a place in world commerce. Further, the navy's own experience was expected to yield dividends applicable to passenger and cargo-carrying airships.[5] In this regard, its contributions proved substantial: helium exploration and conservation; R&D work on impermeable fabrics and aluminum alloys; an improved weather service; studies into aerodynamics; the development of ground handling gear. Further, both organizations retained a nucleus of trained personnel, not to mention hangars and masts applicable to commercial operations.

Five days after the signing, Graf Zeppelin lifted off on her maiden trans-Atlantic run: the first commercial air transit of the North Atlantic. Eckener, again in command, was to hold the headlines whenever and wherever he took "127." This market testing, as it were, had a rousing effect.

As Germany's record mounted, Wall Street and shipping companies took interest; plans for transoceanic air travel blossomed. In charge of commercial affairs at Goodyear-Zeppelin was Jerome C. Hunsaker. He had joined at Litchfield's instigation and now was assisted by Alan MacCracken and Thomas Knowles. "From about

1930 until 1934," MacCracken later recalled, "we were both assistants to . . . Hunsaker, who was Vice President of Goodyear-Zeppelin Corp., and President of International Zeppelin Transport Corp." A "paper" corporation, IZT had been organized to study the feasibility and economic merit of transoceanic operations and explore financing. "Our work for Dr. Hunsaker was concerned with developing information for use in promoting the use of airships in transoceanic service, particularly in an attempt to secure enabling legislation toward mail subsidies." Similarly, the Pacific Zeppelin Transport Company explored trans-Pacific possibilities.[6]

Weather and other investigations confirmed the feasibility of a schedule between Europe and North America. Weekly service connecting California to Hawaii also was projected, to be extended later to Manila, via Guam. A transatlantic line would rely on a joint German-American organization and called for reciprocal use of hangars, masts, ground crews, and support facilities.[7] This coincided with Eckener's international outlook. He knew that intercontinental air travel was not possible without the Americans, so he set out to stimulate U.S. backing and worldwide interest. His aeronautical prowess and gift for public relations would by 1936 bring transoceanic airship service to a practical reality.

Today, it is little appreciated that the large airship was the first commercial carrier to demonstrate—largely through Eckener's genius—the feasibility of transoceanic, intercontinental, nonstop air transport. Until the DC-3, planes of any appreciable payload capacity rarely enjoyed a useful operating radius; airplanes could not yet leap oceans, commercially. The airship had a useful lift measured in tons with a range exceeding six thousand miles at about seventy miles per hour—far faster than surface liners, and with comparable comforts. The potential of this transportation system could not be taken lightly. Prominent bankers, industrialists, and executives of the day saw this and followed the German venture with more than casual interest.[8]

A federal subsidy was crucial to eventual profitability, and so a series of U.S. merchant airship bills were introduced. This would give airships a status as common carriers, lure long-term private investment, and encourage development. (Chief lobbyist and technical spokesman: Jerome Hunsaker.) The airship would repeat the early history of the commercial airplane—relying upon mail and express, not passengers, until reliability and safety had been demonstrated.[9] Against a background of financial collapse, redrafts, crashes, and clever opposition (from the president of Pan American Airways, most damagingly), the matter moved fitfully forward.

Early in 1931, with the form of *ZRS-4* complete and the work of covering the great hull about to begin, draft legislation still awaited lawmakers. "The design practices developed in building the *ZRS-4*," the *Herald Tribune* informed its readership, "are to be followed in the production of merchant airships which are to be operated under the terms of the McNary-Parker airship bill, now before Congress."[10] In the give-and-take of competing interests, enactment teased close—only to skitter beyond grasp. In one instance, the so-called Merchant Airship Act passed the House but, for lack of time in the session, did not get before the Senate. Yet another opportunity arrived in the spring of 1933. As Hunsaker saw it, the likelihood of passage in the new Congress had risen to excellent. Within days, *Akron* foundered—taking the legislation with her.

Disappointments notwithstanding, efforts to bestow status in foreign commerce persisted—along with much lip service. The Navy Department had no particular fondness for lighter-than-air; as the thirties advanced, its (waning) cooperation with private interests in the development of overseas commercial dirigibles would, the navy hoped, relieve it from further obligation. The year 1936 (for one) offered promise; as late as 1939, bills were being introduced to gain consideration for commercial construction. But airships were never opened to subsidies; with the government's attitude unclear, private capital could not be attracted. In the end, mistakes, indifference, opposition, tragedy, and plain bad luck—conjoined with economic malaise and an overlong developmental period—paralyzed planning, leaving only the navy to operate the type. Professed policy notwithstanding, no commercial rigid ever flew the U.S. flag.

Against this backdrop, the engineering for *ZRS-4* and *ZRS-5* opened new design terrain. The navy had insisted that all parts be accessible in flight; the use of inherently stiff, or "deep," rings and the incorporation of *three* keels was Goodyear's answer. The penalty: weight.[11] But the breakup of *ZR-2* and *Shenandoah* had underscored the need for greater strength. The deep-ringed form was very rugged; further, the three-keel arrangement provided exceptional strength and stiffness. One keel ran along the top center; the pair-gangways were placed at forty-five degrees from the hull bottom.

One of the most involved problems was the power plant and its arrangement. Unusual strength plus helium allowed ship's engines—Maybachs were chosen—to be mounted *within* the hull, thereby reducing drag. Drive shafts delivered power to the propellers mounted on outriggers in a four-in-line tandem arrangement. Outriggers made possible swiveling of the props; each could be swung through a vertical ninety-degree arc. Combined with the Maybach's reversibility, this feature offered thrust not only forward and astern but vertically: downward for "heavy" takeoffs, upward during "light" landings.[12] Internal mounting meant that crewmen need no longer climb outside and that the engineering force could have more spacious compartments.[13]

Another feature: *ZRS-4* and *ZRS-5* were expressly designed for the picking up, carrying, and releasing of airplanes. Under navy supervision, the art of hook-on flying was to reach its ultimate degree of development.

In 1926, the department had opened discussions as to the construction of a trapeze to be installed aboard *ZR-3*. On 11 June 1927, the inaugural experiment was flown. After an 0430 takeoff (with two photographers along) the Vought UO-1 biplane rendezvoused for dawn "air exploration tests." Upon signal from Rosendahl, Lt. (j.g.) Delong Mills dropped astern—about eight hundred feet back and about two hundred feet below—then commenced his approach. The upper wing soon blacked his view, but *ZR-3* reappeared as he neared its tail. "I then started climbing and passed about thirty feet below number one car." Throttling his engine just enough to maintain control but still climb slowly, he passed the after wing cars, after which "no trouble was had to keep the plane about ten feet below the keel and maneuver it either ahead and back or up and down. . . . It was possible to keep the plane at any desired location relative to the ship and remain there for extended periods of time." In all, five separate approaches and simulated hook-ons were made.[14]

This experience confirmed the validity of the hook-on concept and served to reassure the airship men, some of whom were voicing a curatorial concern as to possible collision. The day's tests done, all engines at "Half," *L.A.* cruised the Delaware-Maryland-Virginia area. After circling Washington, the airship set a course toward Chesapeake Bay, the Delaware River, and offshore Jersey, executing for training a navigation problem. At 1340, *J-3* was sighted off Ocean City.

In October 1928, the navy announced that *Los Angeles* (then barnstorming in Texas) would begin hook-on experiments "soon." The trapeze was not installed until mid-1929, however, followed by various shed tests. Goodyear had built the trapeze, but credit for its subsequent engineering development belongs to a variety of people at the Naval Aircraft Factory, at Lakehurst, and in BuAer. The most vital aspects of this project, indeed, would continue to preoccupy the station's experimental staff, most prominently Cal Bolster, Lakehurst's experiment officer.

During the afternoon watch of 24 June 1929, a trapeze and yoke on an "airplane landing device" were installed at frame 100, beneath the keel, between the four wing cars. Weight: four hundred pounds Two days later, the electric overhead cranes were connected at frames 25 and 175; the ZR unmoored and was raised about twenty-five feet. Overhead suspensions were secured at frames 100, 110, and 115, and static tests conducted on the hook-on gear. Upon completion, the hull was lowered, the cranes and suspensions detached, the ship secured. On 1 July, *ZR-3* was again lifted—civilian workmen and ship's force assisting as needed—to accommodate a "dynamic test of airplane attachment gear simulating actual condition of contact, employing flat car with engaging hook."

Results? Satisfactory, save for minor repairs and adjustments for flight testing.

For the flight phase, the navy had assigned a UO-1 flown by Lt. A. W. "Jake" Gorton. At 0022, 3 July, Wiley cast off from the northeast field for what was to be more than just another flight. A test was run nearby in connection with a dummy water-recovery panel experiment. A quarter-hour of preliminary operational tests were then

A Vought UO-1 approaches during "air exploration tests." In 1926, the Navy Department had opened discussions as to construction of a trapeze installed aboard ZR-3. On 11 June 1927, the inaugural experiment was flown to her. "It was possible," pilot Lieut. (j.g.) Delong Mills reported, "to keep the plane [no hook] at any desired location relative to the ship and remain there for extended periods of time." (National Air and Space Museum, Smithsonian Institution, SI Neg. No. 2000-11188))

Lakehurst's UO-1 equipped for hook-on experiments. Authorized late in 1927, a trapeze was installed beneath the keel of Los Angeles *at Frame 100—ship's center of buoyancy. Following in-hangar tests, the first contact flown to the trapeze on 3 July 1929. (ADC John A. Iannaccone)*

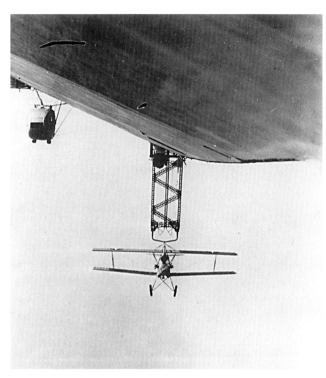

The UO-1 hangs to the bar beneath ZR-3, July–August 1929, during early trapeze tests. The view is aft from about Frame 140—the after end of the passenger car. Power car No. 5 hangs at Frame 115, the centerline car (behind trapeze) at Frame 55. (Charles E. Rosendahl Collection, University of Texas)

run with the "landing device." At 0406, daybreak, number-one engine was stopped and numbers two, three, four, and five were set to "Cruising." Nine minutes thereafter, Wiley ordered the officer of the watch, Lt. George F. Watson (No. 3415), to bring the ship into the wind. (It would soon be realized that because both craft were moving in the same medium, this maneuver was unnecessary.) The test began. At 2,500 feet, Gorton swung in on the tail of *Los Angeles* and lined up a skyhook on his upper wing with the trapeze. After six minutes of "fishing," he touched the trapeze but did not hook it. On his next approach, the plane again touched, remained on the trapeze for a short space then slipped off. Gorton was making his approaches on a course parallel to *ZR-3* and level with the trapeze. To reach the bar, the lieutenant had to fly through turbulence created by her hull, prop wash from power cars four and five, and eddies from the trapeze itself. The exercise was discontinued at 0430. The would-be "carrier" changed altitude, perhaps in search of smoother air.

Within minutes, Wiley again was ready:

> 0451 Plane on trapeze for about 5 seconds but did not hook and slipped off. 0454 Plane on trapeze, did not hook and slipped off. 0500 Test discontinued after total of 15 approaches, four of which resulted in contact with the trapeze.[15]

Gorton was no novice at experimental flying; the problem lay elsewhere. This experience suggested modifications, notably a redesign of the skyhook safety release as well as a change in the pilot's approach.[16] The skyhook was mounted on a spring-loaded tube in such a way that if the pilot struck the bar with too much relative speed, the hook pulled back and released. (Gorton was provided a control by which he could manually release it.) "After extensive 'hook-ons' and releases from the trapeze," Bolster explained, "it became evident that the right way to do this was to use a fixed hook on the airplane with no automatic release, permitting the trapeze bar to swing forward under a gradually increasing braking action when struck by the airplane."

As for the approach, rather than coming in level with the trapeze, Gorton suggested what he called a "reverse carrier approach"—approaching from about a hundred feet below and a hundred behind and climbing to the bar in a stalling attitude, keeping outside the disturbed zone until the hook was very nearly home. A few of the airship men objected—loudly. They feared that Gorton might not come to rest on the trapeze but instead fly on into *ZR-3*—the only such asset they had. But Gorton was not to be deterred by a bunch of "gas bag" sailors. He told Bolster that he thought the existing gear was good. He *could* hook on—but the pressure-release business in the skyhook would have to go.

Given the day's disappointments, the BuAer intervened—in the person of Lt. Cdr. Leslie C. Stevens (CC), who thought that perhaps Bolster and Company had the trapeze idea wrong. Instead of the plane hooking on, maybe the ZR should have the hook and the airplane a simple, trapezelike fitting. As Bolster recalled, "There was much concern particularly in the Bureau as to whether we were on the right track with the trapeze concept, there being some feeling that perhaps the airplane could be flown under the airship and be connected up by some sort of cable device." On 15 August, to test his hypothesis, Stevens was bobbing about in the slipstream of *Los Angeles* in the UO-1, attempting to make contact with a sandbag trailed from frame 100. This proved futile.[17] Some talk was given to the idea of using a bight to engage the plane's skyhook, but it seems that everyone was pleased to retreat to the original concept.

On 20 August, Gorton was back under with the UO-1, trying his new approach. The pressure-release feature had been removed. His first pass realized success. He missed the trapeze on the second, but a third and fourth ended in successful hook-ons. With the plane secured, Gorton killed his engine, and *Los Angeles* circled left, to determine the effects of the added weight and drag on performance.[18] Next day, in rough air, the lieutenant flew to an immediate "landing." As Wiley circled, Gorton turned over the Vought to Lt. Cdr. Charles A. Nicholson (CC), who was handling the bureau end of the project. Nicholson made two approaches and missed before flying his first hook-on. After a minute in contact he "took off," only to make the trapeze again before the exercise was discontinued. Resuming with Stevens at the controls, the UO-1 came up for one hook-on more before landing stations were piped down.

After Nicholson made his report, Moffett pronounced the tests a success. But suggestions that ZRs would be the carrier of the future were, in the admiral's view, exaggerations; the number of planes carried would always be small. Nonetheless, the hook-on development greatly increased the rigid's usefulness to the fleet.[19]

The 1929 National Air Races, held that August in Cleveland, were the year's premier air show. All the great racing pilots of the day attended, lending their glamour. Ordered to Ohio, *Los Angeles* would present her first public performance of the hook-on evolution. ZMC-2 (not yet navy property) would grace the races as well. *Graf Zeppelin* was also slated to pass over, on the last leg of its flight around the world.

Table 5-1 Preparations necessary for flight, USS *Los Angeles*, 1931–32. (George H. Mills Collection, Smithsonian Institution)

1. Place standard amount of drinking and flushing water on board.
2. Remove walkway cover.
3. Check the amount of fuel and oil on board.
4. See that valve covers are off.
5. Warm up engines.
6. Weigh-off ship (same weigh-off sheet as for the morning weigh-off).
7. Put food on board.
8. Place ship on stub or rather the mobile mast with taxi wheel under car No. 1 and handling frames on aft.
9. Attach mechanical handling gear on the sides.
10. Remove all surge tackles and spring balance scales.
11. See that hangar is clear and trolley tracks are clear.
12. Navigator, first lieutenant, and engineering officer report ready.
13. Report to the commanding officer that the ship is ready.

On the 27th, when all was ready for the sortie, landing stations for *ZR-3* were sounded at 1610. Wiley unmoored within twenty minutes. Disposable weights on board: fuel 26,320 pounds, 4,350 pounds of water ballast, oil 1,050 pounds, 561 pounds of food. Accompanying the nine officers and a crew of thirty were a trio of passengers (among them Senator Hiram Bingham, president of the National Aeronautical Association) plus three student officers—total forty-five. The ship rose statically to two hundred feet, whereupon at 1631 her four wing engines were cut in at half-speed and she turned for Philadelphia. Landing stations were piped down in nine minutes.

Not due in Ohio until the following afternoon, the dirigible cruised leisurely westward, changing altitude variously, making about twenty-eight knots ground speed. The Naval Aircraft Factory was below at 1814, Bethlehem almost two hours later, Wilkes-Barre at 2215. Midnight found the airmen under clear skies west of Scranton at 3,500 feet, engines numbers two, three, four, and five at "Cruising." Slightly more than 3,100 pounds of fuel had been consumed. But *ZR-3* had been "making water"; thanks to the recovery gear, the crew had *more* water ballast on hand than at takeoff—1,300 pounds more. At 0745, Niagara Falls foamed below. Here Wiley loitered. "On the way out we went over Niagara Falls," Berry recorded, "and viewed the falls from the air and frankly I have never before seen such a gorgeous sight." The forenoon watch (0810) had Buffalo in view. Following the south shore of Lake Erie, Wiley pressed toward Cleveland. He hung over Erie at 0958, Akron at 1220. Less than three hours later, *L.A.* was holding to various courses south of the Municipal Airport, killing time over Cleveland's suburbs.

On the field and in the stands a hundred thousand persons had collected. Among the honorees: Drs. Eckener and Arnstein, Paul Litchfield, and Hunsaker.

Landing stations were ordered at two thousand feet at 1505. With engines four and five pulling astern at half-speed, the airship lost headway, whereupon "Stop" was rung up on the annunciators for all engines. *Los Angeles* hung like a cloud before Cleveland's assembly. After nearly five minutes, at 1520, engines two, three, four, and five were ordered to "Ahead Half"; two minutes later, landing stations were piped down.

Lieutenant Reichelderfer, officer of the watch, ordered course changed to the airport at 1657. On this new heading, airspeed forty-seven knots, the "airplane landing device" was lowered into position. At 1716, landing stations again were ordered. Flanked and followed by swarms of airplanes plus blimps from Akron, the great floating machine that was *Los Angeles* returned at 1720, increasing speed to fifty-one knots. Circling left, she loitered over the airport, stealing the show.

Gorton had ferried the weary old UO-1 out and now was waiting below. As the rigid plodded down the field under escort, he climbed into the Vought and flew off to meet it. After three approaches and missed landings, Wiley ordered engines four and five cut to half speed; at 1735 Gorton hooked on. Bluejackets cranked at the trapeze-hoisting mechanism until the plane was "housed" against the hull. *ZR-3* came about, to pass in review. At this point, Bolster (without a parachute) "disembarked" onto the Vought's upper wing and swung himself into the front cockpit. The trapeze was lowered. At 1750, Gorton "took off," landing his passenger. (In releasing, the pilot tripped his hook and dropped off in a dive until flying speed was attained.) "It was simple enough," the lieutenant told newsmen. "We would have done it before except that hitherto we haven't been able to pull the plane up close.

Today I merely climbed through the hatch to the top wing, which was only three or four feet below the airship, and that is about all there is to it." Recalling his eccentric boarding, Bolster later said: "This was the first time that a man had transferred from an airship to an airplane in flight and the event received a lot of publicity in the local papers, especially since I was from Ravenna, Ohio, which is near Cleveland."[20]

Wiley now ordered (the log has it) "various courses at various altitudes and speeds preparatory to weighing off for landing." At 1757, engines four and five were signaled stopped, and the airship weighed off. Water recovery was cut out on all engines, and an approach begun. Trail ropes were dropped and caught, spiders attached; at 1817, *ZR-3* touched down. With side handling lines connected at frame 160, *L.A.* was walked to Cleveland's stub mast (shipped out from Lakehurst). Since seasoned men had been left in New Jersey to await *Graf Zeppelin*, a group from Goodyear-Zeppelin augmented navy personnel plus college lads working the races as ushers. Leader of this medley: Zeno Wicks (just resigned from the navy). At 1835, the cone stood locked in the mast cup. A taxiing wheel was secured to the centerline car, a tail drag attached aft, the ship ballasted. Landing stations were piped down and, at 1915, fueling commenced at frames 40, 70, and 85.[21]

At 2330, *Graf Zeppelin* passed over on the last leg of her world flight.

L.A. swung to the mast until the afternoon watch of the 29th; at 1306, she lifted away. Cruising southeast, *ZR-3* crossed Pennsylvania. At 1753, she hung over Cumberland, Maryland, following the Potomac River in the direction of Washington. Two minutes later, number-three engine was stopped to replace an exhaust valve spring. In no hurry, *Los Angeles* loitered in the general area before pushing on via Annapolis, Baltimore, and Atlantic City. Thence came a lengthy jog to the northwest, over Philadelphia to Lancaster, before course was set for home. On the midmorning, over Toms River, a new course: to New York, following the Toms River–Freehold highway. Wiley had the city below at 1145. Here he dallied for forty or so minutes, following the "Eckener parade" up Broadway and Fifth Avenue honoring the circumnavigation. Wiley then steamed west southwest. Over Flemington (west-central Jersey), course was again set for Lakehurst—in sight at 1447, 30 August. Settling into waiting hands, *L.A.* was walked to berth. By 1713, ship and crew ("sore but satisfied," Berry wrote) were being eased through the east door alongside *Graf Zeppelin*. The Germans had circled the globe, Lakehurst to Lakehurst, in twelve days, eleven minutes' flying. Its passengers Richardson (former navigator of *ZR-3*) and Rosendahl returned to duty, "having completed temporary additional duty" in connection with the circumnavigation.

As ship's force conducted routine cleaning and upkeep, crewmen assisted in gassing the zeppelin. On the 31st, Wiley was temporarily relieved; in accordance with a BuNav dispatch, the officer "left the ship this date and reported to the person in charge of Airship *Graf Zeppelin* for passage to Friedrichshafen, Germany." Lt. Roland Mayer held similar orders. In the absence, Rosie assumed temporary command of *ZR-3*. Returned on 23 September, Wiley relieved his former boss in the forenoon watch of the 24th.

The trapeze test program was dormant until 8 October, when approaches were flown without "landings." Next day, Chief Aviation Pilot John J. O'Brien flew aboard; after fourteen minutes in contact, he released. On the 10th, Moffett brought an eminent audience up from Washington, including Secretary of the Navy Charles Francis Adams and his official party as well as Ingalls—nine passengers. As they watched from the control car, fascinated, O'Brien flew the Vought to a hook-on. Five minutes later, the pilot "took off." On 14 October, the Vought was secured after six tries. It unhooked after three minutes in contact, whereupon Wiley ordered altitude changed. He quit Lakehurst airspace for Norfolk and the nation's capital, returning via Philadelphia and New York City. Plane exercises continued. Early on the 15th, having fueled and reprovisioned at the stub mast, Wiley cast off with a dozen student officers and two passengers. At 0940, O'Brien was logged (by Watson) as having "made [an] approach and successful attachment." The year's final hook-on (again, O'Brien) entered the log in the forenoon watch of 20 November—after eight failed tries for the bar.

December saw two sorties, both on the 9th, one with Moffett aboard. Returning from Washington on the 11th, Col. Charles E. Lindbergh landed to the south of the ZR hangar. Lingering three afternoon hours, the famous flier was a dinner guest of Commander Pierce, after which the colonel was shown through *Los Angeles* and inspected *ZMC-2* and the J-ships. With holiday leave granted for most *L.A.* hands, an overhaul commenced. Over the next three weeks experimental cells were installed in the number-two and four positions, and one was purged. (Her gas cells were numbered from 0 to 13, from aft forward.)

The year 1930 witnessed a near-manic schedule of operations. Between 16 January and 5 August, *Los Angeles* logged twenty-five ascents, seven for August alone. The abiding mission: training/indoctrination. On 20 January, for instance, Wiley unmoored for the nation's capital with eight student officers, four officer-passengers, and five student enlisted men. Three days later, he again took his command south, this time to Parris Island with six student officers to test its expeditionary stub mast.

An aging campaigner, *ZR-3* had now flown about one hundred thousand air miles.

Changes were close at hand. In March, as we will see, *Los Angeles* got a new skipper. That summer, the Lakehurst command did also; during July, Class VI graduate Capt. Harry E. Shoemaker (No. 3656) relieved Pierce. "This place has been boiling with activity," Berry (No. 3664) marveled. A comment by Reichelderfer hints of the scope of the bustle. "Little progress has been made on the three dimensional 'structure' of wind shifts and gusts, first because our exposures [for instruments] at Lakehurst are not suitable and cannot be made suitable without erecting structures objectionable as flight hazards, and second, because we have lacked the personnel to go ahead with the proposed installation on radio towers. . . . There has been too much to do in connection with LOS ANGELES operations here for us to follow up . . . as we had hoped."[22]

Among the tests slated for these charged months: a glider drop. Moffett had conceived the idea of launching a glider from *ZR-3*—a way, perhaps, of delivering landing officers to fields unfamiliar with airship handling. The notion was brought to Lt. Ralph S. Barnaby, USN (CC), holder of No. 1 Soaring Certificate (signed by Orville Wright). Reflecting on the question, the lieutenant assured the admiral that it was feasible. Temporary-duty orders were cut and, as Barnaby later remarked, "things started to move immediately." By 18 January, the navy's first glider had reached Lakehurst where, under Barnaby's instruction, it was assembled.[23] A mount and release mechanism were talked over with Cal Bolster, the airship's first lieutenant—the man responsible for fabricating and installing the necessary gear. "The Station is awaiting with keen interest this experiment," the station paper intoned, "and wishes Lieutenant Barnaby the best of luck in his new adventure."

Heavy snow postponed the trial. Next day, general assembly sounded in the clear dawn and cold (sixteen degrees) of 31 January. Delay interposed. Ice jammed the west doors. Then, unmoored from the spring balances,

L.A. overrode the mobile mast, damaging her nose fairing. Repairs ensued. At 0947, general assembly again was announced, and the ZR was made ready for flight. Towing commenced at 1023; in five minutes she stood clear. At the takeoff point the Prüfling glider was reattached to special fittings at frame 100; at 1048, the dirigible unmoored—its 189th ascent. Steering various courses and escorted by *J-3, J-4, ZMC-2,* and planes lofting news photographers, reporters, and observers, Wiley climbed to three thousand feet—Barnaby's requested altitude. Number-one and number-two engines were stopped at 1112; three minutes later, wearing navy blues with a sweater and leather flight jacket, helmet, fleece-lined boots, and seat-pack parachute, Barnaby went aft and climbed (via aluminum ladder) into the slipstream. "Although the ship's engines were idled," he would report, "it was an icy blast which hit me as I climbed down the ladder, removed and passed up to Bolster the control-locking mechanism, folded myself up into the tiny cockpit, secured the cockpit cover around me, and watched the ladder being drawn up."

Wiley ordered airspeed restored—to a predetermined forty knots on Barnaby's meter (to ensure control). Seeing a thumbs-up, Bolster telephoned the bridge, waved goodbye, counted down the seconds, then pulled the release. With spinning propellers so near, the Prüfling's stick was forward. "The glider dropped away like a shot! With a slight backward move on the stick I leveled off

Table 5-2 Duties of man on landing station, Frame 135, USS *Los Angeles,* 1931–32. (George H. Mills Collection, Smithsonian Institution)

PREPARING FOR FLIGHT AND IMMEDIATELY AFTER TAKE OFF:

1. Chief of section see that all landing stations are manned.
2. See that emergency fuel tanks are ready for valving. Report to leading C.P.O. as soon as ready or secured.
3. While weighing off, keep control car notified of remaining water at Frames 140 and 130.
4. Answer telephone.
5. See that cells No. 10 and No. 9 are clear of controls, nettings and structure and straightened out over keel.
6. As soon as ship is underway secure fuel dump tanks at 145 and report same to leading C.P.O.
7. Keep everybody at Landing Stations until "Pipe Down."

LANDING PROCEDURE:

1. Chief of station see that all leading stations are manned.
2. Answer telephone.
3. Report water ballast at Frames 145 and 130 while weighing off.
4. Keep everybody on landing stations until "Pipe Down."
5. Inspect maneuvering valves if ship is valved.
6. Find out from first lieutenant where to put ballast after landing.

AFTER SHIP HAS LANDED AND GOING INTO HANGAR:

1. See that all men remain on landing stations and are carrying out their duties until the word is passed to come forward.

SHIP STAYING MAST AFTER HAVING LANDED:

1. Set mast watch if mooring out.

MOORING TO PATOKA:

1. Supervise attaching of tail drag.
2. Section having watch will rig for mooring. Use off-section if necessary.
3. Section off watch will secure mooring gear after mooring and disembark when ordered by control car.

about a 100 feet below the airship, and let the glider slow down to 30 miles an hour, its normal gliding speed—breathed a sign of relief, and settled back to enjoy the ride." After six and a half minutes, the lieutenant ended his descent by crunching gently onto a snow-covered field. Barnaby was soon at the hangar, drinking welcome coffee and answering questions.[24]

These were the last months of Wiley's tenure. In anticipation of being assigned to *Akron*, he had given up command at his own request; that March, he received orders to sea, serving a year as tactical officer in the battleship USS *Tennessee*. During January, meantime, his command was particularly active—six flights. The year's first took place on the 16th and 17th. In air well below freezing, *L.A.* cruised southeastern Pennsylvania, then returned to loiter in the vicinity of homeport and offshore.[25] Time aloft: thirty hours. On the 20th came the round trip to Washington. Three days later, in twilight chill, *Los Angeles* shoved off for southern airspace. By morning, having circled the Charleston, South Carolina, Navy Yard, the hum of her motors had brought Charleston folk "crowding into the streets." As was the custom, an officer had been dispatched ahead to instruct a ground crew—Scotty Peck plus 250 marines at Marine Base, Parris Island, just north of the Georgia line. The lieutenant had his shipmates secured to the stub mast there at 1752, on the 24th. Exploiting his austral mast-terminus, Wiley let go next day, to follow the coastline as far as Daytona, en route calibrating the St. Augustine radio compass. After loafing along the nighttime Georgia coast, *L.A.* returned to the South Carolina mast (pronounced "satisfactory" by her CO) in the morning watch—at sunrise—of the 26th. Wiley fueled and provisioned. Tarrying less than three hours, the rigid shoved off and set a course for wintry New Jersey. Then had come Barnaby's adventure. A single flight was logged in mid-February[26] and a brief hop late in March—the last of Wiley's command.

On 31 March, with the crew mustered at quarters at the circle, the regulation formalities were held. Wiley read his orders detaching him, and Lt. Cdr. Vincent A. Clarke, Naval Aviator (Airship) No. 3399, read his orders and assumed command. A Naval Academy classmate of Rosendahl and an outstanding submariner, Clarke—solid, square-shouldered, a deep-sea sailor—was a Class III alumnus. "Clarke was given the full course of training," Watson remembered, "and since he had spent most of his time in the Navy in submarines [comparable, as buoyant craft, to airships] he proved to be a very apt pupil. He became an excellent blimp pilot after very brief

Lieut. Cdr. Vincent A. Clarke, Jr., USN, commanding officer USS LOS ANGELES, *31 March 1930 to 21 April 1931. Photo taken April 1928. (National Archives 80-G-460739)*

training and immediately understood how to handle the big airship."[27] Maurice M. "Mike" Bradley (No. 3481) agreed: "He was a good operator." An observer in *Graf Zeppelin* in 1929, Clarke had later served as a technical member of the navy's acceptance board for *ZRS-4*. In over twelve months of operations commanding *ZR-3*, Clarke was to add 963 flight hours. (For his part, Wiley was not detached from ComRATES—now Harry Shoemaker—until May.)[28]

Following the change of command, the flight crew embarked and preparations were effected for getting under way. Unmoored at 0807, *Los Angeles* rose statically to 150 feet before all engines were cut in. Clarke turned coastward. Sailing over Ocean City, the skipper's father waved to the passing airship. Offshore, a navigational problem was run. Throughout this trip (ten hours), five civilians from Lakehurst's Assembly and Repair (A&R) Department measured tensions of "all accessible" head and shear wires—information for the Board of Inspection and Survey. Broken wires had become a problem. Ascrib-

New Jersey astern, L.A. steams past the Hudson River over Manhattan Island, an Omaha-class light cruiser in view, 9 May 1930. Commanding officer: Lieut. Cdr. Vincent A. Clarke, Jr., the fifth of seven skippers for ZR-3. (National Air and Space Museum, Smithsonian Institution (SI Neg. No. 2000-9506))

View forward from power car No. 5, during the May 1930 Presidential Review. Because the control car hung beneath, the ZR platform had a superb field of view. In the interwar period, the naval rigid offered a high-speed solution to the problem of reconnaissance and strategic scouting. (National Air and Space Museum, Smithsonian Institution (SI Neg. No. 2000-9504))

ing these failures to fatigue, BuStandards had advised that an increasing rate could be expected.

Many a training flight would be logged under Clarke. Among these were a half-dozen rendezvous with *Patoka* in Narragansett Bay off Newport. That summer, L.A swung to its masthead more than fifty-nine hours.[29] During April and May, a respectable 165 flight hours were logged—fourteen missions—despite two weeks of high winds. April saw the airship dewinterized; also, pulsating running lights were installed at the stern, and all lights were changed to twenty-one candlepower. Clarke sortied twice to Scranton, made a flying moor to the mobile mast there, and on the 28th he joined a formation of surface units escorting SS *Leviathan*—SecNav Adams aboard—into New York. (A mail bag with newspapers was lowered at the end of a long line. Altitude was reduced to six hundred feet, an approach made, the line cut. It fell over *Leviathan*'s antenna but did no damage.) May was busier yet. On the 9th, for instance, with forty-nine on board, including a student officer and reporters, *L.A.* stood up New York Harbor with *J-3* and *ZMC-2*, passing squadrons of navy planes and the U.S. fleet at anchor in the Hudson. Clarke then steamed southwest, across Jersey, to Philadelphia, before returning to the mobile mast, only to cast off again for another, shorter run to New York—this time with six student passengers.

Eleven days later, *Los Angeles* joined in a presidential review of the fleet—submarines, cruisers, battleships, carriers—twenty-five miles to sea off the Virginia Capes. *Saratoga* then the carrier *Lexington* began to launch aircraft; Nicholson flew the Vought off the *"Lex"* and then climbed to the airship's altitude—the first such rendezvous. As President Hoover looked on from the heavy cruiser *Salt Lake City*, the ZR "carrier" departed her position about five miles to leeward (clear of air operations) and took position to receive the airplane. Crouched over the hatch at frame 100, NBC announcer George Hicks described the pilot's approach: "He did it—no, no, he didn't! Yes, he did! He's on! He's on!" The Vought was still oscillating on its hook when Hicks lowered a microphone to Nicholson, who gave listeners a brief dissertation on the art of hook-on flying.[30] "A good bit in the Sunday papers" resulted.[31]

The last of May's eight ascents again saw a glider. Barnaby was on the sick list, so Settle volunteered to handle the controls. Taking her departure in the dark early hours of the 31st, *Los Angeles* rose statically through an inversion to two hundred feet, then steered eastward. Clarke cruised offshore to Atlantic City then northward, following the beach, awaiting the arrival of *Graf Zeppelin*. At 0555, in the vicinity of Barnegat Light, the *Graf* was sighted on the port quarter. As the zeppelin landed and was housed, Clarke flew his command overhead. In turbulent air not long after, the glider's rudder came free of its chocks; this caused the craft to yaw, damaging its starboard wing. All engines were ordered stopped, the matter assessed; her Maybachs at half-speed, *ZR-3* returned to port, over which the glider was secured by Chief Boatswain's Mate Robert J. Davis (lowered on a bowline). *ZR-3* then cruised on to the Curtiss Marine Trophy races near Washington. Clarke arrived at 1255. Before fifteen thousand spectators, three power cars were stopped; two miles southeast of NAS Anacostia, Settle climbed out. At 1418, he was away for ten minutes of motorless flying. All engines rung up to "Cruising," *Los Angeles* then circled right while Nicholson, in the UO-1, flew two unsuccessful approaches in very bumpy air before latching on.

Besides gliders and hook-ons, scores of training sorties were logged—five with Moffett aboard, at least two with Fulton, and one with Assistant Secretary for Aeronautics Ingalls accompanying.[32] A host of experiments also were run, exploiting *L.A.* as testbed. Commencing that June, the Maybach in number-five car was used for extended tests with a new coolant product, Prestone. In the light of 8 July, the drag of the panel-type condensers planned for *ZRS-4* and *ZRS-5* was investigated via experimental runs at various speeds off the beach between Asbury Park and Seaside Heights. These "water recovery apparatus tests" were simulations: inflatable bags were taped against the outer cover above number-three car and inflated to measure drag. That night, tests on airspeed meters were held. On 6 August, along with four officer and eight enlisted trainees, an RCA representative came along "to conduct test of radio facsimile equipment." Five days thereafter, before proceeding on to Newport, where *Patoka* waited, experiments with aluminum powder bombs were conducted. By September, four of the ship's VL-2s were driving metal propellers, with two and three blades. Only number-one engine, at four hundred horsepower, retained its wooden screw.

As ever, maintenance awaited. Operations were suspended for a week (27 June–3 July) during which cell number ten was purged and helium purity brought up throughout. The water-recovery apparatus was cleaned that week and the experimental air-bag equipment installed.

July saw a casualty. For landing, the captain's main responsibilities were twofold: controlling ground speed and getting the nose mooring line into position for hook-up. Speed had to be very low; too fast, and men could be injured on the ropes. "Vince liked to handle [his airships] at high speed," Watson recalled of Clarke's nervy self-assurance. "He was not reckless but had complete confidence in his ability to handle the ship. But one day this caught up with him." On 3 July, returning after a brief penetration into Pennsylvania, *Los Angeles* hung very light. For landings, normal routine for the last mile was to descend to 250 feet, the two forward engines set to "Ahead Half" and two idling astern—the formula for a slow and careful approach. "But this was not Clarke," Watson continued. With all four wing engines at "Half," he backed down close to the ground crew, settling neatly

Graf Zeppelin *over the field at Lakehurst, 30 May 1930—the fourth arrival. Following a two-day refueling and provisioning, the "Graf" set out for Central Europe and the Friedrichshafen shed. (Lieut. Herbert R. Rowe)*

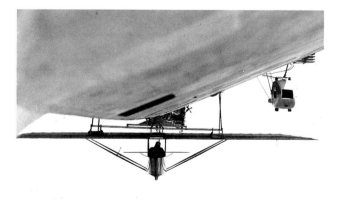

Seated at the controls, Lieut. "Tex" Settle prepares the Prüfling *for launch near Anacostia airfield, Washington, D.C., 31 May 1930. At 1418, Settle was away for ten minutes of motorless flying. This was second—and last—of two such trials. (Vice Admiral T. G. W. Settle)*

in. But too fast—unable to check her, men were taken along with the (rising) airship. A marine, his leg fouled in the port anchor and port spider lines, was dragged up and against the mobile mast. "Clarke instantly ordered, 'Valve her down.' The first lieutenant grabbed a nest of valve controls and opened all valves. This [two minutes' valving] checked the ship and started her down until he could get control by ballast." A bit broken up, the victim settled into waiting arms. The ground crew released a "big cheer, as they knew Clarke had risked his ship to some extent, and his reputation as a skipper, in order to save the man."[33]

Publicity was not neglected. On 19 June (for instance), *Los Angeles* had joined in a welcome for Rear Adm. Richard E. Byrd in New York Harbor. A week later, bow to the wind, the ZR hovered over dedication ceremonies for Floyd Bennett Airport. In August, Clarke hung over City Hall, Philadelphia—participating in the Spanish War veterans parade in that city. Hollywood was hosted: director Frank Capra and stars Jack Holt and Ralph Graves, along with a crew from Columbia Moving Pictures. The project: *Dirigible*, a $650,000 action-adventure film authorized by the Navy Department (a "pep talk" for BuAer, according to a Capra biographer). "The advance guard descended upon us today with a car load of 'properties,'" Berry wrote his wife. "Lt. Cdr. Weed Retired is the technical advisor for the company. Geo. Watson and McCord will be our censors." Granted the run of the place, Columbia made itself at home on station, spending (as cameraman Joe Walker was to recall) "many weeks at Lakehurst, comparatively unhurried, seeking only to get the utmost accuracy and cinematic effectiveness into our scenes."[34]

L.A. also would star. On 19 August, with apparatus installed in the main car, a Columbia crew was flown to a scenic loiter over Manhattan Island. *Los Angeles* circled. Back at the mooring-out site, filming continued but in the keel corridor. Clarke unmoored a second time that day and set course for Philadelphia. The "aeroplane arresting gear" was tested over its Navy Yard, after which, at City Hall, Clarke joined in the veterans parade. Next day, Columbia carried on "making pictures," of airplane hook-ons this time. In the hushed dark of the 21st, further scenes were filmed in the ZR dock.

August's final hook-on proved unusual. Within minutes of unmooring, on the 25th, *L.A.* began steering various courses preparatory to dropping parachutes. Following two practice approaches, three men and seven dummies were dropped over the field. Landing stations were piped down, only to be called away again at 0937 as *ZR-3* began maneuvering for a hook-on—five engines set to "Ahead Cruising." O'Brien landed at 1002. The airplane "housed," Clarke cruised toward New York City, following the coast in a generally northerly direction. Over the inner harbor, he circled in company with *J-3*— for pictures. Early that afternoon, Clarke again had his rudderman circling, over upper New York Bay then Manhattan, as *J-3* took its departure.[35] At 1426, course was set for home. With four engines working half-speed and number one stopped, O'Brien cast off over the reservation at 1536 with a passenger: Harry Shoemaker.

This demonstration was the last hook-on for close to ten months. The basics were down. Yet *Akron*'s operational airplanes (BuAer was still trying to find a model) were procured only in the fall of 1931.

Meantime, with *ZRS-4*, search tactics had assumed some urgency. No ZR had operated with fleet units since 1925, when *Shenandoah* had participated in one exercise with the Scouting Fleet and minor operations with USS *Texas*.[36] As it happened, combined fleet maneuvers were scheduled for February 1931 in the waters off Panama. If information was to be gained before *Akron* flew, however, *Los Angeles*—the only airship to hand—would have to shed her civilian status. At the instigation of Ingalls, a lifting of the Allied-imposed restrictions applied to *ZR-3* was solicited. In its war problem (the Navy Department made clear), the ship would not be armed; instead, *L.A.* was to be simply a commercial craft deployed for scouting.[37] Signatories France and Britain readily agreed; the Japanese government made no reply. Late in November, Henry L. Stimson, secretary of state, was notified of the department's intention to deploy *Los Angeles* as an unarmed observer. Within a week Stimson replied: no objections. In December, the State Department informed Ingalls that it had advised the Japanese ambassador of its intention and had received, as yet, no protest—a silence interpreted as approval.

So a very active year moved toward a close. September had seen a half-dozen sorties plus maintenance work—a cell and engine change, renewal of outer cover, repairs to broken lattices, wires retensioned. One trip had had Moffett aboard: an eighteen-hour run to Syracuse, on 4–5 September. "It was very bumpy," AMM3c John Iannaccone wrote his father, "the wind was blowing strong and it was hard going. We went straight up the Hudson river to Albany and then followed the Erie canal into Syracuse. . . . We read reports the next day that our rudder had been damaged. That was a false rumor. It was a wonderful trip

Table 5-3 Status of *Los Angeles*'s gas cells, 1 October 1930. (National Archives)

Position	Type	Installed	Condition
0	Goodyear skin	18 June 1928	Fair
1	Goodyear skin	18 June 1928	Fair
2	Gutta cellophane	3 January 1930	Fair
3	Goodyear skin	1 March 1929	Excellent
4	Gelatine Latex	6 January 1930	Excellent
5	Gelatine Latex	13 March 1930	Excellent
6	Goodyear skin	12 March 1930	Excellent
7	Goodyear skin	8 February 1929	Excellent
8	Goodyear skin	24 April 1928	Fair
9	Goodyear skin	2 October 1928	Good
10	Goodyear Rubber	28 August 1930	Excellent
11	Goodyear skin	3 October 1928	Good
12	Goodyear skin	10 July 1928	Good
13	Goodyear skin	10 July 1928	Fair

and I hope we have more like them."[38] During October, eighty-five and a half hours were logged under way, including (on the 11th) an exercise with Light Cruiser Division 3 and Destroyer Division 41, in company with *J-4* and *ZMC-2*. That month, winter fronts were installed on the engine cars and other winterizing effected. Three takeoffs entered the log for November, and on the 11th loads were measured during side-handling tests at the circle. A two-hour sally on the 14th (abbreviated due to fog) was the airship's last for 1930.

In anticipation of the Panama operation, *L.A.* was promptly docked along the north side of the shed, her cradles shunted in. A full-scale overhaul would be laid on, with attention to the airframe (e.g., varnishing), engines, and the outer cover. The Maybachs were fully overhauled, about 30 percent of the cover renewed, minor repairs effected throughout.[39] Along with other experimental gear, the trapeze was removed the day before Christmas.

For the upcoming maneuvers, an expeditionary mast would be erected at Guantanamo Bay and *Patoka* would be dispatched to tropical waters. On 22 January (with Freddy Berry along for the passage), the tender stood out from Philadelphia to investigate ports and bays along the Panamanian coast for mooring and servicing *ZR-3* during the problem. (*Patoka* rounded the eastern

L.A. to sea off Barnegat Lightship with J-4 and ZMC-2 in company, 11 October 1930. In formation with light cruisers and destroyers, the trio are maneuvering at various courses and speeds with the surface units. (U.S. Navy Photograph)

end of Cuba on the 28th. "The uniform is whites," Berry recorded, "the sun is hot and yesterday my old mug got its first touch of sunburn.")[40] After transiting the Panama Canal, the surface ship was to await her charge in Dulce Bay, Costa Rica, on the Pacific coast of Central America.

That same month, Clarke notified BuAer that he had been approached by representatives of Paramount Sound News Company seeking approval for a stunt flight: a simulated mooring to the Empire State Building, then nearing its official opening. Having rejected this proposal, Clarke reported that he would recommend against the flight if attempts were made to push it through. The bureau had little difficulty concurring.[41]

Another matter was settled as well. "Clarke left one memento on the *Los Angeles* which was a great service to many of us," Watson volunteered. "Clarke was an inveterate smoker and we had never had a smoking room on the *Los Angeles*. Other skippers had thought this would be a dangerous fire hazard. But not Vince." Concluding rounds of back-and-forth, permission was granted—and a smoking area was installed in the after end of the control car. The Panama trip would confirm its virtues.[42]

Operations for January 1931 involved working up in local airspace, including overflights of New York City, Wilmington (circling *Patoka*), and Philadelphia. The aircraft itself was checked out, her radio compass tested. Under way on the 22nd, the water-recovery condenser to number-four car was disabled when tubes blew out; back in berth, eleven civilian workmen installed another.

February would see the main event—the first use of an airship scout in U.S. fleet maneuvers. In a quickening drumbeat, the ZR concept reentered the headlines: "AIRSHIP TO MEET TEST OF BATTLE IN FLEET GAMES; Craft Nearing Obsolescence to Operate from Limited Base in Defense of Canal." The case for commercial as well as military airships had been much before the public. Yet accomplishment seemed thin, the detractors many. The fleet did not necessarily wish *Los Angeles* well. Nor was the press won over. "Theories," one sheet scoffed, "are all that the best naval authorities have been able to give Congressional committees when they have gone before them in the past seeking government support for an airship program." "It is evident," another announced, "that the *Los Angeles* will be watched as a possibly important factor in the future development of the airship when she lifts her nose from the airship tender *Patoka*. . . . To some extent, the cause of airships, in the United States at least, will succeed or fail according to the success of the performance of the *Los Angeles* in the opening phase of the spring fleet maneuvers. In the Bureau of Aeronautics at Washington during the last several weeks there has been evidence of concern as to how the navy's one dirigible will acquit herself in her first 'war' lesson."[43] This unease was based less on conceptual grounds than on *Los Angeles*'s near obsolescence and the fact that she carried no airplanes. Further, the ZR would be operating for weeks from a temporary base with limited facilities in an unfamiliar climate. Extravagant valving, for instance, could jeopardize the entire operation.

If ever there was a defining moment for *ZR-3*, 1931's full-dress exercise would be it. Aware of the pressure, airship men knew their platform would have to perform.

In the 1200 to 1400 watch, 4 February, ship's force was engaged in preparing for flight. Amid zero-hour excitement, couples embraced and supplies were placed aboard: crates of eggs, beef, oranges, bread. At 1620, *Los Angeles* was unmoored from spring balance scales and moored to the mast. Minutes later, general assembly sounded, all engines warmed, embarkation begun. Declared ready for flight at 1645, *ZR-3* was towed out. Towing stopped at 1703, whereupon the mechanical handling gear was disconnected and *L.A.* allowed to swing to port, into a gusty wind. Towing recommenced eleven minutes thereafter. At the mooring-out circle, final preparations effected, Clarke unmoored at 1726—dusk—with forty-nine persons: thirteen officers, thirty-four crewmen, Lt. Cdr. R. S. Berkey (an Ingalls aide) and a Mr. C. R. Walker, who promised good publicity. Walker had signed up with the *Saturday Evening Post* for an article.

Destination: Panama via the mast at Guantanamo Bay—a three-thousand-mile air leap. Probable absence: six weeks.

"The take-off," Clarke informed Tyler (the trip's press representative) "was entirely satisfactory and was promptly executed. We got away with 5,000 [actually 5,250] pounds of water ballast." Other disposable weights on board: 29,811 pounds of fuel, about five hundred pounds of food, and a bit of mail for sailors already south. Rising statically to three hundred feet, all engines were ordered to "Cruising" speed, and a course was set for Cape Hatteras Light. At 1815, two thousand feet off the water, thousands of brilliant white and colored lights and the straight-line boardwalk at Atlantic City lay abeam. In officers' country, at a table cleated into the bulkhead between bunks, eight men messed—dinner by Peak.

A vital chunk of energy, with red hair and a sailor cap the size of a nickel, emerges from the galley. He carries roast beef in large slices and a stack of paper plates. . . . The next trip to the galley produces potatoes, beans, bread, coffee and more paper plates. . . . After dinner, cigarettes in a fireproof smoker to the rear of the sleeping cabins.[44]

"All hands greatly enjoyed a roast beef supper," reporter Tyler remarked. Cape May—extreme south Jersey—lay abeam at 1900, ten miles to starboard.

A high-pressure area was overspreading the seaboard, so the hours passing Cape Hatteras via the Gulf Stream proved untroubled. The cape stood astern in less than six hours. Pushed by tail winds, the airship was ninety miles east of Florida and 567 miles from Guantanamo the next morning. Clarke, though, was reporting thunderstorms and high winds that had reduced his speed from about seventy to sixty miles per hour (about fifty-five knots). Near the CPO and enlisted men's keelway quarters and mess spaces, a passenger recorded the change of watch.

> Down the four sloping [cross] girder ways come engineers from the power eggs; toward the ship's nose . . . move the new helmsman and elevator man. Poker chips are put away in the mess hall; men roll out of their pup tents, lacing shoes; suddenly a radio, located near the ladder to the control car, bursts into a Mexican tune; the off-watch do a fox trot on the cat walk.[45]

At 1655, on the 5th, the Guantanamo station lay six miles on the starboard beam. Approach commenced twenty minutes later. Having logged 1,535 miles and twenty-four hours, seven minutes aloft, the airship locked its cone into the mast cup at 1733.

Fuel, fresh food supplies, and water were taken on, and all heavy flight clothing was sent ashore. Clarke made an official call on the local commander—standard military etiquette. Cast-off came at 0820 as flying boats maneuvered nearby; *L.A.* rose weightless to three hundred feet, all engines were set to cruising speed, and altitude increased. At thirteen hundred feet, water recovery cut in. Minutes later, though, number one was stopped and four rung up to half-speed, for a fuel economy test. At 0843, Clarke took his departure and set course for Colón, across the Caribbean Sea. Estimated time of arrival: sunset.

Shortly after dark, eighteen vessels of the Scouting Fleet were sighted on the starboard bow, distance ten miles. En route to Panama, these were allies in the coming sham battle. (The "attacking" force, formed from units of the U.S. Battle Fleet, had stood out from San Pedro.) Altering course, the dirigible closed, then loitered, exchanging a series of radio messages with the commander in chief, on board the battleship *Arkansas*. The seven hundred miles to the canal's Atlantic entry were traversed in thirteen hours. At 2035, the Manzanillo Point Light was sighted two points (about twenty-two degrees) on the port bow; by 2116, ZR-3 had gained Colón. Clarke circled. At 2130, USS *Texas*, flagship of the U.S. fleet, lay beneath. Clarke now pressed across the isthmus in exceptionally rough air, following the canal eastward to Balboa. The Gatun Locks stood below at 2136 and at 2215, Balboa. "The dirigible's arrival over Panama City greatly excited the populace, who had never seen such a ship before. They rushed into the streets at the sound of her motors and watched it cruise about, its sides bathed in silvery moonlight."[46]

Instead of crossing land to reach the Gulf of Duce, to the southwest, near the Panamanian–Costa Rican border, the rudderman steered across the Gulf of Panama, to Cape Mala. He followed the beach. A waiting *Patoka* was sighted at 0420, on the 7th; at 0845, Clarke moored. Though her second visit, this was "the first on a war-like mission," the *New York Times* said. "Those who made the trip," the *New York Herald Tribune* announced, "enjoyed stepping down to the *Patoka* for trips to the ship's canteen and to use the laundry. All meals today were taken aboard the mast ship. . . . In the meantime, the *Los Angeles* was supplied with water, food and fuel that it may continue her cruises about Panama, beginning tomorrow morning. . . . The ship will act as a base for the *Los Angeles* throughout the war maneuvers."[47]

Clarke loitered until 0808 on the 8th, when he cast off to retrace his course to Panama Bay. The Atlantic-based Scouting Force was assembling there, so the tender was shifting anchorage. "The top speed of the *Patoka* was eight knots," Watson recalled, "so it took about two days to get from one anchorage to the other. During this time, we had to stay in flight and we did considerable cruising around the offshore islands that are on the west side of Panama and Costa Rica."[48] Having steamed coastwise, sundry points and lights passing to port, Clarke was showing off his command to nighttime Panama City by 2300. Then, well to seaward, various courses were assumed circling the Perlas Islands "for photographic purposes." By 1358 he had reappeared over the city. In the dying hours of the 9th (2034), *Los Angeles*'s cone at last met *Patoka*'s mast cup (after three approaches). Transit time: thirty-six hours, twenty-six minutes.

The mast ship steamed to anchorage. The pair swung to anchor in Panama Roads for over two days as a great naval force came together. By the first watch of the 10th, this concentration comprised sixteen units of the surface fleet. Further ships stood in from sea; by the earliest hours of the 12th, the assemblage included battleships, heavy cruisers, older light cruisers, destroyers, tenders, and submarines, as well as *Saratoga* and *Lexington*—the latter having dropped anchor in the outer harbor during the morning watch.

Table 5-4 Procedure in daily weigh-off, USS *Los Angeles*, 1931–32. (George H. Mills Collection, Smithsonian Institution)

1. At 0700 call the watch—half section of riggers and half section of engineers.
2. If docked in the cradle, the trollies and surge tackles will be set taut.
3. Slack all surge tackles.
4. Have watch stationed at both ends of the ship remove sand until ship comes in equilibrium.
5. Count sand bags and enter weight on chart.
6. Enter on weigh-off sheet amount of fuel at each frame.
7. Enter amount of oil and frame number under miscellaneous weights.
8. Take from rough log an accurate water-ballast reading.
9. Enter flushing and drinking water under other water.
10. Take percentage fullness before weigh-off completed.
11. Enter barometer, temperature of wet and dry bulb, superheat forward and aft, and humidity readings.
12. See if handling frames are on.
13. If docked on spring balance scales, take off sand until ship comes up against scales.

On board airship and tender, the eight-to-twelve watch saw preparations for flight. Clarke unmoored at 0845, 12 February. Seven U.S. and Panamanian guests (including a U.S. envoy extraordinary and minister plenipotentiary, Roy T. Davis, and Dr. J. J. Vallarino, Panama's secretary of foreign relations) enjoyed a ten-hour sight-see. On various courses and dumping water as needed, the views included back-and-forth legs over the canal as well as Panama City, Colón, Penonome (where one guest dropped a message), hometowns, ranches, and the Pearl Islands. Throughout the flight, Clarke's passengers kept the radio engaged with messages. Lunch was served at noon, followed by a short siesta. The afternoon proved especially bumpy. "The ship went approximately 30 miles inland to within ten miles of the high mountains," First Lieutenant George Calnan recorded, "and we took a beating of about the same intensity as what we get over the Pennsylvania mountains."[49] By 1851, *Los Angeles* hung back on *Patoka*. "It was a delightful and never-to-be-forgotten excursion," Dr. Vallarino offered. "At first I thought that ten hours in the air would be boring and tiresome, but once aboard the trip fascinated me and I desired to continue indefinitely. Each time I have the opportunity to make another such trip, I will not lose the chance."[50]

Swinging to equatorial airs, the moor-out in company resumed. To no one's surprise, daytime cycles warmed the ship's helium, topside especially. This bred superheat, and as a result lift bled through the automatics. *L.A.* already had lost two hundred thousand cubic feet of helium, Calnan estimated, by valving and by going over pressure height either in flight or at the mast. "The men who went up the climbing shaft say that you could cook eggs at the top," he told Fulton. "Gas cells and superheat will have to be given some study before any extended tropical operations are contemplated," the skipper later reported. "Have had temperatures up to 130 F, between top outer cover and cells. The skin cells show slight signs of blistering and the gelatin cells take on a glossy surface." Under way, the mechanics endured a hot time of it.[51]

For unsleeping sailors, the midwatch log entry of the 15th recorded a grand aggregation of naval might:

> Moored to U.S.S. PATOKA, in Panama Bay, C.Z., in company with Commander-in-Chief U.S. Fleet (U.S.S. TEXAS); Commander Scouting Fleet (U.S.S. ARKANSAS); Commander Aircraft Squadrons, Battle Fleet, U.S.S. SARATOGA and U.S.S. LEXINGTON; Train Squadrons, Scouting Fleet; Destroyer Squadrons, Scouting Fleet; Submarine Divisions, Scouting Fleet; Aircraft Squadrons, Scouting Fleet; Light Cruiser Divisions, Scouting Fleet; with the following disposable weights on board: fuel 16547 lbs., oil 1187 lbs., water ballast 10400 lbs. Gas cells averaging 91% full. Ballasting as necessary. Sky clear, wind from 330 T, force 10 knots.[52]

Arriving by plane that afternoon, Assistant Secretary Ingalls with Vice Adm. Arthur Lee Willard, Commander Scouting Fleet (ComScoFor), and staff boarded *Patoka*. Willard chose not to linger; Ingalls, though, ducked through the bow hatch and his flag was displayed. Minutes later, at 1629, Clarke unmoored. In darkness, *ZR-3* cruised the bay with its men of war and circled Panama City until, in the morning watch, the dirigible was rigged for mooring. *Los Angeles* resecured at 0650.

TESTBED FOR THE NEW SHIPS

Decks darkened, radios silenced,[53] the defending "Blue" force had stood out late on the 15th. Fleet Problem XII had begun. The battle plan: an inferior surface fleet, assisted by a superior air force, would operate against a greatly superior surface fleet attempting to capture the canal and establish a base in Central America. The defenders would seek to engage the attacking "Black" fleet as far to sea as possible. *Los Angeles* was to assist the search, aiding "Blue" units in gaining command of the sea. ZRs presented an ideal platform from which to surveil wide and uncertain areas—or so they were advertised. Even *ZR-3* could scout a larger sector than any single unit of the defending force, including its cruisers, thereby allowing the defense to keep its fighting force concentrated in the probable area of attack.

In the 1600–1800 dog watch of the 16th, embarkation commenced, Ingalls joining. All engines were warmed. At 1625, landing stations were called away. The pelican hook on the mooring pendant pulled out of shape and failed to release, stressing a compression strut out of line in the nose—minor damage. Clarke sortied at 1635 with thirty-nine on board: eleven officers, a crew of twenty-six, and Cdr. Alger H. Dresel, along as umpire, and Ingalls.[54] *L.A.* loitered on various courses in the vicinity of the fleet anchorage until, at 2210, a dispatch was received from ComScoFor: "Execute Blue Fleet Operation plan number three dash thirty-one." Clarke took his departure for the assigned zone of search. The midwatch of the new day found the ZR pushing southwest en route for "Zone A" in increasing cloudiness, with occasional flashes of lighting. Clarke gained "Point B" at 0955, on the 17th, latitude two degrees, forty minutes north, eighty-six degrees west longitude—some 450 Pacific miles from Costa Rica. Various courses, speeds, and altitudes now were steered. The hunt proved fruitless. Cocas Island was in view at 1515 then, at 1640, the cruiser USS *Trenton*. The dirigible closed to exchange visual signals. A merchant steamer was spotted at 2232, and ten minutes later, two points on the port bow, Burica Point Light—extreme southwestern Panama. The midwatch of the 18th had the rudderman steering coastwise courses, for the Gulf of Dulce and in search of *Patoka*. The tender's flashing-light recognition signal entered the log at 0405; at 0700, the cone locked into the cup. This concluded thirty-eight hours twenty-five minutes under way. Trailing its charge, the tender proceeded to anchorage.

There was no time to relax. So she might continue, *Los Angeles* was refueled and gassed, the former commencing even as *Patoka* steamed. Oil was taken on plus flares

Cdr. Alger H. Dresel. A graduate of Lakehurst Class VI (1929–30), Dresel was Chief Umpire aboard L.A. *for the 1931 fleet problem and, that April, would succeed to the command. The following spring, he received orders to* Akron *as prospective CO and, in 1933, was* Macon's *first commander. Dresel was probably the most popular of the ZR skippers. (U.S. Navy Photograph courtesy Navy Lakehurst Historical Society)*

Los Angeles *steaming the Panama Canal, February 1931. She is steaming tropical airspace as one element of Fleet Problem XII—a defining moment for the ZR concept. The view is forward from No. 2 car, towards No. 4 and the control car. (ADC John A. Iannaccone)*

ZR-3 secured to Patoka, *Canal Zone, 7–20 February 1931. Various units of the surface fleet lie to anchor about the tender and her charge including, well offshore, USS* Saratoga *and USS* Lexington. *(Charles E. Rosendahl Collection, University of Texas)*

(twenty pounds), food, and ice. To replenish the cells, fifty-four thousand cubic feet of helium came aboard during the 1200–1600 watch, during which fueling was finished. Scrambling back in, Ingalls made his way to the bridge. Clarke unmoored at 1835.

As lightning played through the cumulonimbus and thunderstorms roiled the gulf, Clarke cruised the vicinity of Belica Light until 2135, when he took departure for Point A. At 0140, the 19th, the ZR stood out to sea on a track of 200 degrees true, for Rivadeneyra Shoals. With lightning and thunderstorms noted on various bearings, varying courses were steered. On orders from the Blue commander, *Los Angeles* had been directed to search, commencing at daybreak, for "unaccounted [for] enemy vessels."

Another view of the Central American anchorage, with Lexington *and a battleship in company. (U.S. Navy Photograph)*

Though none aboard knew, this was to be *the* day.

At 0545, daybreak, a series of course changes were executed. As per orders, Clarke had begun sweeping a three-hundred-mile-wide arc as, concurrently, patrol planes covered the area of search to the south.[55] At 0950, excitement swept the bridge: a vessel lay in view, bearing 081 degrees. At 1020 this was identified as a carrier, "presumably U.S.S. LEXINGTON." Course was altered so as to shadow the defender. Curiously, neither the watch on *Lexington* nor the pilots patrolling above or about her spotted the dirigible. After more than an hour the horizon became hazy, leaving the carrier barely visible. Then, abruptly, a full-rigged sailing vessel was sighted. Course was changed to intercept. On closing, she proved to be the *Tusitala*, five days overdue from Hawaii.

The task at hand resumed. *L.A.* quickly found action: within sixty minutes (at 1320), her lookouts had a large force in view, bearing 243 degrees, distance thirty miles. "I remember how crowded and excited the control car became in a second," passenger Walker wrote, "and how I ran back to one of the windows in the officer's quarters to get a good look at the enemy. The type and arrangement of warships were now clear to the naked eye."[56]

A contact report sped off. The concentration was promptly identified as the "Black" main body, including USS *Langley*, its only carrier. A further message to the commander of the "Blue" fleet, served to revise the face of battle: "Force previously sighted main body aircraft carrier second line latitude 5 33' longitude 83 30' course 090—1332." Clarke's most effective moment, this was the first reliable report of the whereabouts of the opposing main body received by Admiral Willard from *any* source. The foe having been found, the strategic search had

L.A. *maneuvering with surface units during Fleet Problem XII. This view is from power car No. 5, duty station for AMM3c John A. Iannaccone. (ADC John A. Iannaccone)*

been done; for Clarke, the task now was tactical scouting—to develop the contact, determine composition, course, speed. Altering course to close, exploiting cloud cover so as not to draw protecting planes, Clarke tailed unseen for ten minutes and was able to shadow for thirty before being jumped. As the airship proceeded directly in, its lookouts had failed to notice that the "enemy" was launching planes.

> Then suddenly the attack began. They dove at us from a great height, leveling off just as they reached the back of the ship, or zoomed past our lookout windows at what seemed arm's length. All this was only a maneuver, but it had moments of realism, as when one fighter dived under the ship and "destroyed" our antennae. No one found it hard to visualize a duel to the death in the next war.[57]

The "attack" was effective, as the airship's log records, "1350 Sighted three enemy planes in formation on starboard quarter, distance one mile. 1355 One enemy plane made a bombing attack approach, as result of this attack Chief Umpire, Cdr. Dresel, ruled this vessel destroyed, secured radio"—in the middle of a message. Capt. Ernest J. King commanded a force from the *Langley*. In May 1933, he would succeed Moffett as chief. Just now, his attackers had dispatched the surveilling gas bag (a luscious target) "with singular pleasure."[58]

Los Angeles (a nonparticipant now) passed over two of the attacking vessels, after which a "Black" plane actually carried away her aerial—a further rub. At 1408, Clarke received a visual message from Adm. William V. Pratt, CNO and chief umpire: "You are sunk," followed

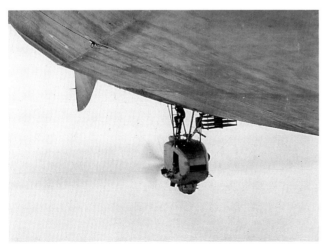

AMMC2c *Frank T. Karpinski clambers into No. 1 car during the 28-day mission into the tropics. (ADC John A. Iannaccone)*

The control car vanes in over the mooring deck of Patoka *during a war-game loiter, Panama Bay, 20–22 February. In the forenoon watch of the 22nd, Vince Clarke cast-off for the expeditionary masts at Guantanamo Bay and Parris Island, South Carolina—and home. (ADC John A. Iannaccone)*

by "pleasant voyage." At 1523 another message, this time from Admiral Schofield, C-in-C, U.S. Fleet, advised: "Consider *Los Angeles* out of action during remainder of problem direct *Patoka* proceed Panama Bay and anchor approximately 3000 yards south of entrance buoys Panama Canal clear of fairway and await further instructions." Course was set for Jicaron Island.

In such exercises, customarily, "sunk" assets were soon declared back in action. (In 1934 games, the "shot down" *Macon* continued as "ZRS-6.") "But for some reason," Watson told the writer, "we were not given this privilege. We were 'shot down,' and that was the end of us."[59] *ZR-3* was not resurrected.

It is to be regretted that Clarke had had no plan as to how he would develop his contact while concealing himself at a distance or in cloud, let alone engineer an escape. Yet his sighting, along with two radio reports, allowed his side to launch a highly successful attack. Wide-flung aerial and surface sweeps had succeeded in locating the advancing attackers in two main detachments. (Finding nothing, it should be noted, was significant in itself: it meant no "enemy" in the sectors searched.) Concentrating its attention on known threats, the defenders attacked in force, so crippling the "invaders" that a successful attack upon the canal was rendered impractical.

The war game concluded at 1800 on the 20th. First among the defenders "quayside," *ZR-3* reengaged the mast at 0643, ending a protracted sortie totaling thirty-six hours, eight minutes. Ingalls disembarked for good. In less than two hours, refueled, the airmen were again aloft—a ten-hour photographic survey over the Perlas Islands.

His fall in "battle" notwithstanding, Clarke pronounced himself satisfied. "All the Admirals and staffs, lukewarm at the start, have daily seemed to be more enthusiastic about the *Los Angeles* and L.T.A.," he informed Moffett. "We will get out as soon as we can and leave that favorable impression." Fred Berry, liaison officer aboard *Patoka*, agreed. "The *L.A.* created at least a favorable impression in some quarters and an enthusiastic one in others. On the whole, lighter-than-air has materially benefited from her association with the fleet." In his remarks to reporters, Clarke was decidedly upbeat: the successors to *ZR-3*, he opined, would be able to search an ocean area of six thousand square miles each hour. As to his own command, "In two flights the ship cruised 2,315 nautical miles, searching an area during daylight hours averaging 3,000 square miles an hour. This figure far exceeds that which can be covered by any type of surface craft." Ingalls, for his part, was no less pleased. Described by Clarke as

having "had the time of his life and . . . [having been more] pleased and more interested than ever," the secretary underlined the positives of *ZR-3*'s performance. "I am exceedingly gratified," he radioed via the Universal Service correspondent (Tyler) "at the marked success attendant upon the use of the *Los Angeles* in connection with the fleet maneuvers. . . . Although the ship was employed only in a scouting capacity and therefore no defense of her was attempted, one experience has conclusively established that such type ships are of material value in naval operations at sea in defense of our coast. . . . The repeated contacts with theoretically friendly ships and her final discovery of the main body of the theoretically hostile fleet attacking the Panama Canal definitely established the advisability or, rather, the necessity, of continued development and maintenance of lighter-than-air craft by the United States Navy." No less a personage than Admiral Pratt remarked that one of the outstanding lessons of the fleet problem was "the justification, under favorable circumstances, of the lighter-than-air craft, as shown by the use of the *Los Angeles*."[60]

Ingalls, though, was soon to vanish; in June 1932, his resignation was accepted by the president. (As an economy measure, the office of the assistant secretary of the navy for aeronautics stayed vacant until 1941.) Moffett would be lost as well, with *Akron*.

Meantime, labeling the airship's mock destruction as "immaterial," Rosendahl (still an indefatigable cheerleader) professed his own satisfaction. *L.A.*, he said—slow, obsolete, of small range—had been unarmed, without camouflage, and had made no attempt to withdraw following its contact reports "but continued on into the battle."[61] This emphasis upon the airship as scout, and too little regard for its vulnerability to airplanes, was the old story. One 1931 lesson lay unlearned: a ZR "carrier" had to use its own planes to search ahead and hang back itself in relative safety. But this stood unappreciated during *Akron*'s career and, aboard *Macon*, received emphasis only in 1934. In short, no truly pertinent conclusions were drawn from the Panama experience.

Admiring applause from airship men and ephemeral political heads was one thing, rousing the topside leadership another. There, doubt had hardened to contempt. Pratt's remark notwithstanding, "we did not get a very favorable report from the Commander of the Scouting Force, who was our boss for the exercise," Watson noted. "We never understood quite why, and maybe we were too biased to understand, but we felt that he had prejudged the airship as not being a very valuable member of the fleet anyway. . . . The Scouting Fleet didn't think much of airships to start with."⁶² Panama left Watson with no illusions. At briefings and critiques attending the exercise, the mood, he later remembered, was one of barely concealed hostility. "It was my impression that the Fleet commanders could have cared less whether we were there [off Panama] or not."

Had participation been fundamentally futile? Renewed discussion was provoked, certainly. For most critics, *L.A.* merely confirmed preconceived opinions. Undervalued by most naval aviators and at-sea commanders, it had remained in an experimental status for what was proving a maddeningly protracted period. The ZR was seldom seen by fleet forces, let alone exploited, and it seemed to offer little practical good when its services *were* presented. Clarke, in short, had failed to win the fleet's regard. Watson may have put a finger on the problem. "The relations between the Navy high command and the top LTA people were always puzzling—and still are. In retrospect, it seems to me to have been mostly a matter of funds. The development of HTA, carriers etc., was just beginning and BuAer needed every penny it could get. So when Moffett reached in and grabbed the money to build *ZRS-4&5*, it was a tragedy for BuAer."⁶³ Fleet Problem XII was emblematic of such emotions, especially as regarded carriers, to which the navy was ceding primacy. Appropriations insecurities added a further knot. Changing political and economic conditions—and faint praise—could not disentangle the ZR from competing interests favoring heavier-than-air. Big airships, in sum, continued politically and militarily peripheral.

There is rough justice in the criticisms presented by policy makers and fleet commanders. Given its chronic experimental/marginal status during 1923–35, one is left to wonder whether the ZR was ever *able* to perform, truly to help itself.

Meantime, the issue again jumped into the press, certain elements of which proved ferociously anti-airship. The *New York Sun* had waged a particularly virulent campaign. *ZR-3*, the paper declared, had "failed to live up to its advance notices." Further, "the big balloon was of little use even for scouting" and "observers generally agree that the gas bag *Los Angeles* completely failed to justify confidence in dirigibles as engines of war." Exasperated, Rosendahl prepared a rejoinder for Moffett who, in turn, directed a letter against this broadside.⁶⁴ The editors, doubtless, could not have cared less. "I do not ask for the airship," Rosie complained, "anything but fair, open-minded consideration." This was fantasy, an exercise in optimism. "As a result of years of effort, the Navy will soon be in a position with modern ships and bases to give the airship project a practical trial with our naval forces. Until fair conclusions as to modern airships in a modern Navy can be thus achieved, it would seem unjust to condemn this type of aircraft of such great potentialities, solely on isolated performance of a single, obsolete craft."⁶⁵

Armed with a weather map suitable for the jump, Clarke cast off for Cuba in the forenoon watch of the 22nd. (Detached from *Arkansas*, Berry was on board for the return.) Once again, rough air met *ZR-3* over the isthmus; it brought four head wires broken and small holes in cells five, eight, and eleven. The return proved protracted—eight calendar days. Contesting a stiff head wind above the Caribbean, *Los Angeles* got quite light. "Nobody liked to valve helium very much," Bradley remembered. "When the ship was so light you had to keep the nose down, to keep her from going up. And it got to the point where we were going along with about ten degrees down-inclination—and the Captain didn't like that very much. We had quite a to-do about whether we ought to valve a little helium." Clarke prevailed—sixty seconds from all maneuvering valves. "We rode a little more comfortable thereafter."⁶⁶

At 1215, the island's southern coast in sight, bearing north, course was set for Guantanamo. Meeting very turbulent, bumpy air on arrival, Clarke elected to abort his approach. Landing stations were piped down—and three and a half hours more expended aloft. The ship's cone locked to the expeditionary mast at 2125. Aft, the stern carriage and tail drag were secured. Ballasting as needed (often in gusts and abruptly changing superheat), winter flight gear plus three hundred pounds of alcohol were received back on board, six cells were gassed, the aircraft fueled, and just prior to landing stations, drinking water and food replenished. The layover stretched to 84.5 hours (23–27 February), due partly to head-wire replacements plus repairs to leaks in the radiator on number-four car. Farther on, at Parris Island, a 38.5-hour stay was logged (28 February–1 March).

Clarke had homeport beneath at 1720, on the 2nd. At 1833, battle weary, their vessel under roof, the flight crew commenced disembarking. Time away: twenty-seven days.⁶⁷

In physical terms, *Los Angeles*'s performance, and that of captain and crew, had been impressive; all had passed the test admirably. Operating solely from mooring masts, *ZR-3* had traveled 14,445 statute miles altogether in 274 flight hours. This old campaigner had pressed nearly to the equator (two degrees, thirty minutes north) and as far west as ninety degrees west longitude. Over the course of ten moors, the operation had added 352 hours at masts: 214 with *Patoka* (seven moorings); ninety-nine hours at Guantanamo; thirty-eight hours at Parris Island. "We had thought," Watson observed, "that we did a rather credible job of operating so far from home with no regular base under conditions of weather that we were not familiar with," especially superheating, high humidity, and possible effects of heat on the cells.[68] Among other recommendations, Clarke requested that ZRs be allowed to operate more often and for longer periods away from the hangar. Secretary Ingalls asked BuAer to consider immediate arrangements for combat exercises for *L.A.* in the Atlantic, in conjunction with *Langley* and the Scouting Force.

Such were never ordered. Further, save for a trio of missions later in the year—one a Navy Day flight to Atlanta thence to the Parris Island mast, another to Boston early in 1932—no more long flights would be logged.

That spring (April), the first of the year's detailed quarterly reports regarding *Los Angeles*'s material condition was forwarded. By July 1931, Moffett had in hand the latest test results from the Bureau of Standards. At midmonth, these were routed to the Board of Inspection and Survey. In its view, "These data . . . are considered to show nothing to cause alarm as to the general condition" of the airship, "and continuing her in active service under the limitations now in effect is considered to be fully warranted." An equally comprehensive report for the quarter ending March 1932 was anticipated, but a more extensive structural sampling for 1932 was recommended. The importance of this sampling and testing program is hard to overstate; by 1931, if not earlier, it had become apparent that the condition of the duralumin structure (coupled with the attendant system of wires) would be the limiting factor in defining the craft's useful life.[69]

Closing a frenetic twelve months, *Los Angeles* had logged more than seven years service. The quarterly report ending 31 December found her condition essentially unchanged. Upon reviewing the report the next April, the chief of the bureau found it "gratifying" to note the ship's overall condition but emphasized the necessity for continued exercise of care and judgment in dealing with problems concerning material condition.

The personnel needs of *Akron* were generating new orders. That April, fourteen *Los Angeles* enlisted men (Patzig among them) were transferred to the air station for duty in connection with ZRS-4. In the forenoon watch, the 21st, Clarke read his orders detaching him to temporary duty in BuAer, and Alger Dresel read his orders directing him to report to ComRATES for duty in command of *ZR-3*.

The sixth of the seven commanding officers *ZR-3* was to have, Dresel was to be skipper for eleven months. A graduate of Class VI (1929–30), Dresel—Naval Aviator (Airship) No. 3665—was one of the relatively senior officers brought in for grooming for future command in new billets to be created by expansion, including a West Coast base.[70] Enthusiastic but cautious, reserved in style and a polar opposite of Rosendahl, Dresel was not dictatorial; he was simply in charge. A pleasant personality, he was good at setting an informal yet professional tone. Dresel was perhaps the most popular of the big-ship commanders. "Commander Dresel was my favorite skipper," Bolster confided. "He was a man who inspired loyalty and determination in his crew and yet was stern and honest in his manner. I always felt that Commander Dresel had an ideal personality for an airship skipper."[71] Under his command, the airship's log recorded the mix of missions characteristic of this period: radio calibration, training and experimental flights, publicity hops, and intensive trapeze work.

Late in 1930 BuNav had announced that billets were open in the area of hook-on flying. As one outcome, in February 1931, Lt. Daniel Ward Harrigan—the charter member of *Akron*'s hook-on unit—reported to Lakehurst.[72] *Los Angeles* was then in Panama and her trapeze at the Naval Aircraft Factory, for modifications. Within weeks, the first (of six) hook-equipped Consolidated N2Y-1 trainers also reached the station, displacing the UO-1. But the trapeze was not reinstalled until May. For want of anything to hook onto, Harrigan went through the lighter-than-air course, qualifying as a balloon and blimp pilot (No. 3909).

The forenoon watch of 26 May found a half-dozen civilian workmen installing the trapeze; next day, O'Brien flew practice approaches. A week later, the device was modified. On the 9th, the first hook-on under Dresel was logged—O'Brien again at the controls. A dozen guests witnessed the demonstration (which was filmed for training use), among them Ingalls, Moffett, and newspapermen. On 17 June, as *ZR-3* again circled the field, Harrigan flew his first approach "and had few difficulties in running through ten hook landings. He discovered that, after only a little practice, it was easier to hang the N2Y onto the trapeze than it was to set a plane down on a concrete runway, much less on a carrier's flight deck."[73] Securing from hook-on practice, Dresel crossed the coast south of Beach Haven to cruise offshore. Having set course for Ambrose Light, *Los Angeles* intercepted SS *Leviathan*, then escorted

Looking aft from the mobile mast in position on the West Field, 27 July 1931. As a mast watch looks on, ship's crew stands to attention for His Majesty King Prajadhipok of Siam and the royal party. (National Air and Space Museum, Smithsonian Institution (SI Neg. No. 2000-9500))

SS *France*. Eight days later, the dirigible appeared over Wilkes-Barre—a nod to the National Convention of Disabled War Veterans under way in that city.

On 24 June, with power supplied from the hangar, experiments with ship's radio were under way. At 1130, the watch noted a thread of curling smoke, port side. By the time an extinguisher had been brought to bear, flames were licking the cover above the control car. The fire was subdued, but not before "assembly" had summoned the entire station. Perhaps ten square feet of cover had burned. To effect repairs, lower panels between frames 145 to 165 would have to be peeled. Apparent cause: a short circuit between a live wire and girder. Dresel dismissed the damage as "slight." Acting in Shoemaker's absence, Berry predicted that *Los Angeles* "will be ready for flight tomorrow and will fly tomorrow if the weather is favorable."[74] He was correct; ship's force and civilian workmen were through by noon on the 25th. A generator test was run, and at 1530 *L.A.* unmoored from the spring balance scales, then married to the mobile mast.

During July, Dresel and Company both were treated royally and entertained royalty. Concluding an easy, loitering run, *L.A.* joined *Patoka* in Fort Pond Bay, Long Island, on the 12th. She hung to her tether for seventy-five hours, during which time ship's officers—and the later-arriving Moffett, Ingalls, and other notables—were entertained ashore. Congressman Fred A. Britten, chairman of the Committee on Naval Affairs, and his wife gave a dinner in honor of the admiral. Members of the Easthampton and Southampton summer colonies then joined with Montauk residents in feting the visiting officers. (The entertainments for enlisted personnel passed unrecorded by the press.)

The 15th saw *Los Angeles* back home.

Late that month, His Majesty King Prajadhipok of Siam, accompanied by Queen Rambaibarni and their party, were embarked as special guests. Boarding on the 24th, the sovereigns had their ride postponed due to a thunderstorm threat. Three days later, at the mooring circle, the royal party—thirteen men and *ZR-3*'s first women passengers—again climbed the stairs. *Los Angeles* unmoored at 1030. (To deal with stifling temperatures, disposable weights included 250 pounds of ice for 281 pounds of food.) At three hundred feet "Cruising" was rung up to the power cars. Off the beach en route to New York, Harrigan hooked on in Lakehurst's N2Y at 1110; after five minutes, he released. Lunch by Peak, served as the rudderman steered south off Jersey's coastal edge consisted of veal cutlets, mashed potatoes, cucumber salad, a desert, and lemonade. Upon landing (which was delayed two hours until breezes moderated the heat), the king spoke enthusiastically, calling his excursion "the most marvelous experience of all the experiences which I have had on this visit to the United States." He added, "I found it quite warm. It was all wonderful, but I think the most wonderful sight was at the start, the flying over New York City."[75]

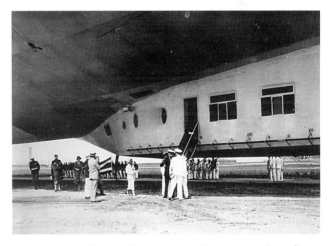

Queen Rambaibarni about to board—Los Angeles's first woman passenger, 27 July 1931. Unmoored at 1030, L.A. logged a seven-hour-fifteen-minute sortie. Off the beach en route to New York City, Lieut. Harrigan hooked on with Lakehurst's N2Y; after five minutes, he released. (National Air and Space Museum, Smithsonian Institution (SI Neg. No. 2000-9501))

This same summer, Harrigan was joined by Lt. Howard L. Young, Naval Aviator No. 3298. On 18 August, both were maneuvering over Barnegat Inlet at 1,800 feet, each in an N2Y. Around 0933, the pair commenced taking turns at the trapeze as *L.A.* steered various courses paralleling the beach two or three miles off. Harrigan swung on three times. This was Young's introduction to the maneuver; he experienced no difficulty in running through eight landings.

In the forenoon watch of 8 September, off Brigantine, hook-on operations were again practiced as Dresel followed the coast, nine student-officers observing. The public watched as well; at Atlantic City, the midair contacts offshore "thrilled thousands of Boardwalk strollers." A total of sixty-four landings were completed between 0905 and 1054. ZR-3 then steamed to *Patoka*, off Newport. In less than two hours Dresel again was under way, destination Lakehurst. There also he loitered but briefly, fueling and gassing at the mobile mast before getting away at 1048 on the 9th—now with ten student officers. The beach was crossed at Bay Head. On northward courses following the shore, the pilots were again drilled: when the exercise ended, at 1140, three landings had been made. (Minutes later, three miles east of Long Branch, number-three engine had to be stopped—a

Two to three miles off the beach, at 1,800 feet, a Consolidated N2Y-1 climbs to the bar during "hook-on tests," 18 August 1931. Lieuts. D. Ward Harrigan and Howard L. "Brigham" Young, the second member of the HTA Unit slated for ZRS-4, are taking turns at the trapeze. This was Young's introduction to the maneuver, but he had no difficulty in running through eight "landings." (National Air and Space Museum, Smithsonian Institution (SI Neg. No. 2000-9514))

ZR-3 with a Mark VI water-recovery condenser mounted above No. 3 car, 4–5 August 1931. Dresel is commanding, Lieut. George F. Watson watch officer, Admiral Moffett guest-passenger. The test was held to sea off Long Island. Light, rugged and compact, the Mark IV collector comprised four rows of aluminum tubes extending aft and upward for fifty feet. Unlike units mounted open-air, an external condenser close to the hull reduced drag. (Fred Tupper/Hepburn Walker Collection)

U-bolt strap had carried away, allowing a counterweight to be thrown through the crankcase.) On the morning of the 18th, hook-on work was repeated—this time over the station. In a stiff northeast wind, and with engines two, three, four, and five at cruising speed and number one at 1,175 rpm, three hook-ons were completed at an airspeed of forty-six knots. By 0959, when practice concluded, forty contacts more had been flown to the bar.

By month's close, Harrigan and Young had decided that it was time to find out how hooking-on worked at night. With a dozen officers and a crew of thirty-one (plus seven student officers and a dozen passengers), *Los Angeles* rose statically at 0824, 29 September. She cruised southwestward, passing over Philadelphia, Wilmington, and Baltimore before laying a course for Washington. At

A N2Y-1 drops away, 18 August 1931. Commanding Officer, USS Los Angeles: Commander Alger H. Dresel, Naval Aviator (Airship) No. 3665. (National Air and Space Museum, Smithsonian Institution (SI Neg. No. 2000-9513))

Los Angeles is hauled in—forward and down—by the main winch on the mobile mast, summer 1931. The bow of ZRS-4, then under construction, was built strong enough to take a lateral force of 40,000 lbs. Note the airplane trapeze, here retracted against the hull. (Fred Tupper/Hepburn Walker Collection)

Port and starboard yaw lines taut, her main wire dropped and coupled, L.A. floats like a cloud during haul-down to the old "iron mike" crawler mast, summer 1931. (Fred Tupper/Hepburn Walker Collection)

L.A. secured to the mobile mast. Note the Jacob's ladder and the trapeze. CNO orders to decommission would be issued the following May. ZR-3 is the only Navy aircraft to have observed all the naval traditions of decommissioning. (James R. Shock Collection)

Swinging to Patoka's *mast at Newport, 8 September 1931. This was the 10th moor to the floating base that year and* L.A.*'s final hours with the tender. Career total at the floating masthead: 718 hours, 34 minutes. (Warren M. Anderson)*

1145, the old Navy Building on Constitution Avenue stood below. Course was set for Annapolis, where *L.A.* arrived at 1225. Dover, Delaware was logged within ten minutes. At 1428, Dresel crossed the coastline at Jersey's Sea Isle City. On northerly courses following the shore, Atlantic City lay abeam to port at 1450. At Bay Head, the ship's bow was pointed inland; *L.A.* moored at 1723. She loitered at the mast for sixty-seven minutes, disembarking passengers and taking on water ballast and fuel. Airborne once again with eight student officers, *ZR-3* pounded eastward—to meet her planes over the sea, in the area of Barnegat. At 1940, the first N2Y commenced its approach. To illuminate the trapeze, the mechs in number-two and number-three cars were given flashlights to aim forward at the yoke. Except for the crude lighting, the two pilots discovered that night trapeze operations were somewhat more agreeable than during the day—the calmer air made for easier flying.[76]

The operation concluded at 2030, a dozen landings effected.

Next day's flight was odd: sound pictures over New York City plus a smoke screen laid about the dirigible by airplane. Smoke-related particles descended to the streets, stinging the eyes of city spectators. Two weeks later, on 13 October, over the Jersey coastline, Young again flew an N2Y to the trapeze, completing twenty-eight hook-on tests.

This same month Harrigan took delivery of a Curtiss XF9C-1 that had finally been equipped with a skyhook. In the early hours of 23 October, *Los Angeles* unmoored from the mobile mast and steamed east. Shortly after landing stations were piped down, over Barnegat Bay, Harrigan and Young put the fighter through its paces on the trapeze. Just before 0900, Harrigan met the airship preparatory to making a series of practice hook-ons; at 0926 at fifty-nine knots, he hooked on—his first of five. At 1011, Young joined in. On his third hook-on, however, Young found that he could not get the plane *off* the trapeze; his hook refused to release. The XF9C was equipped with the same system as the N2Y trainers, but it was a heavier machine; its weight had stretched the cable between the pilot's lever and the releasing gear in the hook.

Dresel could not land with a plane hanging below the keel. With the airship at half-speed and following various courses in the vicinity of the station, Young struggled with the mechanism from the cockpit as much as he dared. Lieutenant Calnan decided that the only way to release the plane was to climb down and knock it off himself. So, with number four and number five engines killed and numbers one, two, and three slowed to cruising, the officer clambered down the trapeze girders to a position on the yoke. At 1122, after ten minutes of pounding with a wrench, the hook flipped open, and the plane fell free.

A new release wire was installed, and Harrigan took off to meet *ZR-3* over Jersey City's municipal airport, where the navy had scheduled a trapeze exhibition. At 1335, the pilot began hooking on. Following a ten-minute demonstration, the maneuver was discontinued; the rigid steamed south to Lakehurst, where it arrived at 1555. Landing stations sounded, the platform weighed off at a thousand feet. Dresel commenced his approach for mooring, valving for sixty seconds; his trail ropes and main cable were then dropped. The latter was connected at 1639, ship's cone locked in the mast cup eleven minutes thereafter. By 1735, *ZR-3* was through the west doors, unshipped from the mast, and moored to spring balance scales in the north berth.

Three days later, Harrigan was in the air once more, pushing the XF9C toward western Georgia to rendezvous with Dresel for a Navy Day exhibition over Atlanta. During the 1200–1600 watch on the 27th, over College Park, he made a pair of hook-ons at full speed. These concluded at 1225, when—as Lt. Ben May recorded in the log—course was set for Athens, Georgia. After thirty-four hours and ten minutes under way, *Los Angeles* settled to the mast and circle at Parris Island; the main wire was connected at 1828, the cone secured at 1840.

The three-hundredth flight had ended.

No one could know, but the final hook-on to *ZR-3* was now in the log. The just-completed hook-ons were to be the last ones anywhere for seven months. The trapeze was now cannibalized to provide parts for *ZRS-4*. *Akron* did not have its trapeze installed until January 1932, however, and she received no planes on board until May.

For *Los Angeles*, the autumn and winter months of 1931–32 were a turning point. On 23 September, *ZRS-4* lifted off for its first trial. Commanding officer: Lt. Cdr. Charles E. Rosendahl; executive officer, Lt. Cdr. Herbert V. Wiley. Passengers embarked (among them Secretary of the Navy Charles F. Adams and former secretary Ingalls, Admiral Moffett, Commander Fulton, Dr. Arnstein, C. P. Burgess, Commander Weyerbacher), the dirigible was weighed off. Upon Rosendahl's order "Up ship!" *Akron* was cast off from the mast. Pushed aloft by the ground crew, she rose statically to two hundred feet, whereupon her eight Maybachs were telegraphed to life. The great ship moved off on her own power.[77]

The first of ten performance trials was under way. Pressure-distribution measurements, deceleration tests, turning trials, and rates of ascent and descent (among other data) were compiled.[78] Concluding her trials series, the navy's "newest and largest recruit" steamed into Lakehurst at sunrise, 22 October. Before about a thousand spectators, *Akron* was shunted under roof using rail-mounted ground-handling gear. First, locked in the cup of the rail mast in a "mooring out circle," *ZRS-4* was shunted to the center of a similar layout, called a "hauling up circle." Here rode the stern-handling beam, or "Bolster beam"—named for its inventor. Resembling a long flatcar, the device was attached by bridle arrangement to the lower fin. It lay broadside to the ship, thus complementing the mast by providing mechanical control of the stern. Once the rigid was secured, a small locomotive towed it around parallel to the wind, after which the beam was transferred from the circle's tracks to those of the dock.[79] Gripped at both ends, she was hustled in—safely delivered alongside *Los Angeles*. This ground system, though cumbersome, eliminated the small army of ground personnel and provided security against severe side loads in gusts or high winds.[80]

Driven down from New York, Dr. Eckener and von Meister were honored guests for the occasion. The airman "expressed his admiration for the big airship and for the manner in which it was handled by . . . Rosendahl and his crew of 60 men."[81] One week later, *ZRS-4* was commissioned. In Ohio, construction had commenced on the sister ship.

To get a second chance, a first one must be taken. *ZR-1* and *ZR-3* had introduced the type: a record of performance had been shaped, a bit of history written. Now airship men had a platform designed expressly for fleet operations. Further, *Akron* was lofting U.S. hopes for a commercial service. The bitter pioneer years were astern, valuable experience logged, obstacles and hostility met and (seemingly) overcome.[82] The program was alive with excitement and portent.

These were the busiest years, and among the happiest—in retrospect, the national apogee for the ZR. The new ships represented a huge emotional and financial investment. At Lakehurst, training soared to a peacetime high, new construction was authorized, and civilian support

Commissioning-Day ceremonies, 27 October 1931, NAS Lakehurst. Capt. Harry H. Shoemaker (No. 3656) reads orders placing Akron *into commission as Rosendahl, Wiley and ship's crew (out of view) stand to attention. Rosendahl formally accepted command, after which Wiley was ordered to set the watch. A bosun's pipe shrilled—and her crew marched aboard. (U.S. Navy Photograph courtesy Navy Lakehurst Historical Society)*

Akron's first sortie as a commissioned naval vessel, 2 November 1931. Cast off the railroad mast at 0657, she is rising statically on four propellers tilted for vertical thrust. Following rendezvous with L.A. over the nation's capital, the pair cruised northward in company to New York City—a joint operation unique in U.S. aeronautics. (Acme photo)

was added. In terms of personnel, the air-station command climbed to about forty officers and nearly three hundred men. *Los Angeles*'s complement added a dozen officers more and seventy enlisteds, *Akron* another sixteen and seventy-five, respectively. Contingents of student officers and enlisted aviator trainees further swelled the on-board population.[83]

The fervor did not end at the perimeter fence. "Ashore," neighborly feelings were strong. Civilian employees, aviators, officers, sailors, and marines, and their dependents, had adopted surrounding communities. Also, the base meant jobs. To locals, the flight crews were "our boys," their machines a matter of intense personal pride—and concern. In sum, the air station was woven deep into the regional fabric.

On 2 November, *Akron* made its first flight as a commissioned vessel. With Moffett and a party of aviation journalists and NBC men along, a leisurely ground speed was logged coastwise, to the Washington area. In a radio address of greeting, Ernest Lee Jahncke, assistant navy secretary, declared that *ZRS-4* had been built not only for naval purposes but also for "a new and more rapid means of international communication and commerce." A fleet of merchant airships would follow *ZRS-5*, he predicted. Moffett and Rosendahl were congratulated, then "a word of credit" proffered to Goodyear-Zeppelin.[84] Having rendezvoused over the nation's capital, *Akron* and *Los Angeles* cruised in formation—the sole episode of its kind in U.S. skies. As the forenoon watch recorded,

Admiral William A. Moffett, Chief, BuAer, on the field following a ten-hour sortie, 25 March 1932. Here the admiral is flanked by two of his three Navy sons, Ensign William Jr. (right) and George, a lieutenant (j.g.). (Navy Lakehurst Historical Society)

U.S.S. AKRON in sight on starboard quarter. . . . 0930 Joined company with U.S.S. AKRON carrying flag of Rear Admiral W. A. Moffett, USN. 0940 Various speeds and courses following movements of flagship in maneuvers over Washington D.C. 1030 Following astern flagship en route Washington, D.C. to Baltimore, Md., northerly courses, various speeds. 1035 Circling over Baltimore, Md., astern of U.S.S. AKRON.[85]

Taking course for Philadelphia, the pair continued in company, arriving in the afternoon watch. There they circled. At 1250, after twenty-nine minutes, navigators set course for Trenton, where the pair again circled. The convoy of two then pounded northeast, for New York. Bathed in bright afternoon sunshine, it gained the harbor's Lower Bay at 1420. "1422 Various speeds following U.S.S. AKRON over New York City," Tyler recorded. With *L.A.* astern (and a blimp farther aft), *Akron* loitered, thrilling the metropolis: "Thousands of New Yorkers swarmed to the streets or took up vantage points on high buildings and in windows," the *Daily Mirror* declared, "as the country's mistress of the air soared above the Hudson River, across to the East River and then swept over the heart of the world's largest city before its bow was turned to its home hangar at Lakehurst, N.J."[86] After maneuvering with the larger ship for forty minutes, Dresel ordered a northeast course.

Overflying Connecticut, Rhode Island, and southeastern Massachusetts, *L.A.* was circling the Boston Navy Yard at 1909. Dresel then pressed south, toward Block Island, the ocean, and home. At 1109, 3 November, he regained Lakehurst but loitered in the general area. At 1321, over Barnegat Inlet, *Akron* was sighted bearing 180 degrees, distance fifteen miles. An hour later, inland, while cruising

Ground handling was the bete noir of the large airship. On 2 February 1932, while undocking at Lakehurst, Akron broke free of the "Bolster beam," a device used to secure her stern. Weathervaning about the nose-hold, Akron banged her lower fin until she lay parallel to the wind. Operating budgets—always problematic—were ravaged to effect repairs. (National Air and Space Museum, Smithsonian Institution (SI Neg. No. 77-8198))

Table 5-5 Comparative factors of strength and performance in ZR-1, ZR-3, and ZRS-4 (Akron). (Design Memorandum 102)

Name of Airship		Shenandoah	Los Angeles	Akron
Gross lift	lbs.	124,500	153,000	403,000
Useful lift	lbs.	44,100	62,600	170,000
Useful/gross		0.354	0.410	0.422
Fuel load	lbs.	26,000	40,000	127,000
Fuel per hour at 50 kts.	lbs.	580	510	705
Endurance at 50 kts.	hrs.	45	78.5	180
Propulsive coeff. at 50 kts.		40	44	60
Specific fuel consumption,	lbs./hp/hr.	0.55	0.47	0.47
Horse-power at 50 kts.		1,050	1,090	1,500
Air volume	ft.^{3}	2,300,000	2,800,000	7,250,000
Designed speed	kts.	50	68	72.8
" "	ft./sec.	85	115	123
Max. velocity of sudden gust,	ft./sec.	17.0	20.6	56.0
Bending moment due to gust,	ft./lbs.	650,000	1,300,000	9,750,000
Max. stress due to gust,	lbs./in.^{2}	3,800	5,000	10,800
Static and secondary stresses	" "	9,700	9,900	5,300
Max. total stress	" "	13,500	14,900	16,100
Ult. strength of longitudinals	" "	26,000	33,000	37,000
Factor of safety		1.9	2.2	2.3

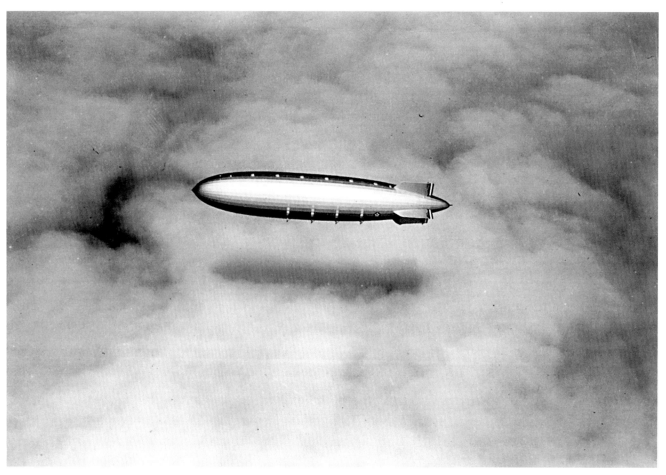

Akron skimming a cloud deck. Because of their clean hull lines, ZRS-4&5 were the most striking rigid airships ever designed. (San Diego Aerospace Museum)

to the northwestward of base, a sudden atmospheric change caused *ZR-3* to rise—on an even keel—eleven hundred feet. An increase in cruising speed from four engines was telegraphed to check the ascent. Maximum estimated rate of rise: twelve meters per second. The next watch had *L.A.* circling the field. Minutes shy of thirty-six hours under way, at 1727, the bow again locked into the cup of the mobile mooring mast, after which *ZR-3* was towed toward the hangar. Mechanical handling gear connected aft, the ship's stern was now hauled up. Towing recommenced at 1820; *L.A.* docked (1824), unmoored from the mast, and moored to the spring balance scales (1840). At 1840, Lt. Clinton S. Rounds closed out the watch.

"Up Ship!" was heard next on Armistice Day. Under power nearly a day and a half, Dresel penetrated to the Hampton Roads–Norfolk area, after which he loitered to seaward of Jersey and Delaware—familiar airspace. The five officer and three enlisted students accompanying chalked up a lengthy indoctrination. A half-day sortie was made on 23–24 November, during which Hoovenair Sound equipment and a General Electric sonic altimeter were tested. Student officers numbered four, passengers six. During December, five training and experimental flights more—forty-eight hours—entered the log.

The year's flying had concluded. As Dresel sat at his desk at year's end scheduling operations for the first months of 1932, *Los Angeles* and the officers and crew with orders to her had but six months flying left to them.

CHAPTER 6

Grounded

It is recommended that the airship be docked in the shed at Lakehurst in such a manner . . . that she can be recommissioned at some future date.

—Adm. William A. Moffett, USN (May 1932)

At the close of 1930, the number of U.S. jobless hovered near six million. Following faint signs of improvement, the decline renewed its toboggan run. Production ebbed, stock prices faded, failures multiplied. The Great Depression endured.

With *Akron* operational, further officers received orders to her tiny (six pilots) HTA Unit. The first of six Curtiss F9C-2s were delivered; the first "squadron" F9C was flown to *Akron*'s trapeze in June 1932. Hook-on flying embarked upon its operational phase, though experimentation continued. Bolster had redesigned the skyhook so as to give the pilot maximum vertical tolerance during the approach. "In developing the hook-on equipment for the airplane," he recalled,

the hook itself became a key problem. Our original idea of a good hook was shaped something like a shepherd's crook. Actually this hook required almost a perfect approach by the pilot in order for the trapeze bar to enter the hook opening and be latched in place. One day while worrying about this I tried to figure out how to give the pilot the greatest possible vertical tolerance. . . . This resulted in the design of a hook so shaped that a relatively large vertical opening lead the trapeze bar into a space under an inclined flat surface where gravity and forward momentum would force the bar into the latch. . . . Successful hook-ons now became pretty much a matter of routine.[1]

The Akron would never carry more than three planes; *Macon* could handle five, but the fifth plane was seldom carried. Among the pilots, it came to be the consensus that hook-on "landings" were easy compared to conventional runway approaches. Lt. Harold B. "Min" Miller, who reported aboard *ZRS-4* that July, became the senior aviator of *Macon*'s HTA Unit. "I might say that landing aboard [an airship] trapeze was a relatively simple thing after one got the hang of it."[2]

For *Los Angeles*, the year's operations opened during the eight-to-twelve watch of 7 January. Gassing commenced, after which water came aboard, then fuel, then food. Cast-off came at 1624 for the first of four sorties that month. The longest—twenty-five hours six minutes—had her roving New England as far as Boston and Cape Cod. At Nantucket Island, the Surfside radio-compass station was calibrated. On the 25th, *L.A.* lifted away on Dresel's final sortie as skipper. Having run facsimile tests with the base, he proceeded to Cape Henry Light and NAS Hampton Roads, returning via Richmond,

Ship's officers and crew pose beneath ZR-3, February 1932. (Elizabeth Tobin)

Washington, Dover. Back over south Jersey, facsimile tests resumed.

An accident interposed.

Inspections were not confined to the keel corridor; riggers walked the girders as well—pushing out and up as far as ship's equator—longitudinal number three. "The only time you got away from the keel proper was if you were making an inspection of a cell," Monty Rowe explained. "And then you had to walk on one of the transverse rings and get out behind a cell, and then you walked a longitudinal girder which would run parallel to the keel. Other than that, all the traffic was in the keel or to the engine cars." "Up in there," the former rigger continued, "the cell would fit right up tight against the ship, so you had to actually force your way in between the cell and the outer cover, or the Ramie Cord netting, which took the pressure of the cell onto the framework."

In the 12 to 16 watch of the 26th, while pushing through at longitudinal number two (port side) inspecting cell nine, Lieutenant Calnan, overcome by helium, fell into the cell. A. M. Oliver, Seaman First Class, at the keel, noticed its corridor section twisting about. Pulling slack fabric over, he felt a body inside. Help (always within hailing distance) was promptly on scene: Reginald H. "Ducky" Ward, CBM, and Coxswain Malvin M. Hill. While effecting a rescue, Oliver also was rendered unconscious. The extrication badly ripped number nine—about seventy-five feet horizontally, fifteen feet vertically. There were maintenance kits with cements, fabrics, needles, thread, scissors, brushes, dope, tapes, and safety pins, for quick mends (particularly bottom tears), after which the temporary patch normally was cemented over. "That time, they didn't even patch it," Rowe added. "All they did was just pin it up with safety pins"—and head for port, with engines telegraphed to "Cruising."³ Failure of a gas cell in flight was primarily a trimming problem, to attain as nearly as possible a horizontal trim. A ton of water was let go near the location (to compensate for loss of lift), speed was

Dolled up in gaudy chrome-yellow "visibility markings," L.A. rises statically for a training flight—among her last, spring 1932. (National Archives)

Table 6-1 Cost to the Navy of maintenance and operation of USS *Los Angeles*, 1 July 1931 to 30 June 1932. (Garland Fulton Collection, Smithsonian Institution)

1. Repairs:

Labor	$16,719.89
Material	$65,624.58
	$82,344.47

2. Maintenance of shop facilities, overhaul etc. directly connected with repairs, upkeep, and operation:

Maintenance of hangar and shop facilities	$13,220.31
Maintenance of mooring mast and stub mast	$9,450.97
Leave and holiday pay (estimated)	$2,527.22
Heat, light, and power (estimated)	$4,645.51
	$29,844.01

3. Operating expenses:

Total helium receipts	$77,547.18
Value of helium in ship 30 June 1932	$21,639.79
Total direct helium expenditures	$55,907.39
Chargeable losses in handling, purification etc.	$7,517.00
Total helium costs	$63,424.39

4. | | |
   |---|---|
   | Value of gasoline | $7,187.22 |
   | Value of oil | $277.48 |
   | Miscellaneous charges to operation | $12,529.78 |
   | Total operating expenses | $83,418.87 |
   | Grand Total maintenance and operation | $195,607.35 |
   | Cost of operation and maintenance per flight hour | $260.67 |
   | Cost of operation and maintenance per air mile | $5.15 |
   | Man miles flown | 2,106,190 |
   | Cost of operation and maintenance per man mile flown | $0.093 |

reduced, and weight was shifted away from the casualty—all to ease strain on ship's structure. (The most severe loading situation for the main frames was a deflated cell adjacent to a full one.)

Dresel was back on deck by 1441. Upon docking, preparations were made to remove the injured cell. Replacements for it and for numbers seven and eleven were installed.

That same month, the Committee on Naval Affairs instigated an investigation of reported defects in *Akron*. While rebutting these charges, Fulton read a letter from Admiral Moffett to Secretary Adams suggesting that the ZRS-5 design be increased in size to 7,500,000 cubic feet—larger than the yet-unnamed *Hindenburg*. "In view of the fact that treaties forbid us to use the *Los Angeles* as a warship, wouldn't it be advisable to sell her and use the money for enlarging the *ZRS-5*?" Fulton was asked. "That would be a very good trade," the commander replied. "We suggested it" to Goodyear-Zeppelin, Fulton added.[4] Representative Fred Britten (R-Illinois) remarked that he did not desire merely to "swap" her for the extra work. "I would suggest the company increase the size and also pay $200,000 for the *Los Angeles*." Dr. Hunsaker, vice president of the firm, then conferred with Britten. Goodyear-Zeppelin, he said later, was "seriously considering" taking ZR-3 in payment for enlarging *Akron*'s sister, and he disclosed that congressional leaders were launching a move to commercialize *L.A.*, perhaps as a sightseeing "bus" for the World's Fair. Offers to purchase *Los Angeles* now reached the papers. Howard Hughes, for one, let it be known he wanted her for his newest thriller, *Sky Devils*.[5]

Late in the forenoon watch on 1 February 1932, Lt. Cdr. Fred T. Berry read his orders to assume command. A Naval Academy graduate (Class of 1909) Berry had a distinguished record in destroyers. At Lakehurst, he had qualified with Class VI as Naval Aviator (Airship) No. 3664.[6] Upon designation, "Freddy" (or "F.T.") had received orders to ComRATES, for duty involving flying and for additional duty as the naval air station's executive officer. In August 1931, he reported to *Los Angeles*'s skipper as his relief. By the change of command, Berry had logged fifty ascents in *ZR-3*, nearly 1,030 hours.

Short in stature, Berry was "of a genial and friendly nature," George Watson later recalled. The appointment was "very popular with us. He was the kind of skipper that knew and called his crew by their first names. Berry was a competent ship handler but relied heavily on the veteran crew of the *L.A.*—they responded loyally, for all liked Berry."[7]

Berry and his crew were to fly together mere months; he would append 301 hours, seventeen minutes to ship's log during twenty-three sorties under his command. (In December 1932, as relief for ComRATES, Berry would also assume command of the naval air station. The following April, he would be lost with *Akron*.) Their longest excursion occurred that April—a flight deep across New England, into Maine. The earliest hours of 19 April found *Los Angeles* tracking across the nightglow of Biddleford, Portland, and Brunswick before pushing inland, up the Kennebec River, until Berry had Augusta below.[8] There course was altered. New Hampshire, Massachusetts, and Connecticut were overflown en route to the Jersey coast, off which *L.A.* roved through the afternoon watch. Time aloft: eleven minutes shy of one full day.

Meantime, within days of his assuming command, press reports had Berry and Company visiting the winter Olympic Games at Lake Placid. On 9 February, *L.A.* unmoored at 1700, heading not at first for the Hudson Valley, thence the games, but well to the south, to Cape May, then reversing course and following the beach. By 2245 the Hudson's silver path glimmered beneath. Albany gained the log at 0100—the wind rising. Berry pressed as far as Troy (and circled it) before turning southerly to regain the Hudson River. Wind had increased to forty knots. Near Newburgh water recovery was cut out on all engines and altitude twice increased (perhaps seeking warmer air)—the airship hung heavy due to snowfall. New York City was logged at 0620; the approach for mooring at Lakehurst begun at 0845. The cone was secured within thirty-one minutes, the west door cleared about 1010. Air time: sixteen hours, sixteen minutes.

Three missions more were flown that month, four in March—the last (on the 24th) with Moffett accompanying.

On 22 February, having just cleared the hangar, *Akron* broke free of her stern-beam restraints. Her tail soared upward before crunching down, after which it banged its way about the mast-hold until she lay parallel to a six-to-fourteen-knot blow. Repairs ravished the budget, obliging a new policy. Heretofore, all wire and metal work on *ZR-3* had been done by civilians. To reduce operating expenses and free civilian workers to nurse *Akron*, crewmen would replace their own broken lattices, netting, and brace wires as well as repair walkway girders.

This March, *Los Angeles* took on a distinct look. Complaints had been lodged that planes found it difficult to see the big airships in low visibility. On 3 February, the bureau proposed an experimental marking, to make *L.A.* more visible in daylight. It was "desirable in peace times," the CNO was advised, "to forestall possible criticism and accident by giving to large airships a more conspicuous marking than they now carry." A modification was therefore proposed; if it was found satisfactory in service, a similar arrangement for *Akron* and *ZRS-5* might be recommended:[9] chrome yellow applied to the main car, fins, and number-one car *(L.A.)*, a star-shaped figure on the bow, and star-in-circle insignia forward as well as topside.[10]

Meantime, determined to conserve the treasury, Congress—and, soon, the newly elected Franklin D. Roosevelt—were drawing the purse strings. When the Naval Subcommittee of the House met to consider fiscal year 1933 appropriations, the requested $270,000 for maintenance and operation of *ZR-3* represented a target.[11] *Akron* could carry on training and experimentation; accordingly, the committee unanimously recommended that *Los Angeles* not be operated after 1 July 1932—and the new fiscal year. By spring, with *Macon* nearing completion, there was consensus in the Appropriations Committee; in an April report to the House explaining the 1933 naval appropriations bill (HR 11452), it urged that *L.A.* be removed from the fiscal 1933 budget. During hearings on the bill, Moffett expressed the intention of laying up the ship as soon as *ZRS-5* entered service. One representative, for his part, suggested that unless the airship was to be sold (as advised repeatedly by Congress), *Los Angeles* should be salvaged.[12]

Correspondence ensued. On 22 April, less than three months into Berry's tenure as skipper, Secretary Adams dispatched a letter to the chairman of the House Appropriations Committee, James W. Byrnes, urging reconsideration. It was "extremely important" to keep *L.A.* in service, Adams argued (echoing Moffett), into the next fiscal year, until *Macon* became operational; an abrupt grounding would interrupt development projects. Further, an uneconomical utilization of material and personnel would inevitably follow. Perhaps, the secretary speculated, the funding that had been allowed would be sufficient to meet the expense. Adams and Moffett had misjudged the mood; Byrnes was having none of this. In a strong reply, he not only agreed with the Naval Subcommittee on withholding the $270,000 but threatened to impound the funds the navy proposed to divert from other purposes to support *Los Angeles:*

> The bill, of course, does not specifically provide that the *Los Angeles* shall not be operated, but if there is a real desire on the part of the Navy Department and the Administration to avoid other than absolutely necessary expenditures, I am sure this possible saving of at least a quarter of a million dollars will be made.[13]

It was decided to operate only two ZRs; on 9 May, Moffett advised that *Los Angeles* be placed on an inactive status on 30 June. "It is recommended," the admiral continued, "the airship be docked in the shed at Lakehurst in such a manner and preservation of material be arranged to the end, that she can be recommissioned at some future date."

Maneuvers for recommissioning would continue. "The recommendation as to possible recommissioning at a late date," Moffett elaborated, "is made because: (1) the material condition of the LOS ANGELES is generally good and scrapping her is not warranted at this time; (2) prior to scrapping her it is desired to carry out certain destruction tests that will yield valuable scientific data, and there is not sufficient time to carry out those tests prior to 30 June; and (3) there have been conversations regarding a possible sale of the LOS ANGELES, in 1933, which may culminate in a definite offer to purchase."

Inactive status would have unfortunate effects, the admiral warned—not least for the training of officers and men. Further, "serious interference" to a number of development projects would attend a decommissioning, the results of which were expected to have "important bearing on the future of airships." One possible casualty would be service testing of improved cells and cover materials; also, the trial installation of an American-built airship engine would have to be postponed or abandoned, likewise new conceptions of plane landing and stowage arrangements. "By careful planning," Moffett concluded, "and transfer of projects wherever practicable, the Bureau of Aeronautics will endeavor to salvage a maximum of information and thereby escape complete vitiation of certain projects."[14]

Within forty-eight hours, Berry was warned of the prospective action.

The decision was grounded on certain realities. Funds were short. Due to attrition, moreover, the number of trained officers and men available was barely enough for two crews. And *L.A. was* getting on. There was no way of knowing, with assurance, how much her hull structure had been strained by years of service. *Akron*—with its vaulting performance (and expectations)—was operational. So, rather than resisting the retirement, most airmen probably were philosophical about it.

Lighter-than-air advocates had long been "selling" the project. "This was part of the mission of practically everyone in airships," Reichelderfer later affirmed, "because it was a small service. It was a service that people had many questions and doubts about, so that you were automatically—if you believed in airships—cast in the role of being a salesman whenever you talked with people who were interested in knowing more about them."[15] *Los Angeles,* in sum, had lived a good, long life—for the type. Now, with *ZRS-4* and *ZRS-5,* airship men meant to prove the worth of their vision—militarily as well as commercially.

> For trans-Atlantic air transport service to Europe, greater cruising range and carrying capacity are required than can be efficiently provided by heavier-than-air craft at the present stage of aeronautical development. Rigid airships at this time offer a prospect for air passenger service to Europe. They are already being used by a European nation in providing regular air passenger service across the South Atlantic. The Congress has done much to establish a rigid-airship industry in this country by appropriations for the construction of the naval airships *Akron* and *Macon*. The navy's experience with lighter-than-air craft has provided valuable information and data for commercial airship development and operation. The committee believes that the United States should continue to encourage the development and use of rigid airships as a means of ocean transportation.[16]

Berry (on his own initiative) submitted an airworthiness report on his command. Structural condition, he believed, was good to excellent; with careful maintenance, *ZR-3* would remain sound for an indefinite period. The 1930 survey, the lieutenant commander noted, had estimated the airship's useful life at from two to four years; the lower estimate had elapsed, but the structure, due to extensive overhaul over the top of the hull, was now actually in better condition than when the original projection had been made. When the exhaustive quarterly report ending in March was forwarded to the Board of Inspection and Survey, Moffett included Berry's assessment and his own appraisal:

> From a study [of the reports], it is the opinion of the Bureau of Aeronautics that the material condition of the U.S.S. LOS ANGELES is generally good. Nothing has been disclosed from a material standpoint that would warrant discontinuing operations, and in so far as can be judged her material condition can be considered satisfactory for several years longer.[17]

When these latest reports were routed up the line, the Board of Inspection and Survey concurred with BuAer and summarized to the CNO:

> After almost two years of operating the U.S.S. LOS ANGELES under the restrictions imposed in the Board of Inspection and Survey Material Inspection Report of 3 September 1930, combined with a very careful and thorough system of maintenance and overhaul, the airship appears to be in a highly satisfactory material condition. It now appears that the factors governing the limit of apparent usefulness of the U.S.S. LOS ANGELES are the maintenance and overhaul costs and the increasing obsolescence of the airship rather than deterioration of essential parts of the hull structure.[18]

Within twelve months, the board would adopt a distinctly different point of view. However, at the close of May 1932, it appears, all concerned with airworthiness were in accord: *L.A.*'s structural condition was good and, with careful maintenance, could remain so for years, if not indefinitely.

On 13 May 1932, the CNO ordered Lakehurst to decommission *ZR-3* no later than 30 June; within hours, the impending action reached the press. This was the first instance of a U.S. rigid's being placed out of commission for an extended period, so the bureau emphasized a careful program of preservation. Berry was advised to assume an indefinite grounding, certainly not less than a year. Recommissioning was to be made practicable on thirty days' notice, however, and with a minimum of funds and effort. On the 27th, a detailed plan was submitted for bureau approval (granted two weeks later). For inspection and upkeep, a nucleus crew of fifteen experienced men—boatswains, coxswains, seamen, aviation machinist's mates from the ship's crew and from station personnel—was authorized.

That June, Shoemaker (then ComRATES) recommended that the enlisted force of *ZR-3* (less the maintenance gang) be transferred to the air-station command for reassignment.

> A crew for the U.S.S. MACON, now under construction at Akron, Ohio, will be required on or about 1 November, 1932. It is proposed to form a crew for the MACON from the present crew of the LOS ANGELES, in order that this crew may be trained as a unit. It is desired to organize this crew as soon as possible after the decommissioning of the LOS ANGELES. In order that this crew be thoroughly familiar with the MACON it is proposed to fly each section of the MACON crew on the AKRON with the AKRON crew. The two ships are similar in almost every respect and this plan will enable the crew of the MACON to be entirely familiar with the ship prior to any trial flight.

Accordingly, fifty sets of flight orders were prepared for the enlisted men of *ZRS-5*. (At that moment, the base was allotted forty-three men, with twenty deemed minimum for operating requirements.)[19]

Los Angeles's operations for May–June featured practice with the mechanical handling gear as well as student indoctrination. In all, eleven sorties entered her log—hours flavored with regret, affection, nostalgia. June saw a half-dozen flying moors to the mobile mast and seventy-seven hours, forty-one minutes flying altogether—nearly thirty-eight hundred air miles.

On 23 June, Berry and Company were undocked; within a half-hour, at 0618, he unmoored with fifty-two persons on board. Crossing the surf at Asbury Park, the rigid began a training flight following the shore line on various northerly and southerly courses. Off Seaside Heights at 1950, having cruised through three watches, Berry changed course for the naval air station. The approach began at 2100, the cone again locked in the cup of the mobile mast at 2205. Kept on the field at the circle, warming all engines, *ZR-3* was reballasted, fueled, and watered.

Next day, the 24th, Freddy Berry unmoored with forty-nine crew and passengers. Time: 0750. Rising to fifteen hundred feet, *L.A.* pushed for the coast with all engines at 1,175 rpm as Lt. (j.g.) Ben May closed out the dawn watch. A quarter-hour into the eight-to-twelve watch, the beach was crossed at Sea Girt. Having steered various northerly courses, the ZR swung round over Spring Lake and steered various southerly courses to follow the beach along Jersey, Delaware, and Maryland. Over Assateague Light, course was ordered reversed. Beneath a clear sky in excellent visibility, *Los Angeles* flew toward New York. The 1600-to-1800 watch found little to note, at least in the log: "Cruising as before. 1700 Over Scotland Lightship. 1750 Vicinity of New York Harbor."

At 1850, the coast again was crossed, this time at Seaside. Having hung a half-day aloft, Berry had the airfield below him at 1935. Landing stations were piped, the approach commenced on course 260. The trail ropes and main wire were dropped; at 2037, the cone stood locked in the cup of the mobile mast. The stay proved brief. After fueling then weighing off, the skipper unmoored for a night training flight. At 2135, the final flight was under way.

The aircraft rose statically to three hundred feet, all engines at half-speed. A pair of youngsters were on board: Berry's son and Tyler's boy. "Once aloft," Owen Tyler recalls, "Fred and I were permitted to go into the control car and Dad turned over the rudder wheel to me for a short time—what an unforgettable thrill for a 9-year-old!" At 2330, the elder Berry had Philadelphia 2,500 feet below the keel. The dirigible circled on various courses three miles off the New York–Camden Airway. *Los Angeles* pressed northeast in ideal conditions to cross New Jersey's narrow waist, passing east of Princeton. Destination: New York City. Coastal Keyport sparkled beneath at 0042, New York at 0115. In no haste to return, Berry ran various courses following the coastline south. At Barnegat Inlet, he reversed his heading before, finally, telling the rudderman to take her inland. "At daylight," Tyler continues, "prior to landing, I grabbed at the chance to trail behind Dad along the catwalk as he made a routine inspection of the ship fore and aft, and understandably, didn't have to be reminded to keep a very tight grip on the safety cable that ran along the length of the catwalk!"[20]

Landing stations were piped at 0430; five minutes thereafter, Berry commenced his approach. The maneuvering valves were opened for ninety seconds, water recovery cut out on all engines. At 0448, the trail ropes and main cable dropped to the deck. Waiting hands connected the latter; at 0511, ship's cone was locked in and the water line connected. Docking was immediate. At 0523, ground personnel commenced towing *ZR-3* to the docking rails; when in position, the handling gear aft hauled up the stern. Towing commenced after twenty minutes. The sill was crossed within three, and at 0552 *ZR-3* was disconnected from the mast and moored to the spring balance scales.

Los Angeles never flew again.

The crew was mustered at quarters at 0855 (no absentees), after which it engaged in postflight inspection, cleaning, and upkeep. Five days later, the midwatch found the aircraft to be 3,500 pounds heavy in berth, with 33,680 pounds of disposable weights aboard. Average gas-cell fullness: 90 percent. The morning watch of 30 June recorded ship's last operational entry: "0700 Weighed off ship in equilibrium with disposable weights as before except water ballast 26000 lbs, sand and iron weights 6770 lbs."

During the next watch, the ship's company duly assembled in dress whites, and USS *Los Angeles* was formally decommissioned. The concluding entry of the airship's log records this sad duty, unique in naval aeronautics:

> 0815 Mustered crew at quarters—no absentees for decommissioning ceremony under Commander F. T. Berry, USN, read his orders BuNav let. 7105-140 of 15 June, 1932, detaching him as Commanding Officer, U.S.S. LOS ANGELES to duty as Executive Officer of the Naval Air Station, Lakehurst, N.J. Capt. H. E. Shoemaker, USN, read his orders BuNav let. 5132-176 of 25 June, 1932 detaching him from the LOS ANGELES to the U.S.S. AKRON for duty involving flying as Commander, Rigid Airship Training and Experimental Squadron. The Commanding Officer, Naval Air Station, Lakehurst, N.J., read Chief of Naval Operations order ZR3/46-1 (320513) of 13 May, 1932 directing that the U.S.S. LOS ANGELES be decommissioned not later than 30 June, 1932 and docked in the shed at Lakehurst in the custody of the Commanding Officer, Naval Air Station, Lakehurst, N.J. The ship was declared decommissioned by Captain H. E. Shoemaker, USN, and the colors and commission pennant lowered.[21]

Berry (George Watson later recounted) had "turned to the officer of the watch (me) and said, 'Haul down the Commission Pennant.' There was not a dry eye in the house."[22]

Emotions rippled far away. "My reaction, of course, when I heard about the *Los Angeles* was one of deep loss," Reichelderfer (who had been in Norway at the time) later observed. "I loved the ship and had many interesting flights on her. And I had the same attachment that most sailormen have with respect to the ship they have served on."[23] As for her complement, most were seasoned campaigners; their attachments were strong. Among the longtime hands, Martin O. "Mo" Miller, ACMM, and Paul A. Jandick, AMM1c had served aboard since commissioning. Officers, for their part, had shifted from billet to billet—from sea back to shore (often Lakehurst) and back to sea. "We were accustomed to moving; change didn't seem to bother us very much," Mike Bradley recalled. Tyler, for one, had had several tours—in all, five years of flying orders to *L.A.* Certain shipmates—Buckley, Thurman, Settle, Wiley, McCord, Rodgers, Peck, Dennett, and Watson among them—had long and ably attended her operational career.

Now came new orders. Ten officers and fifty-five men were reassigned to *Akron* or *Macon* billets, to the naval air station, to sea. Frank McCord, who in 1931 had replaced Peck as exec, received orders to *ZRS-4*, "to duty involving flying under instruction as prospective Commanding Officer." Calnan went to the air station. Watson got orders to sea duty, aboard the battleship *Mississippi.* For enlisted personnel, the immediate orders were to the ComRATES organization.

June brought happier news from Washington; the House passed a bill authorizing the transport of mail overseas by airship. Goodyear-Zeppelin and others, it seemed, could look forward to the construction of airships for commerce. In Akron, the Daniel Guggenheim Airship Institute, with a unique vertical wind tunnel, was dedicated. Its director of research: the esteemed Dr. Theodore von Karman of Cal Tech. Facilities now were to hand with which to advance the art, to conduct special research in the fundamental sciences pertaining to the construction and operation of lighter-than-air craft. "It is hoped," Arnstein remarked, "that the new Guggenheim Airship Institute will combine its efforts with the N.A.C.A., the U.S. Army and Navy, the Bureau of Standards, and other governmental agencies which have been of such valuable assistance in the past."[24]

In New Jersey, the new ship would weather fiscal storms of its own. Formally detached from fleet duty on 1 July, *Akron* was assigned to ComRATES, thereby replacing *ZR-3* in training the precommissioning detail of her prospective sister ship.[25] As for development and experimental work, *ZRS-4* logged extended flights to sea. Her lookouts were drilled, hooking-on was intensively practiced, a new method of navigating her planes beyond the horizon tested. Also, the first steps were taken toward improving radio communications and homing gear for the ZR "carrier" and its search planes.

On 3 January 1933, McCord took command of *ZRS-4*; Dresel had orders to Ohio, as prospective CO of *ZRS-5*. Certain officers and men of the *Macon* detail also stood detached at this time. Within hours of the change of command, *Akron* was on her way to the expeditionary mast at the Naval Aviation Reserve Base at Opa-Locka, near Miami—which was to be a winter airship base. From there, she jumped off for Cuba.[26] McCord appreciated the added flying time that Florida and the Caribbean afforded, well away from the miserable northeastern winter.

Inside, *L.A.* was suspended high, docked on shores and cradles near one of the doors, close to the north wall. This arrangement would permit men to work beneath it, and, at one end, allow small ships to be housed. *Los Angeles*'s cells were partially deflated (later they were removed). In the power cars, each Maybach was unbolted, lifted off, and wheeled into the shops, where they received storage-related attention. In the berthing space, tarpaulins closed off each car, and the ship's instruments, equipment, storage batteries, generators, papers, and other items were sent to storage ashore. Along the keel corridor, ballast bags and fuel tanks were unshipped, cleaned, then reinstalled. The fuel and water lines were drained. Throughout the airship's hull, essentials not removed and stored were cleaned, lubricated, greased, lashed down, and sealed as required.[27]

As the decommissioning orders had stated, and pending recommissioning, *Los Angeles* was now the responsibility of Lakehurst's commanding officer. Though funds for maintenance and upkeep proved lean, the station would dote upon her. *L.A.* would lie silent and snug in berth for the next eighteen months.

Admiral Moffett completed his third tour in the Bureau of Aeronautics in March of 1933. His final weeks saw the christening of *Macon*—and the admiral's last public speech on lighter-than-air matters. It included this paragraph:

> The Navy's constant object has been to build the AKRON and MACON not only for strictly Naval purposes but to show the way for the commercial uses of airships. We have done the experimental work, developed new types of mooring masts, trained personnel, and shown the way for commercial use. We feel we have done our part, and that it is up to Congress to pass legislation that will make it fundamentally practicable to build and operate commercial airships.

Table 6-2 Comparison of fleet-type airship scout *(ZRS-5)* with the new 10,000-ton-treaty surface cruiser, about 1933. Restricted in the thirties as to number and size, cruisers were the conventional scouts for the fleet. (Garland Fulton Collection, Smithsonian Institution)

	Macon	Surface Ship
Cost of construction	$2.45 million	$18 million
Cost Maintenance/Operation	$400,000/year	$1 million/year
Pay of Crew	$225,000	$650,000
Top Speed	72 knots	32.5 knots
Cruising Speed	50–60 knots	18–22 knots
Complement	65 men/15 officers	550 men/55 officers
Cruising Range at Cruising Speed	9,000 nu. miles	10,000 nu. miles
12-hour Daylight Scouting Area	26,400 sq. mi. (20-mile visibility)	4,800 sq. mi. (10-mile visibility)
Estimated Life	10–15 years	20–25 years
Planes/Armament	4 airplanes/50-cal machine guns (defensive only)	4 airplanes/8-in and smaller

In interview, he implored Congress to fund two ZRs more, the construction of which would be "a powerful and direct means of unemployment relief." (The Navy Department actually had no interest in such matters.) "We need more [rigid] dirigibles," Moffett opined. "I hope the President will include at least two more of them in his public works construction program. But the deep-sea admirals in the navy don't believe in them. They're suspicious. They have an idea they're not safe."[28]

In the failing light of 3 April, *Akron* cast off on a routine flight. Rising statically, her silver bulk dissolved into a three-hundred-foot ceiling plus fog. A hook-on drill, McCord radioed back, was canceled.[29] *Akron* proceeded to sea with no planes on board. Neither the airship, her four passenger-guests, nor her complement (save three) were ever seen again.[30]

Offshore, conditions underwent a radical change. Unknown to McCord and Aerology, *Akron* was flying ahead of an exceedingly violent front. She was soon enveloped. Shortly after midnight, the air became extremely turbulent; severe downdrafts swept her dangerously low. On a second descent, in a nose-up attitude, the lower fin of the 785-foot-long hull struck the sea. Laboring to free the ship, the eight engines could only drag her through the water and pull her nose up grotesquely—until *Akron* stalled and surrendered.[31]

Of those who escaped the wreck, most succumbed to the frigid water. Seventy-three officers and men were lost, among them Admiral Moffett, Cdr. Fred Berry, and an official from IZT—the International Zeppelin Transport

4 April 1933: Lakehurst and Los Angeles *maintain a futile vigil for the ship and aircrew that never came home.* Akron's *foundering was the beginning of the end for the rigid airship in the United States. (Associated Press photo courtesy Thomas Raub.)*

Seventy-three officers and men were lost with Akron, nearly all of whom had had duty aboard ZR-3. Navy Secretary Claude A. Swanson termed the crash "one of the greatest peace-time blows the Navy has ever experienced." (Lynn Berry)

Columbus (OH) Dispatch, 5 April 1933

ZRS-5 under construction inside the Goodyear Airdock at Akron, Ohio. (Wide World Photo courtesy Kevin Pace)

Company, chartered to study airship travel over the Atlantic. Three lived, including Wiley—the sole officer.[32] Tragedy was compounded when on 4 April *J-3* went into the water while assisting in the search for *Akron* survivors. Two men more were gone. Once again, the nation's headlines shrilled. Franklin Roosevelt issued a statement calling the loss "a national disaster;" the Secretary of the Navy, Claude A. Swanson, termed the crash "one of the greatest peace-time blows the Navy has ever experienced."

In Ohio, the finishing touches were being applied to *ZRS-5*. The ship's first lieutenant, Cal Bolster, recalling the event decades after, remarked, "It is difficult to describe the impact of this sad news on those of us in Akron. The overall personnel serving in rigid airships at that time were a relatively small, closely knit group. The sudden word that all but three of the eighty-one officers and men on board had been lost, many of whom were close friends, had an effect that is difficult to describe."[33] At Lakehurst, grief-stricken wives, mothers, sisters, and sweethearts waited for word. Consoled by the chaplain and aided by the station's physician, one group huddled anxiously in the Welfare Building. "A radio was tuned in on the first floor. . . . Each time some news of the disaster was broadcast there would be a rush to the top of the stairs. White-faced, grim-lipped women would wait breathlessly for news and then return to their anguished vigil."[34] The lost fliers' automobiles stood lined up in *Akron*'s berth, to be claimed by relatives. Children at the village school, many now fatherless, were photographed saluting a half-masted flag.

Offshore, some bodies were recovered, plus wreckage. Hope faded for the missing. "The happy, normal social community . . . at Lakehurst . . . was no more. Their bridge clubs, their golf foursomes, their dances, and all their healthy diversions have been wiped out. . . . Empty houses, empty cars and empty hearts are here in their stead."[35]

On board the battleship *West Virginia*, Rosendahl defended his former command. "Through my personal experience I know the *Akron* previously survived the severest tests of any airship in history. The *Akron*'s loss baffles me," he opined.[36] From Berlin, Eckener offered his "deep regret," expressing hope that the majority of the crew might yet be saved. "I am unable to comment on possible causes of the crash until I learn more details," he said.[37]

The world at large had become none too friendly toward airships. Within weeks of the accident, Rear Adm. Ernest J. King, Moffett's successor as chief of the bureau (*and* in the NACA), tried to obtain a replacement as well as a "new training airship," using funds provided under the Industrial Recovery Act.[38] King's mind was open. Despite the sorry picture, there *was* congressional sentiment for this. Outrage prevailed, however; the collective reaction was nearly hysterical. Clamoring for an investigation, lawmakers demanded that no more dirigibles be constructed and that current building plans be curtailed—"for the present at least." While Congress wrangled, the press had a field day. Abetting the lurid headlines and acres of newsprint, editorial writers scolded, wringing their hands over lost the taxpayers' dollars. "Whether the dirigible is doomed henceforth for use in peace or war is a question for the experts, after due consideration, to decide," the *New York Evening Journal* opined. ". . . Certainly the loss of Akron, of the *J-3*, of the *Shenandoah* and of large dirigibles abroad places the burden of proof on the advocates of these types of craft. . . . Informed opinion differs widely as to their usefulness. . . . All that the layman can conclude is that they have not yet reached a stage of development which fulfills the expectations of their sponsors." This was a balanced judgment. Among opinion makers and personnel topside, however, headwinds ran strong; the ZR's obituary was already being written. "You can take this from me," Representative Carl Vinson, chairman of the House Committee on Naval Affairs, predicted. "There won't be any more big airships built. We have built three—and lost two."[39]

With *Macon* not yet in service, the navy had little immediate need for more helium. In the interests of econ-

USS Macon *berthed next to* Los Angeles, *24 June 1933. Following official trials,* ZRS-5 *took leave for her new West Coast homeport. The naval air station never again handled a commissioned ZR; an era of naval aeronautics had ended at Lakehurst.* (National Archives)

omy, Amarillo's production and its operating staff were slashed by half.⁴⁰

On Capitol Hill, a joint investigation by a special committee convened to probe not only the loss of *ZRS-4* but *all* dirigible disasters. In mid-June 1933, the postmortem—944 pages—was presented. It was a clean bill of health for the airship. The immediate question—should the Navy Department abandon its investment?—was rejoined in the negative: the ZR retained a naval mission. Noting that the airplane-carrying airship lessened the scouting obligations of surface ships, it summed up the matter tersely: "It supplants nothing. It supplements all." But if the lessons of the past were to be turned to profit, continuity of experience, training, and knowledge was vital. Accordingly, construction to replace *Akron* and procure a ZR "training airship" were recommended. Further, pending completion of the latter, *Los Angeles* should be recommissioned. But the joint committee was not Congress. The future was to provide little in the way of funds; indeed, neither a fleet-type replacement to fulfill the requirements of Moffett's 1926 Five-Year Aircraft Program nor the trainer was ever built.

The crash wrought incalculable damage, in public opinion as well as Congress. "I can find no justification," Secretary Swanson advised the president, "for recommending the construction of additional lighter-than-air ships so long as the MACON is available for carrying on the experimental work, or until such operations of this ship clearly indicate that further construction is desirable."⁴¹ To her credit, *Macon* generated a doctrine of operation useful to the fleet. Swanson, though, despised airships; as far as he was concerned, building more ZRs was never to be "desirable."

For the rigid type in the United States, it was the beginning of the end.

On 15 June, the bureau recommended that the Board of Inspection and Survey evaluate the material condition of *ZR-3*. So it was that a BIS met at Lakehurst that July "to determine the material condition of the U.S.S. LOS ANGELES and to report on her readiness to resume operations for training service." President of the board: Rear Adm. George C. Day. ⁴² Among its assessments and inspections was a test on the bulkheads of number-ten cell, to determine whether any serious deterioration had taken place in the head wiring. Within weeks (the survey yet incomplete), BuAer urged that *L.A.* be placed in service, as recommended by the Joint Committee. Swanson, though, needed no report to condemn the idea. "Some scientists of the National Research Council wanted us to take the *Los Angeles* to Akron, Ohio, for experimentation," the secretary announced. "After the inspection of the airship showed such deterioration from old age as to render it unsafe to make such a flight we told the scientists there was no way of getting the *Los Angeles* to Ohio intact."⁴³

The board's findings were submitted that September. Though the inspection failed to detect any specific evidence of structural deterioration and loss of strength, the report was generally unfavorable to the resumption of operations. For about fifteen thousand dollars, *ZR-3* could be made reasonably safe for restricted flying service. After citing the limited returns that might be expected from her use merely as trainer, the board concluded that further expenditures were not justified. Disposal by sale⁴⁴ or by scrapping was therefore recommended.⁴⁵

Predictably, to officers familiar with *Los Angeles,* these judgments seemed unduly harsh—and wholly unsubstantiated. BuAer geared up for damage control. In a memorandum for Rear Admiral King, Fulton took exception to the general tenor of the report:

> The Board seems to have lost sight of the fact that the LOS ANGELES was operating successfully up until June 30, 1932, when she was laid up carefully with a view to her later recommissioning. There is not one iota of evidence to show that her condition today is any worse than when she was laid up. Such proof tests as were made recently at the Board's direction, and such sampling of lattices, girders, and wires as was made recently did not disclose evidence of advancing deterioration.⁴⁶

The commander went on to cite the March 1932 quarterly report, Berry's independent evaluation of hull condition a month prior to layup, and the board's own evaluation of condition based on that information. During August–September, Fulton's views were echoed by King to Secretary Swanson. The "outstanding lesson" of *Akron*'s loss (King wrote) was "the necessity for better trained and more experienced personnel to handle the Navy's airships." King urged that Lakehurst be kept open and that officer and enlisted men classes be ordered for instruction. So important did he regard the matter that on his own authority "funds would be diverted from other programs in order to carry out" his recommendations. Within weeks, BuAer reiterated its desire to recommission: "The Board's adverse findings appear to be based on a fear of what might occur rather than upon any actual evidence of a seriously deteriorated condition." The board to the contrary, he argued, "The training of personnel for airship work is especially important at this time." As for the risk or danger to personnel, the ship's structure (King opined) would give ample warning, via an inordinate amount of required upkeep and maintenance. To date, *Los Angeles* had given no such indication. The admiral concluded: "The Bureau considers that the LOS ANGELES can be placed in readiness for limited service similar to that prevailing in January–July, 1932, for a sum approximating $7,500, and does not expect that maintenance costs thereafter will prove to be abnormal."[47]

The dustup played out. With benefit of hindsight, it is possible to see the board's report as an instrument to help sideline naval airships. BuAer waged its campaign, taking issue on nearly every point and insisting that *ZR-3* resume flight service. The Navy Department, indifferent when not actively hostile, continued to beg off, citing the 1933 findings regarding her material condition and usefulness. A memorandum endorsement for the assistant CNO evinces departmental sentiments on the matter. Dated 23 September, it reads in part: "If the placing of this airship in service depends solely on her material condition, efforts should be made to reconcile the opposing opinions of the Bureau of Aeronautics and the Board of Inspection and Survey as to her material fitness. If, as a matter of policy, it is not desired to put the vessel in commission, such procedure will not be necessary."[48]

Time would show that it was not.

On 3 October, Adm. William H. Standley, Chief of Naval Operations (and an airship skeptic), endorsed the board's negative report. Owing to the restricted appropriations for aeronautical equipment, the availability of *Macon* to meet reasonably well the immediate demands of experimentation and training, "and the doubtful material condition of the LOS ANGELES, the CNO considers that placing the airship in active service is not warranted, and, consequently, has no plans to that end." A week later, Swanson approved the CNO's position—and so advised King: Inasmuch as further expenditure on *ZR-3* was "not justified," as stated by the board, "this airship is hereby placed on the list of naval vessels to be disposed of by sale or by scrapping, as recommended by the Board of Inspection and Survey, after the removal of such machinery and equipment as may be directed by the Bureaus concerned."[49]

Time, it seemed, had run out. But *Los Angeles* possessed remarkable powers of survival; even in dotage, *ZR-3* survived destruction for six years more.

In its significant particulars—length, volume, power plant—*Macon* was a copy of *ZRS-4*; the external differences were very few. Alterations had been worked in, however, to enhance efficiency and save weight. Indeed, *ZRS-5* was approximately eight thousand pounds lighter and, thanks to cleaner hull lines and metal propellers, three knots faster. The ship's trapeze was an advance over that on *Akron*, and, unlike her sister, *Macon* was complete when she left the Airdock.

In the predawn of 21 April, *ZRS-5* was unhoused into a chill breeze. At 0602, as reporters, photographers, news broadcasters, Goodyear publicists, and well-wishers looked on, Dresel ordered, "Up Ship!" "Then," a reporter recounted, "came the roar of four motors and triumphantly, gloriously, U.S.S. *Macon* rose, soaring away from the earth in a movement so swift and sure that it brought exultation to every heart. As it turned and headed east, the sun came from behind the clouds and shone brilliantly, making the beautiful ship resplendent silver."[50] "The ship handled well underway," Dresel told newsmen. "The maiden flight," Arnstein beamed, "was a complete success. Not only were we able to conduct a great number of tests [e.g., speed, deceleration], but we also found that the ship performed better than we hoped."[51]

For more than decade as chief of the Bureau of Aeronautics, Admiral Moffett had defended airships as naval instruments. King, new to the office, knew that the rest of the navy was watching him and his attitude toward airships. He therefore had made it a point to be on board during *Macon*'s delivery. Neither an apologist nor a debunker, he could not dismiss the ZR's potential as the only available platform for very-long-range reconnaissance.

At 0320, 24 April, *Macon* arrived over Lakehurst. Landing stations sounded over the howlers, and the approach begun. Once down, she was gripped by the same handling gear used by *Akron* then towed astern into the south berth, alongside the mothballed *ZR-3*. As crew disembarked a water line was snaked to her, the dock's electrical mains connected. Four thousand pounds of liquid ballast were pumped aboard, and the ship's generators were secured. At 0747, *ZRS-5* was unmoored from the rail mast and secured to spring balances fore and aft.

The career of *ZRS-5* proved happier than that of her sister, and more fruitful. Precisely because *Akron* had preceded her, *Macon* was more ready to meet the challenges of 1930s naval warfare. However, *Akron* had participated in but two fleet exercises, neither including aircraft carriers. As one consequence, certain hard lessons had yet to be confronted, let alone digested and mastered. In short, the military test of the lighter-than-air "carrier" had been left to *Macon*.

While the Navy Department pondered *Los Angeles*'s final disposition, BuAer (through King) continued to press for renewed operations. A handwritten note to Admiral Standley, dated 2 February 1934, is emblematic. "Strongly renew my former recommendation [he wrote] that LOS ANGELES, if not to fly, be made available for experimental development, as at high mast at Lakehurst, *not* including flight, etc." That same month, SecNav Swanson and the CNO revised their position—somewhat. "After further considering this question authority is hereby granted to use the LOS ANGELES for experimental development to as great an extent as is practicable in a nonflight status."[52]

Although this revision was hardly all he wanted, Admiral King, wasting no time, proceeded to contact selected authorities, requesting their suggestions as to experimental uses. A heap of recommendations and suggestions resulted.[53] BuAer could not hope to execute them all; the process, therefore, was one of elimination. On 20 June, the responses were directed to the Lakehurst station's new commanding officer: Lt. Cdr. Charles E. Rosendahl. The enclosures, King advised, were to be studied with the view of recommending "a program that is at once practical, economical, can be accomplished within a reasonable time, and one that will be a real value to airship development."[54]

Rosie immersed himself, studying and correlating the responses. In a bulging July memo, he outlined a program. A tireless campaigner, this officer used the opportunity to revisit the 1933 material inspection, deploring the "apparent misunderstanding" as to what the Board of Inspection and Survey had actually found. In his estimate,

> The conclusion to be reached after a study of the Board's report is that the Board did not condemn the LOS ANGELES as unfit for further service but did actually find that the ship can at an estimated cost of about $15,000 be made reasonably safe for restricted flying service in good weather, and at airspeeds not in excess of 50 knots. This station concurs in such findings as a result of a study of both the Board's report and the actual condition of the ship today.

Subsequent hull-board reports (Rosendahl continued) had revealed no fundamental change from the conditions reported the year before. In his estimate, further, the rate of corrosion was so slow as to be considered negligible as far as serious loss of strength was concerned.[55] Unfortunately, however, maintenance personnel as well as funds had been slashed and were now insufficient to ensure indefinite preservation. Nor did matters improve; to the end, the shortage of both personnel and money for maintenance was to be a litany repeated in every hull-board report prepared for the bureau.[56]

On 28 July, King endorsed Lakehurst's program of experiments and recommended flight resumption. In his view, *Los Angeles*'s material condition continued to justify fully the flying and mooring-out program now proposed. It would ensure maximum scientific return and achieve appreciable training and experience. The experimental work, added to the training that would naturally follow flight, had not been considered in 1933—thus placing "a new aspect" on the question. In conclusion, King called for reconsideration and looked forward to operating *ZR-3* in order to realize full benefit from the experimental program already approved. Cost for one year's activities: less than $85,000.[57]

To King's regret and that of like-minded operators, the Navy Department that summer decided against the recommendation, primarily for reasons of "the unjustifiable hazard entailed," as per the much-quoted 1933 report.

Nonetheless, a program *was* on the books. On 17 August, a project order for reconditioning *Los Angeles* for experiments not involving flight was issued. To help execute it, a detail of seasoned hands received orders to Lakehurst. Lieutenant Tyler (for one) was detached from duty on board the light cruiser USS *Richmond* (then in Cuban waters); at month's end, he reported to Rosendahl. Appointed as officer in charge, Tyler relieved the naval air station's operations officer of upkeep duties on the retired aircraft. "Your duties," read his 31 August orders, "include the coordination of plans, supervision and inspection of work connected with the reconditioning and experimental program of the *Los Angeles* prior to her complete readiness for mooring out." (Responsible directly to the commanding officer, the lieutenant would keep Rosie informed.) For material procurement and shop work, a small committee of officers were assigned: Lt. C. V. S. "Connie" Knox (CC, No. 5024), assembly and repair work; Lt. Mike Bradley, engineering; Lt. (j.g.) Alexander MacIntyre (No. 3541), communications and commissary; and Lt. Cdr. Griffen (CEC), for work under the Public Works shops.[58]

To augment this core experience, students from the Enlistedmen School were drafted. These men "will supply," BuNav was advised, "a very essential part of their airship training in the necessary phases of upkeep, maintenance, and mast routine, and this duty is considered as a portion of their airship schooling and training." Practical experience, in other words, would now displace the classroom.

Kickoff work was under way by early September. Hull condition was examined and assessed in every component, after which appropriate repairs or replacements were effected. For example, many of girder lattices were found to be bent or broken. Laboring throughout the hull spaces, Aviation Metalsmith (3c) William "Bill" MacDonald and shipmates replaced hundreds of them. (During flight, these damaged "spoons" were tagged when found, and replaced when opportunity afforded.) Tellingly, none were found broken as a result of corrosion or "working" of ship's structure. *ZR-3*, MacDonald remembered, was "as sound as a dollar."[59]

To realize maximum benefit, reinflation was essential. Shored and empty of helium, *Los Angeles*'s hull was merely a complex of voids, compartments, and spaces with wire bracing. Hence, the magnitude and distribution of test stresses would be markedly different from those affecting a gas-filled airframe. "The general opinion here," Lt. Cdr. Jesse L. Kenworthy (No. 3911) advised Fulton, "is that not much information of value to operators can be had in tests with the ship fully deflated." Inflated, moreover, *ZR-3* could be undocked and moored out.

As these matters resolved, *Macon* took departure for a far-off homeport: U.S. Naval Air Station Sunnyvale, Mountain View, California.[60] On 12 October, the crew busied itself preparing for flight. The cells were topped off to 97 percent fullness, and flight luggage was stowed along with a full load of fuel (more than forty-six tons), six tons of ballast, and 883 pounds of fresh provisions. Eight Maybach VL-IIs warmed, the airship was weighed off in equilibrium. At 1620, general assembly sounded. *Macon* was placed on the self-propelled rail mast, after which the flight crew embarked. Towing commenced at 1702. On the

U.S. Naval Air Station, Sunnyvale, California. Its function—a complete serving facility for rigid airships—would last but sixteen months. (Mr. and Mrs. Joseph T. N. Suarez Collection)

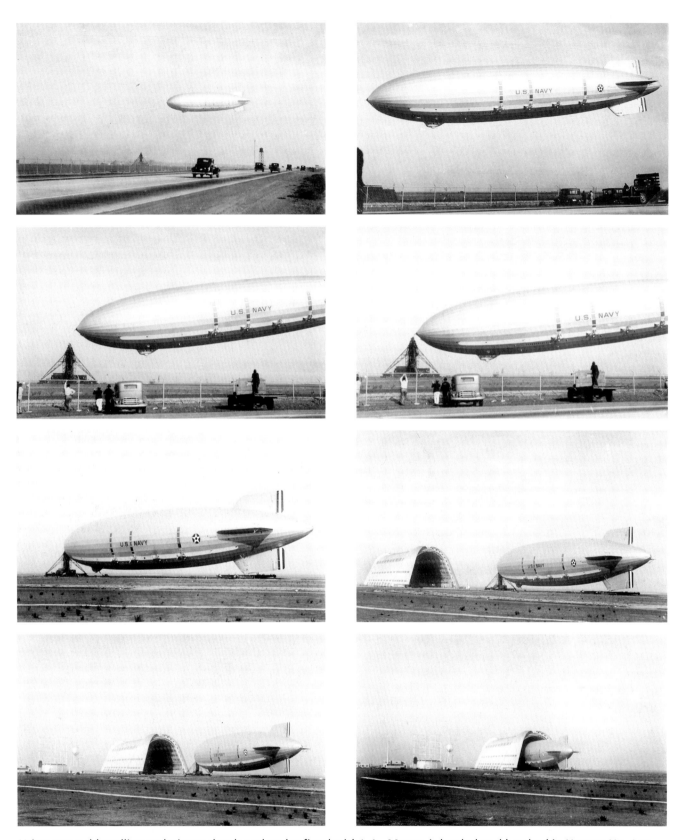

Using ground-handling techniques developed and refined with L.A., Macon is landed and berthed in Hangar No. 1 at Sunnyvale. This sequence is published here for the first time. (Joseph Billiams)

hauling-up circle, she was allowed to swing to the wind; twenty minutes later, the mast stood centered on the mooring out circle.

ZRS-5 cast off at 1805. With four propellers tilted, she rose statically to seven hundred feet. The annunciators rang up standard speed, and *Macon* climbed to cruising altitude. On the landing field, the drone faded on the crisp evening air. The air station would never again handle a commissioned ZR of the U.S. Navy. An era of naval aviation had ended at Lakehurst.

Its future clouded, the base was assigned "restricted" status and expenses were cut to bare-bones minimums. Ever the backwater, Lakehurst descended into somnolence. The quiet of a boom town "gone flat" lay over the air station, with *L.A.* the sole, barren ZR—"antiquated and forgotten." Compounding the gloom, local news sheets predicted the demise of big-ship operations, characterizing the base as "a monument . . . of an experiment that cost upwards of $100,000,000 and more than 135 lives."[61]

Lt. Cdr. Edwin F. Cockrane (No. 3933), acting skipper, espoused optimism. "No one knows anything for certain yet," he commented. "But the station is not going to close. No one knows anything about weather [sic] or not a new ship will be built, and no one knows if the *Los Angeles* will be recommissioned to be used as a training ship, but if orders came for this we could recommission her in two months, and we keep enough helium on hand to fill her at any time. . . . But even without a new ship or the *Los Angeles*, we would need a ground crew and men to service the two nonrigid aircraft stationed here. They are smaller but they need gasoline, helium and a ground crew. This base will be kept open as a training field."[62] Cockrane was correct; though slowed, flight operations persisted—with *K-1, ZMC-2,* and balloons. (*J-4* was at Sunnyvale.) This scanty fleet did not increase one cubic foot until *G-1* arrived from Goodyear in the fall of 1935.

Meantime, Class IX reported aboard (in 1934), as did a class of enlisteds. For the latter, training commenced that March. Tenant activities—the Aerographer's School, Parachute School, the (homing) Pigeon Loft—soldiered on. *Macon* had barely vanished west, however, before the station's authorized officer and enlisted complements were slashed from about four hundred to roughly half that. The number held at about two hundred at least through 1936; in October of that year, the authorized staff for restricted operations stood at twenty-nine officers, the balance enlisted men—navy and marine[63]

The naval air station was barely operating.

Civilian staffing also suffered. Civilians were (and remain) integral to naval shore commands and operations: in this case, manning the various shops as well as the helium and power plants, and operating much of the heavy equipment (e.g., the overhead cranes). Certain specialists worked closely with naval personnel, providing a host of skills essential to running the reservation and keeping its aircraft repaired, maintained, and flying. Inescapably, their number fluctuated with the see-saw fortunes of the base itself. In April 1925 (a comparatively heady time), 216 permanent civil employees labored; by 1933, to conserve funds, civilian support was slashed to less than a hundred men and women. Recalls from furlough were occasionally made; in the uncertain economic recovery of 1933–37, however, these proved erratic. One general helper from the Wire and Netting Shop was discharged in March 1936 only to be reinstated five months later. Placed on "indefinite furlough without pay" that December, the man was returned to

Men from Lakehurst's Marine detachment on the landing field, 1934–35. Along with security-related duties, Marine personnel were sometimes called out to assist in handling the nonrigids aboard—"a real pain" in Northeastern chill. (Don Brandemeuhl)

Another view of the steel hangar and its reigning "hangar queen," summer 1934. (Don Brandemuhl)

Major C. L. Fordney, USMCR (l.), technical observer, and Lieut. Cdr. T. G. W. Settle, pilot, hand their barograph to a National Aeronautical Association representative near Bridgeton, NJ—landing site for their stratosphere balloon flight, 21 November 1933. Altitude attained: 61,236.691 feet. (Vice Admiral T. G. W. Settle)

duty in March 1937, then furloughed and recalled three times more over the next twelve months. Maximum per diem (1937): $5.47 per day.[64]

Civilian support bottomed out in 1936, at about eighty. It edged upward, then soared during the frenetic days preceding Pearl Harbor. More than a few civilian employees devoted their entire working careers to the naval air station.

In the rigid hangar, meantime, the work list proved a long one. *ZR-3* was a complicated structure; still its cables and wiring, controls, equipment, lines, and myriad systems received alert examination. Dust and dirt had accumulated throughout *L.A.*, so, as far as practicable, the hull structure was inspected, cleaned, and varnished (a task facilitated by the absence of cells). Eventually, the bulkhead wiring was tested by inflating alternate cells to 100 percent. Prior to installation, the maneuvering valves were overhauled, new rubber seats installed, the seats for the automatic valves (the German originals) varnished. The major and minor controls were checked and overhauled as needed, including the clutch mechanisms in the control car. The engine telegraph cables were inspected, ballast bags examined, the water-ballast and fuel-dump controls checked out for serviceability. Save for the chronometer, all fixed instruments required for flight were reinstalled. (Somewhat later, a new automatic telephone system would be fitted as well.) During September–November,

Maybach VL-1s were retrieved from storage; upon examination and reconditioning in the hangar shops, the power plants were shunted to their pods and rebolted onto their seats.

These weeks witnessed a restoration of *Los Angeles* to virtual operating status. Though desirable in terms of scientific return from the program of experiments, the refurbishing was not without calculation: the bureau had not given up on flying *ZR-3* once again, at least for limited training service. The reconditioning was such that the various systems would be satisfactory for flight, were it ever authorized.

This was a hope never realized.

The 1934 reconditioning and reinflation came to approximately $35,000, including all helium charges. Work stood virtually complete by 14 December. Four days later came a relaunching—the first walk-out in nearly two and a half years. Following a six-hour mooring-equipment test, *Los Angeles* was shunted back inside. Her complexion seemed odd: thanks to scaling paint and removal or renewal of certain cover panels, the bright-yellow "visibility" scheme had deteriorated. "The Commanding Officer considers that these markings detract from the general appearance of the ship," Rosie advised Washington, "and are entirely unnecessary for the approved employment of the ship. Accordingly, authority is requested in repainting the required portions of the hull, to remove or obliterate the additional experi-

Reconditioned and reinflated, L.A. is undocked for the first time in thirty months using the light railway mast, 18 December 1934. Purpose: a six-hour moorings equipment test. In March, an experimental moor-out program began in earnest. (Author)

Mooring-out circle No. 1, 1935. When this site was modified to receive Hindenburg *(DLZ-129), another circle and rail connection were installed for the moor-out program. Note the surfaced runways, to accommodate American Airlines' DC-3 shuttle connection to Newark for zeppelin travelers. (Mrs. Charles M. Ruth)*

mental markings that were required [in 1932] and to apply in place thereof the standard aluminum color."[65]

The day after Christmas, BuAer approved. The ship's earlier appearance was restored, with silver paint.

Commencing that March and well into 1937, *L.A.* was shunted from berth to Mooring Site One, on the West Field, for various experiments, tests, and measurements. The light railway mast had been retailored for her—that is, the mast-cup height had been lowered. "It will be the first time," Rosendahl remarked, "that we've ever been able to conduct a study of this type with a full-sized ship. In the past we've been obliged to use models. Naturally some of our conclusions were based on scale calculations which we will now be able to check."[66] Nose to the cup and number-one car affixed to a ride-out track, *Los Angeles* assumed a unique function: as a laboratory, a full-scale model for solving the problems of future airships. Although retired and deprived of authority to fly under any conditions, *ZR-3* was "back to duty."

These ride-outs served also to keep ground crews familiar with big-ship handling. Yet the inaugural moor proved inauspicious. It commenced at 1720, 9 March. Midday of the 13th found nine men aboard—one to each station. *L.A.* herself was ballasted approximately 4,000 pounds heavy aft, 1,500 pounds heavy forward—to keep her from becoming light due to gusts. At 1300, as the stern answered the wind, a severe gust from starboard halted the swing. An upward force caused the (shifting) stern to rise, lifting the *track* two feet up, as well as mastward, and jamming the gripper-rollers of the riding-out carriage. "Preventer cables" had been installed, intended to destroy important structural members at frames 55–65, precluding *Los Angeles* from becoming a free-ballooning derelict should she escape her moorings. Though it took a damaging strain on the carriage, this system failed.

Freed from the carriage (which remained attached to the rails), ship's after section lifted about thirty feet and swung to port. Number-one car hit hard, dragging along the track before control was restored and another riding-out car shunted under.

Riding-out scheme for Los Angeles. "A" is the cradle for her riding-out car; "B" weigh-off springs of the riding-out carriage; "D" the quick-release hook for a "rip" or breakaway cable from the ship; "E" a bridle connection to that cable. Note the A-frame struts. (National Air and Space Museum, Smithsonian Institution (SI Neg. No. 2000-9241))

The impact had torn the A-frames from the hull, pulled three suspension cables free, and caused the car's struts to fail. The rip cables' progressive action on the hull structure augmented the injuries. The damage, however, Fulton wrote, following his own inspection, "is not great and involves replacement of four or five girders in the airship, minor repairs to the aft engine car, replacing struts between aft engine car and airship, and rebuilding the weighing-off platform. The estimated cost is five to seven thousand dollars. Due to the depleted condition of Lakehurst's force, work will necessarily proceed slowly and may occupy four to six weeks' time."[67]

The epicenter of the navy's airship community, LTA operations were but a pale phantom of those of remembered days. "We do not have any new airships yet," Rosendahl explained to a friend at the German Air Ministry, "but I am optimistic over our prospects for resuming building. Meanwhile we are doing interesting experimental work with the *Los Angeles* moored out."[68]

Attention had shifted westward. In July 1934, "Doc" Wiley had relieved Dresel as skipper of *ZRS-5*. (That month also, the General Board began to consider revisions in the department's airship policy.) Under Dresel, valuable—if tentative—search tactics had been studied. Several search methods exploiting her planes were tried, the first efforts were explored to *control* them from the "carrier," and experiments begun with a true radio-direction finding device. In the months left to *Macon*, Wiley fused these and other innovations into a platform for very-long-range search.[69] Wiley and the pilots of the HTA Unit (but few others)

New wounds: 13 March 1935. Reacting to the vertical force of a gust, L.A. lifted the riding-out car and trackage. Excess of weight over bouyancy: approximately 19,000 lbs. The gripper-rollers had jammed; with the weighted car jammed on the rails and the airship fixed in place, her stern broke free. No. 1 car came down hard then dragged over the track. The damage was serious enough to cost $5–7,000. (National Air and Space Museum, Smithsonian Institution (SI Neg. No. 2000-9239))

saw *Macon* for what she was—the means of extending the range of its airplanes. For example, development of a low-frequency homing device proved to be the answer to most of the communication and navigation problems for *ZRS-5* and her F9Cs. Soon the hook-on pilots were operating at the limit of their radius, well out of sight of the lumbering ZR. By the dawn of 1935, indeed, *Macon* was performing as a true lighter-than-air carrier. With her planes stationed sixty miles on each beam, she could advance a high-speed, high-endurance scouting front approaching two hundred miles—a fantastic feat for the time. There was not a military airplane in the world that, in 1935, offered comparable performance.

Unfortunately for *Macon* and airships, the fleet exercises of 1934 were framed with the purpose of developing the offensive qualities of carrier aviation and gave no consideration to the uses of an airship, much less an aerial "carrier." Instead, the trials obliged the undefended *Macon* to get in close to "enemy" formations, to be a *tactical* scout. The unique, very-long-range surveillance capability of this expensive machine was never tested

A jury-rigged taxi wheel for No. 1 power car, shown here bandaged up, 13 March 1935. (National Air and Space Museum, Smithsonian Institution (SI Neg. No. 2000-9240))

Charles S. "Chick" Solar, ACMM, at one of Macon's dial-system telephones, port lower keel. A seasoned airman, chance had kept Solar from making the fatal test of ZR-2. Standing by are Sea1c William H. Herndon and AMM1c Edward R. Morris (standing). No. 8 gas cell bulges overhead. At right is a 2,000-lb.-capacity fuel tank and, forward, a cell repair kit. Following the gangway at upper left is the water line and water-ballast control leads. Note the daylight illumination through the translucent (unpigmented) outer cover. (Lieut. Cdr. Leonard E. Schellberg)

against a Fleet Problem. And *Macon* did not survive to be tried in strategic patrols out of Hawaii.

Nevertheless, as the only commissioned rigid in the navy, the future of large U.S. airships was riding with her.

On the morning of 11 February 1935, at 0600, the siren for general assembly sounded. Men tumbled from bunks, donned flight gear, proceeded to the sleek Sunnyvale hangar, now ablaze with light. Inside, ship's engines were warmed. At 0620, the hangar doors rolled aside. "All departments ready for flight, captain," the officer of the deck announced. Wiley nodded to the mooring officer, who in turn blew his whistle and called: "Walk the ship out!" At 0630, two hundred men and some four hundred tons of mooring gear towed *ZRS-5* through the south doors and into the circle.

This takeoff was to be the last ever by a ZR.

Swung into the wind, the hatches were closed for takeoff. "Get ready aft! Get ready forward!," Wiley commanded. "Heavy aft!" reported the officer in charge there. "Two thousand pounds heavy forward!" called out one of the mooring party from atop the mast. In response, the first lieutenant jerked the toggles—and water gushed forth forward and astern. "Two thousand pounds light forward, sir!" "Two thousand pounds light aft, sir!" *Macon* was ready, buoyant. Wiley glanced quickly round; every man was at his station—the rudderman at his wheel, the engineering officer at the telegraphs, the elevatorman port side, wheel firmly in hand, the first lieutenant alongside. The officer of the deck was bending to his log, the radio man at his phone.

"Up ship!"

It was 0710. At bow and stern, holding-down bolts and the securing lugs released. Forward was heard a distinct metallic click as the nose jerked free. "Four engines, up standard speed," Wiley ordered. Engine room bells respond to the bridge's signal; the three-bladed props (each more than sixteen feet in diameter) spun into service. "Rising two hundred feet per minute," the first lieutenant reported. "Increasing, sir."[70]

Macon pushed to two thousand feet and set a course over the Santa Clara Valley, for sea. Having waited until she was about fifty miles out, a pair of F9C-2s zeroed in with their homing gear and raced for the trapeze. Ordered to stand by to receive aircraft, the compartment opened its doors, lowered the trapeze. The fighters commenced their approach. The hangar crew watched as each, with expert nonchalance, climbed to the bar: "The pilot grins up at his cheering section. This is old stuff to him also. He bends to his job, jazzes the engine a trifle. Man and engine are working as one complete living unit now. There is a metallic clang, the pilot relaxes, looks up and nods. Someone says, 'He's hooked.'"

On signal, the winch operator swung the trapeze up and in; as the plane passed through into its tiny hangar, its pilot cut his engine. His F9C stowed, the airman scrambled nimbly along girders and out. By 0748, both airplanes hung secured. As the HTA boys headed for a smoke or coffee, the hangar crew removed the F9Cs' landing gear, installed auxiliary fuel tanks, topped up their fuel.

During 11–12 February, the fleet was moving from San Diego and Long Beach northward to San Francisco.

Macon *landing at Sunnyvale's south circle, 15 October 1933. From the day of her arrival, U.S. Naval Air Station, Sunnyvale—named Moffett Field that spring—had only sixteen months in which to serve its planned function: a complete servicing facility for rigid airships. (Peter M. Bowers Collection)*

Though not a direct participant, *Macon* had orders to use this movement for training in strategic scouting. Wiley and his officers put in a highly credible performance. Between 1010 and 1023, the pair of F9Cs were swung out, to sweep ahead; five minutes later, another two, fresh from Sunnyvale, hooked on, then were swung up into waiting hands.

At 1310, the 12th, Commander in Chief, U.S. Fleet (CinCUS) released *Macon* from the exercise, whereupon Wiley steamed for Monterey Bay, the Santa Clara Valley, and the barn. At 1705, *ZRS-5* was about two miles off Point Sur, near 1,250 feet, eight engines at standard speed. The captain was at the conn, with Scotty Peck assisting the navigator of the watch, Lt. Cdr. Cochrane. Lt. (j.g.) George W. Campbell (No. 3812) had the deck. The executive officer of *Macon*, Lt. Cdr. Jesse Kenworthy, was on the bridge as well, observing the elevatorman, as was the Construction and Repair Officer, Cal Bolster. Just then a side gust seemed to strike the port side. The ship abruptly lurched to starboard and downward, shaking the control car violently. This "gust" in fact was the leading edge of the upper fin lifting away from the hull; immediately afterward the fin itself blew apart, puncturing the after cells. Within minutes, the emergency signal sounded; men donned life jackets.

Steersmen in Macon's *auxiliary control room, in the lower fin. (Goodyear Photograph courtesy Kevin Pace)*

At CinCUS, a signal was received from the stricken dirigible: "WE HAVE BAD CASUALTY 1715" and, not long after, "MACON WILL ABANDON SHIP ON ACCOUNT SERIOUS CASUALTY," followed by her approximate position.

The tail kissed the sea at 1739; soon rafts were launched aft.[71] Darkness settled. As crewmen watched, *Macon* broke up progressively from aft, frame by frame. By 1800, the hull was submerged to the bridge, bow vertical. Scrambling down a line from the bow, Coulter landed on the curved front of the car "and for a brief moment I tried to reach and yank off the commission pennant which was secured to a stay just forward of the car. I missed it and slid into the water."[72]

At 1820, *Macon* was seen to sink.

An immediate search had been ordered; at 1918, the *Richmond* had two boats in sight. Motor launches soon were retrieving survivors. In all, two men were lost.

It is commonly assumed that rigid airship development in the United States ceased with *Macon*. Much of the remaining hope for a future for large airships, indeed, had foundered with her. *Macon* was the last of her kind commissioned in the United States and the last to be built there. Nor did a ZR ever again take to the air to explore further the uses of airplane-carrying airships for strategic scouting. The political climate had soured, certainly. Within days of the foundering, President Roosevelt said he would not recommend further expenditure for such projects. Carl Vinson, whose home city had given *Macon* its name, sounded the death knell for new construction. "Frankly," Swanson remarked, "I do not know whether lighter-than-air craft justify expenses and accidents. We need other things worse, ships and air-planes, for instance."[73] The collective weight of such opinion would prevail.

The drama though was not quite done; backstage, several acts had yet to be played out. The navy's relationship with large airships—not least in commercial connections—persisted to 1940. More immediately, a court of inquiry was held, as well as a broader investigation initiated by the Special Committee on Airships of the Science Advisory Board at Secretary Swanson's request—the so-called Durand Committee.[74] During 1935–37, this panel examined a good deal of information, including general design data on airships, materials, and methods of construction. At the request of BuAer on behalf of the special committee, NACA obtained load measurements on a large model of *Akron* in its wind tunnel.[75] Some of the committee's reports and recommendations bore on desirable strength factors for future ships, if built. Concerning ZR-3, physical and metallurgical test data aplenty were to hand. In May 1935, at the panel's request, a summary of BuStandards' results were compiled relative to *Los Angeles*'s girders, wire terminals, and test specimens. Among the data were 1933 reports to the Board of Inspection and Survey.

Twenty-foot (1/40 scale) model of Akron *in NACA's full-scale wind tunnel at Langley Field, 1935–1936. Design data pertaining to airships remained largely empirical, especially with respect to gusts. Knowledge of the aerodynamic forces imposed and their distribution helped designers calculate the resulting stresses in structural hull-members. Object of this Navy-initiated study: to determine the forces on the airship when handled near the ground. (National Air and Space Museum, Smithsonian Institution (SI Neg. No. 7A-46950))*

As the committee labored, voices urged that *L.A.* be recommissioned for limited duty or that a trainer be authorized, so as to maintain proficiency and train new personnel. (Similar recommendations had followed loss of *Akron*.) Rosie's was one. *Los Angeles*, he asserted, was "practically in flying condition." Further, the navy needed a big airship for "experimental and training purposes." In his view, she was capable of such a program "on a conservative basis" for another two years. "A fair trial" was called for before any conclusion was reached as to the airship's usefulness.[76] All this presupposed further construction. In a waning cause, the situation had become extreme. Airshipmen had no standby, no platform in commission flying.

The Durand Committee proved unanimous in its opinion: given the lessons from casualties, there was no reason why safe rigids could not be built. "We said," von Karman, a member, remarked, "[that] it was our general feeling that the navy and others had been too complacent about the airworthiness of the ships and that they were given their regular naval assignments too soon without treating them as full-scale models for further study." It therefore urged upon the navy a "continuing program" of construction and use.

> We further recommend most strongly that the first large airship built under such a program should, at least for a time, be considered not an adjunct to the fleet but rather a flying laboratory or flying training ship, not only for extensive technical observations of the structure under operating conditions, but also for enlarging our knowledge regarding the best conditions of service for such vessels, and, as well, for giving opportunity for the training of officers and crew in the technique of handling airships under all conditions of weather and service.[77]

A series of studies and proposals were put forth, involving military airships as well as commercial versions for overseas transport. Policy and preparedness obligated the Navy Department to cooperate in development of commercial dirigibles. Privately, however, support was fast ebbing. Behind the politics lay genuine contentious issues: would these platforms (or their services) meet demonstrated needs at a the price the government (or the market) would accept? In the naval realm, ZRs had been thrown into tactical exercises, seldom deployed *strategically*—and with inconclusive results. Was further development justified? Or affordable? As the 1930s waned, these questions faded, unresolved. In the meanwhile, Admiral King pressed the case for large airships, as a means of holding and perfecting the gains already realized. He even took such practical steps as assigning officers to transoceanic crossings flown by *Hindenburg*.

For its part, *ZR-3* was yet capable—structurally—of a return to operations, albeit restricted ones. Proponents would finally opt for a new trainer; one was in fact authorized and funded in 1938. "That training ship," Capt. Mike Bradley conceded, "was bandied about for several years, as I recall it. I don't think any thought was given very much to flying the *Los Angeles* again. Unless there was going to be another rigid airship program, there wasn't any point in it."[78]

Among the might-have-beens was a "cut-down" version of *ZR-3*, based on new work on wind loading. Wind shifts and gusts had long troubled designers (witness the 1925–26 NACA investigations). Studies relating to maximum sudden changes in velocity, the characteristics of gusts especially, were therefore ongoing.[79] Objective: to analyze wind—everything from gusts to the turbulence in thunderstorms—in three dimensions and, in particular, to dissect its strong discontinuities. Aerologists did not know enough to meet the requirements of designers, who as yet had to *assume* aerodynamic-force conditions. "It was a matter of life and death," Reichelderfer insisted, "to know more about them. I just can't emphasize too much the way this idea, and this need, permeated the whole airship world."[80]

Enough *had* been learned to revise strength criteria—upward. Wind and water-tunnel research as well as flight tests had shown, for instance, that one of the governing cases for longitudinal strength was gusting. Yet gust loading had not been considered when *ZR-3* was designed. "The *Los Angeles* is not, never was, strong enough to encounter extreme squall conditions such as were considered in the strength calculations of the *ZRS-4*," Burgess had advised when *L.A.* was still operational. "Such squalls must be avoided."[81] The Durand Committee's deliberations focused attention on the fact that the inherent strength built into airships had risen. It followed that *Los Angeles*, although stronger then any predecessor, did not embody the latest engineering opinion.

In January 1936 Lakehurst's Engineering Division suggested that her three largest bays—frames 100, 115, and 130—be cut away. This would reduce volume and length and, in consequence, the gust bending moment (force multiplied by length of moment arm), thereby increasing the margin of safety. The reduction in lift would be partly compensated for (it was calculated) by eliminating the forward power cars and by replacing the main car with a *Macon*-type (smaller) control car. Further weight savings would come by replacing the Maybach in number-one car with an Allison V-1710-4 reversible engine developed for *ZRS-5*. If tests proved satisfactory, the remaining two Maybachs could be replaced as well—subtracting yet more weight.[82] Estimated cost for the changes: two hundred thousand dollars.[83] The conversion offered a quicker and cheaper way to resume training than building a new ZR. Redesigned, *Los Angeles* would retain sufficient range and lift for normal needs of training and, further, be more economical to operate than it had been. "Above all," one officer commented (others at Lakehurst agreed), "it affords a chance of putting a rigid airship in the air at an early date." The impact upon morale, publicity, and, perhaps, the CNO was obvious.

Not all favored the modernization, however. One fear: if the suggested remodeling were to be rejected by the department, the likelihood of flying *ZR-3* unmodified was nil. Also, several officers took issue with deadweight. Tyler, for one, flatly opposed the "so-called modernization;" the blueprint-given figure for deadweight (ship's weight, as distinct from that of its contents, loaded); and the numbers claimed for speed: a maximum of seventy-one knots. "The cutting out of a section from the center would," the lieutenant argued, "have very little effect on the resistance of the hull to the airstream. During five years of duty on the *Los Angeles* I have never seen a speed greater than 62 knots attained with five engines at approximately 1600 rpm each." In his view, modification would produce an "antiquated airship whose performance characteristics would be far below those outlined on blueprint." The lieutenant wanted *ZR-3* back aloft—but unmutated. "It is believed," Tyler concluded, "that the estimated cost of $200,000 is [unrealistically] low and even if the conversion would not exceed this amount, the results would definitely not warrant this expenditure. For a fraction of this amount, the ship could be placed on a sound flight status and would be ideally equipped as a training ship."[84]

The modification proved to be a nonstarter, probably because it was deemed too costly for the results anticipated.

As for commercial dirigibles, the domestic airlines were themselves planning transoceanic routes. Powerfully self-interested, their lobbyists intrigued against legislation supportive of any such government venture.[85] The loss of *Akron* had shaken public confidence, undercutting support. Now *Macon* was gone. In short, nothing flying remained in inventory that could conceivably continue airship research and development. Among the commercial proposals was a rigid of 8,700,000 cubic feet. Prepared by Goodyear-Zeppelin, a 1939 prospectus called for a maximum speed of ninety miles per hour, a cruising speed of seventy-five. With a commercial load of fifty passengers and eighteen thousand pounds of mail and freight, a cruising range in still air of 6,750 miles was projected. The design rather resembled *ZRS-4* and *ZRS-5*.[86]

It never left the drawing board.

Meanwhile, navy LTA operations persisted, using what few nonrigids were to hand. In September 1935, *G-1* was procured from the Goodyear fleet with seventy-five thousand dollars from the 1936 appropriation. Delivered that October, *G-1* was impressed into training duties plus utility work. Two weeks later, Lakehurst staged a mass operation. Under the command of Lt. Cdr. George H. Mills (No. 3925), *K-1, G-1, J-4,* and *ZMC-2*—the full inventory—flew in procession to Philadelphia to welcome delegates visiting in honor of Air Navigation Week. Upon the return and a change of crews, the squadron streamed northward, up the coast, Lieutenant Commander Peck in charge. Over New York City that afternoon, the four were joined by a commercial blimp. "It's the first time we've had four ships here at one time and we wanted to give the New Yorkers a good look at them," Rosendahl commented.[87]

At the naval air station, the workload had hardly slackened. "Lakehurst is just as busy with our various experiments and nonrigid projects as it was in the days of [navy] rigid airships," Reichelderfer wrote a colleague, "and our study [of a gust-research report] has been subject to numerous interruptions."[88] Among the programs: exploitation of *ZR-3*. *Los Angeles* was towed to her test track four times during 1935. Extended moor-outs took place between 11 to 20 June, 11 July to 2 August, 17 to 26 September, and—the year's last appearance—on 19–23 November.

While masted, *L.A.* carried a riding-out watch—a skeleton crew of officers and men. One section was on board in favorable weather, more as conditions demanded primarily to adjust ballast, compensating for changes in lift resulting from superheat changes, rain load, and so on, and to ensure that nothing interfered with free swinging in the horizontal. (The stern was kept from moving vertically by a weighted car or taxi-wheel carriage or by rail clamps.) The customary records were maintained as, on the bridge, the watch operated the water-ballast controls and the hand-pulls

A kite balloon and K-1 on the West Field, summer 1934. Shored and deflated in the north berth, a laid-up Los Angeles *is being reconditioned for "experimental development" in a non-flight status. (Kevin Pace)*

for the maneuvering valves. (To reduce helium leakage and entry of air into the cells, "jam-pot" covers had been fitted onto the automatics.)

For winter riding-out, the matter of snow and icing stood unsolved. Under way, normal practice called for seeking out a warmer layer. Absent the dynamic effect of the engines and wind (as when masted), snow and ice buildup could impose excessive loads on all frames—the fins and control surfaces particularly. "The problem of preventing accumulation of large quantities of snow on a moored airship is recognized as

Works Project Administration hirees attend to mooring-out circle No. 2, constructed for the experimental moor-out program exploiting L.A. (Charles E. Rosendahl Collection, University of Texas)

being an important one," the bureau had advised. "In fact, the problem will probably be insoluble by the application of heat alone and brushes or sweeps in some form will have to be used."

Accordingly, *L.A.* was inducted into snow and ice-removal experiments.

To attack the problem, in February 1935 a project order in the amount of $1,500 was issued. "It is recommended," the bureau wrote, "that part of this sum be used for the procurement of a large size conveniently handled oil-fired unit heater of commercial type which can be experimented with to determine its efficacy when heat is directed to the fins, etc., (and also to determine data in regard to creation of artificial superheat while moored)."[89] In the cold of 1935–36, tests were conducted relative to snow removal on the ship's horizontal fins and top side—applying heat to the fins and testing mechanical means of removal:

> Nets [Rosendahl wrote] have been rigged over the cover which can be moved along the cover to scrape off snow. Snow has been swept off fins and pushed off control surfaces by long-handled brooms. Recently a system of flaps have been made in top side of the horizontal fin covering to permit men to work on the fins without actually going outside on this rather dangerous platform. Experiment is being made with a heating unit riding on a flat car attached to stern riding-out car which will supply heat from a coke-oven through fabric sleeves to the horizontal fins.[90]

Winter ballast problems were also assessed. The ballast bags—original and *Akron-Macon* spares—were furnished with canvas jackets lined with felt. For the tests, these were half-emptied and refilled with hot water, realizing a temperature of from fifty to fifty-five degrees.

On station, *L.A.* recalled a happier, less dubious time. Grounded status notwithstanding, she fed a (faltering) sense of possibility. To believers, her presence on the field—her sheer continuance—represented a modicum of reassurance for a struggling, compromised program. But as far as the department and most politicos were concerned, the loss of *ZRS-5* had brought the issue to closure.

Lakehurst, however, was still in the big-ship business. Arrangements for the reception and handling of *Hindenburg (LZ 129)* had commenced late in 1934. That November, Rosie hosted King and Eckener (*Herr Doktor* escorted by von Meister).[91] The party, along with heads of naval station departments, watched a demonstration of the mobile telescopic mast, the stern beam, and its locomotive and other mechanical handling gear, and then discussed docking and servicing arrangements.[92]

Hindenburg would begin a series of demonstration trips across the North Atlantic, made possible by leasing terminal facilities at Lakehurst and the U.S. Naval Reserve Aviation Base near Miami. Privately owned U.S. terminals were nonexistent, so Eckener had written President Roosevelt to argue the benefits inherent "in demonstrating airship transport and placing it on a firm basis." Germany, he reassured Roosevelt, was not out to dominate the field. "My proposal is not intended to interfere with any commercial airship effort in the United States, but rather to stimulate such effort."[93] Fortunately for Berlin, authority and precedent allowed private operators to use government aviation facilities—with restrictions. Washington assumed no liability or expense whatsoever; use was entirely at the risk of the operators. Further, domestic resentment against the Nazi regime dictated that the Germans pay their way in every respect. Plainly, though, a demonstration service to the East Coast lay in the public interest.[94]

The Navy kept a hand in. Scotty Peck was detailed to Europe, to help keep the bureau advised of developments. Reaching Berlin in mid-February, he reported to the naval attache and had meetings with German navy and air ministry bigwigs. "The impression I received from the various officials that I talked with is that airships are an accepted vehicle of transoceanic transportation and are no longer considered experimental." In Friedrichshafen, he found, the plans and schedule of construction also reflected this view. Granted a tour of the ship's passenger spaces and (a portion of) the keel corridor, the officer was elated by what he saw:

> The lounging spaces for the passengers can be called no less than luxurious. The interior appointments compare favorably with the better class of ocean liner. Every conceivable convenience has been provided. There is no swimming pool or gymnasium but everything else is there. I may appear over enthusiastic but to me, it would stretch the imagination to the breaking point to believe such comfort and luxury possible on an airship.[95]

The maiden flight occurred on 4 March 1936. "The LZ people all seem delighted with their ship," Peck wrote as the trials proceeded, "and the old Doctor fairly glows with elation." Back home, there was a new splash of zeppelin copy; "NEW ZEPPELIN IS DUE AT LAKEHURST MAY 9," one headline intoned. "The coming of the new dirigible is the first promise of a variance from

En route Lakehurst: Hindenburg's *inaugural North Atlantic crossing, 6–9 May 1936. Here Dr. Eckener (second from r.) takes his ease with unidentified fellow passengers. (Kevin Pace courtesy Navy Lakehurst Historical Society)*

On watch in the control room of Hindenburg. *The view is forward, from the plotting room of the control car. The elevatorman is nearest the camera, the rudderman beyond. (Kevin Pace courtesy Navy Lakehurst Historical Society)*

the routine which has settled down here since the navy had no more large dirigibles with which to experiment. Officials at this station have lamented this deprivation and done all they could to keep their personnel acquainted with the latest developments in the flying and handling of the great bags."[96]

At dawn, 31 March, *LZ-129* eased aloft on her first round trip to Rio de Janeiro. That May, the zeppelin was scheduled to begin the first in a series of ten round trips to the United States; on the 6th, *Hindenburg* lifted off for America. On board were fifty passengers plus fifty-five crewmen, including three U.S. observers. Stowed aft were 2,255 pounds of mail and 295 pounds of freight. Total payload: 13,550 pounds. Pushing westward, Capt. Ernst Lehmann selected his course to take full advantage of the available winds—estimating the situation (Peck was to marvel) "perfectly."

By the early hours of the 9th (more than two days out), and thanks to a tailwind, *Hindenburg* was approaching the eastern seaboard. At the naval reservation, word began to circulate that the visitor would appear at about sunrise. At 0350, Lt. George Watson, communications officer, posted a bulletin in a very crowded press room. "The Station has received word that the *Hindenburg* will be over New York between 4 and 5 A.M., EST, and will land at Lakehurst about 5:30 A.M., EST. Lakehurst will be ready. Sunrise is 4:48 EST. The Station is *now* closed to the general public until 4 A.M. EST."

At about 0440, the east just silvering, the zeppelin was sighted at the far end of the field. It pounded over, circled once to "feel" out conditions, and started in toward the mobile rail mast—the station's second (and heavier) railroad mooring mast, completed in 1933. *Macon* had used its mast cup; now it would help ground handle the *"129"* for two dockings—and receive each of its eleven moors that year.

As per German practice, three senior watch officers controlled the various functions during landings and takeoffs. The commanding officer was on hand, certainly, but in a supervisory capacity. "The fact that

Los Angeles *is undocked for a mooring-out, 8 May (?) 1936. Among the experiments for 1936–1937: hauling-up tests, pressure-distribution measurements, checking superheat cycles, studying camouflage effects, and measurements of gust forces on a full-scale airship. (National Archives)*

scarcely a word is spoken during these periods," Tyler reported of a later crossing, "is an illustration of the splendid organization and fine spirit of cooperation which exists aboard the *Hindenburg*. Years of close association in airship operation on the part of most members of the crew and officers has developed this organization into a smooth working machine where the casual observer is impressed by the apparent smoothness and lack of effort with which the entire unit operates." "Along with several other Navy officers," Louise Tyler Rixey remembers, "all of them friends of their German counterparts, Dad flew on the last trip of the '36 season and returned with glowing reports to us of the flight."[97] Contrasts were inescapable. "Comparing the *Hindenburg* with American airships viz the *Shenandoah*, the *Los Angeles* and the *Akron* and *Macon*," Tyler continued, "I was impressed with the relatively small amount of vibration present, even during periods of passing through turbulent air."[98]

Upon securing, "Aerology was the headquarters for the *Hindenburg* people," Reichelderfer said with satisfaction. "They spent rather little time at the hangar—only what they had to do for operations and much of the time they spent in the quarters, where they were guests."[99] Meantime, beneath the great aircraft, the passengers disembarked.[100] Not far off, an American Airlines DC-3 idled, waiting to carry them on to Newark.

Planning was meticulous. To clear the big shed, *Los*

GROUNDED

Lieut. Raymond F. Tyler's ticket for passage—the ninth North Atlantic crossing by Hindenburg, *eastbound. (Navy Lakehurst Historical Society)*

Gripped by Lakehurst's stern beam and line handlers, Hindenburg *is shunted under roof, 9 May 1936. Well away, unheralded, ZR-3 rides to the mast at mooring-out circle No. 2. (Airship International Press)*

Angeles had been shunted out; bow secured to the mobile mast, she was riding a newly graded permanent circle (designated number two) well to the westward. Rails had been extended from the shed through the hauling-up circle to its center. Eight hundred and seventy-four feet in diameter, these circle-tracks held a ride-out car. *Hindenburg*, though, would exploit the ZR hangar but twice. Zeppelin officials viewed the docking equipment as too heavy by far, making for slow handling. With each stay necessarily brief, "they rather prefer to stay out on the mooring mast as that will reasonably assure their timetable which is one of their most important concerns."[101]

The zeppelin's outsized, sprawling bulk—the first "live" rigid here since *Macon*—seemed to fill the great space: eleven *inches* separated bow and stern from the door leafs. In deference to seven million cubic feet of hydrogen, and to prevent stray accumulations, the door leafs were "cracked" at each end and the skylights kept open. Guards stood at the elevators and stairways to ensure that only authorized personnel climbed topside.[102]

During an open house, an estimated 75,000 persons "walked, motored, flew or crawled" to the naval reservation. The Navy Relief Society sold hot dogs, soft drinks, and the like. A temporary post office did a brisk business from philatelists, who prepared mail for Europe. Away to the west, largely ignored, a geriatric *Los Angeles* swung to her mast. "She was decommissioned on June 30, 1932, for reasons of economy as only the AKRON and MACON were to be operated," a fact sheet advised visitors. "Late in 1934, she was again inflated and although

Night view: Hindenburg *on the mast at Lakehurst. The zeppelin's presence announced unmistakably the arrival of transoceanic air travel. (Wide World Photo courtesy Navy Lakehurst Historical Society)*

in flying condition, is being used for experimental development mainly while at a mooring mast under widely varying weather conditions."

Seasoned hands were struck by this newest model of German know-how—the world's largest, most luxurious flying machine. The children were awed as well. "As an impressionable 11-year-old in 1936," Louise Tyler Rixey reminisced, "I was thrilled when the *Hindenburg* crossings began and those of us in quarters [on-base housing] were allowed to be on the field to watch the ship take off and to greet the passengers on each incoming flight. I always had my autograph book in hand and particularly remember meeting Douglas Fairbanks, Sr. and his then new bride, Lady Sylvia Ashley, Dorothy Kilgallen, and the very dashing Leslie Charteris with his monocle—I treasured for years afterward the stick figure he hastily drew for me of his literary character, 'The Saint.'"[103]

The zeppelin announced unmistakably the arrival of transoceanic air travel. Lakehurst, though, was but a brief port of call. In Reichelderfer's phrase, "They were all business, all operations." Eckener nonetheless found time to view his former command. "After that inspection," Congressman William H. Sitphen advised BuNav, "Dr. Eckener states that the *Los Angeles* was in perfect flying condition, and in his judgment, absolutely safe to recondition and put into operation immediately." Likewise, Dr. Sachs, director of the Duerener Aluminum Works, Germany, expressed "complete satisfaction" with the condition of the aluminum alloy structure, "stating that he believed it is as serviceable today as it was when it was first built." Ever at the pulpit, "C.E.R." had chimed in. "Commander Rosendahl, whom I [Sitphen] believe is the acknowledged leader in lighter-than-air in this Country, has also declared that he is convinced the ship is in satisfactory condition for flying, and has urged that course." With this information, the good congressman urged an immediate commissioning. "Surely," he closed, "we cannot admit failure by the American Navy when foreigners are able to operate these ships over a long period successfully."[104]

The buzz for commercial dirigibles had resumed. As one and all admired the airship and commentators generated excited copy as to the implications, Navy personnel began cleaning, servicing, and loading *Hindenburg* for the eastbound run. Reserve hydrogen had been cached—about one-half million cubic feet in the station gasometer, plus four tank cars. Nearly thirteen thousand gallons of diesel fuel went aboard, plus lubricating oil and provisions. Off-loaded trash and garbage was burned under Department of Agriculture supervision. (Any contraband and undeclared merchandise confiscated by customs also was destroyed.) A navy detail emptied the sanitary waste system. As for the German crew, the off-watch officers were berthed in the station BOQ, while Eckener shared Rosendahl's quarters.

The prime objective this transatlantic season: to win support for a joint American-German venture, building upon the record of *Graf Zeppelin*. "The main point to the establishment of a transoceanic airship corporation," Eckener remarked, "is to re-establish public confidence on this side of the water, which was lost when the *Akron* crashed. President Roosevelt is interested, but wanted the feasibility of regular schedules tried out." The airman expressed hope for a rebirth of the large naval airship as well. Indeed, as the season progressed, *Hindenburg* served as a platform to help train personnel—navigation, ground handling, weather, and general big-ship operation.[105]

The aeronautical question of the hour: Was it possible to inaugurate regular air traffic across the North Atlantic? At 2227 EST, 11 May, *Hindenburg* lifted off, destination Frankfurt. After detouring over New York City, Lehmann steamed out over Long Island Sound, toward Cape Cod and the great circle route. On board, the ocean weather map—based upon surface-ship reports and land-station readings—was maintained and analyzed by Lehmann and his team, who set courses accordingly. Lakehurst, for its part, was drawing four complete maps daily and dispatching planes to seventeen thousand feet to collect upper-air data. Its daily forecasts of flying conditions, including special summaries, were (of course) available to fellow airmen.

The continent-to-continent leaps of *129* (fifty-six that year) aroused international notice as well as investor interest. Berlin seemed resolved. That June, design number *LZ-130* was laid down by the Luftschiffbau. Intended for passenger-mail-freight service between Europe and the Americas, the new rigid's construction followed conven-

Table 6-3 Commercial record for *Graf Zeppelin (LZ-127)*, 1928–1937.

First Flight	18 September 1928
Last Flight	19 June 1937
Dismantled (along with *LZ-130*)	Spring 1940
Total No. Flights	590
Total Flight Time	17,177 hours (716 days)
Miles Flown	1,053,391
Mail	100,500 lbs.
Freight	134,800 lbs.
Passengers	13,110
No. Ocean Crossings	144
North Atlantic	7
South Atlantic	136
Pacific	1
Visits to NAS Lakehurst:	
First Westbound Crossing	15–28 October 1928
Round-The-World Flight (start, end)	5–7 August 1929
	29–31 August 1929
"Triangle Flight" from South America	30 May–2 June 1930

The keel corridor of Hindenburg. *(Kevin Pace courtesy Navy Lakehurst Historical Society)*

tional zeppelin practice—a streamlined hull composed of main rings braced radially with wires, intermediate rings, and longitudinal girders. A catwalk or axial corridor ran along ship's centerline from nose to tail. The gas space was divided into sixteen cells, each held in place by wire netting. Intended lifting medium: hydrogen.[106] The date set for the first flight was August 1937, with an inaugural crossing to Rio late in October.

As for the U.S Navy, Class X had reported for duty under instruction. That summer, Lieutenants Richard S. Andrews (No. 6749) and Richard N. Antrim (No. 6750) soloed in free balloons and logged nonrigid hours. "When we were not flying or assisting in ground operations with other aircraft," Andrews said, "we attended ground school." The navy had no flight-status ZRs so, by arrangement, its trainees had flight hours on *Hindenburg*.[107] That October, Andrew and Antrim had joined VIP guests for an exhibition tour over New England. "We were taken throughout the ship by various members of the officer crew and shown every possible aspect of the ship," Andrews later enthused. "As a special treat, we were allowed to go up the vertical shaft from the central corridor and look out on the top of the *Hindenburg*. [Also,] one at a time we visited the engine gondolas and tried to understand in part what our German engineer was trying to tell us."

Both were slated to make a round trip—Lakehurst to Lakehurst—as observers in 1937.

By the fifth eastbound farewell (in mid-July), the 1936 season seemed destined on success. Commercially speaking, things were looking up. The Business Advisory Council made a report to the secretary of commerce recommending a U.S. airship policy and program. Within months, an extension of the Maritime Act of 1936 to transoceanic airships was in planning. This would attract private capital. As for Germany, *LZ-131* was being projected; with it and the *129*, the Deutsche Zeppelin Reed-erei (an operating company set up by Air Minister Hermann Goering) could offer weekly sailings from Europe and the States.

For *Los Angeles,* the year's program had commenced on 8 May—the day prior to *Hindenburg*'s first berthing. Shunted out, *L.A.* rode to her circle for a week "with grace and distinction." The next moor-out occurred on 17–24 July. Purposes of this and later series: hauling-up tests, checks of superheat cycles, and studies of camouflage effects on airships. (For testing visibility, a section of the cover from frame 145 to frame 160 was coated with cold-water paint; this scheme was changed as the months ensued. During September, the envelope of *J-4* was given a coat as well.) By August, planning was under way for hauling-up tests and pressure-distribution measurements. Sufficient experience in riding out, the bureau advised, had been gained to warrant keeping *ZR-3* out in strong winds. As soon as certain preparatory tests were completed, Lakehurst replied, it would do so. For his part, Rosendahl voiced "confidence in the ability of this airship to successfully withstand any but very exceptional weather conditions at the mast. Heavy snow, hail and icing conditions at the mast have been the subject of study and experimentation, but the problems involved have not yet been fully solved."[108]

Field trials resumed on 9 September: a two-day moor-out. From 30 September to 12 November, though, *ZR-3* was shored up in order to have four cells changed. (Retailored *Akron-Macon* spares, three new cells had been installed that March.) The year's final moor took place from 19 to 23 November. As yet, the aerodynamic forces (and resulting stresses) acting on airships subjected to side winds near the ground were not well enough understood to define the wind-velocity limits for ground handling in such conditions. The purpose of hauling-up tests: to obtain data on the horizontal and vertical forces acting on the hull for definite angles of yaw.

Table 6-4 Performance summary for *Hindenburg, (DLZ–129)*, 1936 North Atlantic season.[1]

Total Flight Time[2]	Actual Distance Covered	Best Ave. Speed, Flight Time	No. Passengers	Mail	Freight[3]	No. Observers[4]
Eastbound (to Frankfurt)	35,410 nu. mi.	84.7–42.9 hr. (6th)	528	4962 lbs.	6644 lbs.	40
Westbound (to Lakehurst)	38,293 nu. mi.	71.3–51.4 hr. (4th)	485	4780 lbs.	3601 lbs.	40

1. Four round trips also were made to South America between March and August 1936, for an additional 702.3 flight hours, 309 passengers, 2329 pounds of mail, and 7938 pounds of freight.
2. Another 11.9 hours were lost to non-operational causes, such as non-availability of ground crew, government officials, and circling cities.
3. Includes freight and express.
4. U.S. Navy and Goodyear-Zeppelin representatives.

Hindenburg *over lower Manhattan, 19 August 1936—ship's seventh crossing, westbound. Among the passengers: Cdr. Garland Fulton, Lieut. Cdr. Anton Heinen, Karl Lange (No. 547 and pilot with Goodyear-Zeppelin), and Lieut. Cdr. George H. Mills. (Elsie C. Harwood)*

At 1600 hours on the 19th, *L.A.* was undocked and secured to the rail mast at the mooring site. Next day, a preliminary test was held to check equipment and procedures. But reasonably favorable conditions—a steady wind—did not occur until the afternoon of the 23rd; at 1430 on that date it was decided to commence as soon as possible. Even before the hauling-up could begin, however, the wind dropped below the desired velocity and continued to fall. With tow cables attached to her starboard side, *ZR-3* was allowed to swing; the mean position of the stern carriage was taken as a "zero angle" position. She then was hauled up to starboard (via tractor and spider groups) by ten-degree increments until ninety degrees from "zero" was reached. At each ten-degree position, *L.A.* was held for two minutes, during which observers recorded data.

Results were inconclusive. Data from follow-on trials would be required as to the magnitude and distribution of forces. Accordingly, "from time to time," Rosendahl reported, when conditions permitted, similar full-scale tests would be conducted.[109]

That August, hearings were held before the House Naval Affairs Committee on a bill to replace *ZR-3* with a similar-sized ship. Despite favorable testimony, no action was taken. "The *Los Angeles*" in this period, Andrews recalled, "was no museum piece but rather a living reminder of what she once was and what she might be again should she be allowed to take off. She also represented the fact that

Lieut. Cdr. George H. Mills (l.) and Cdr. Garland Fulton on the bridge of Hindenburg. (Langdon H. Fulton)

time required by fast steamers to the English or French coasts. North Atlantic passengers carried (eastward and westward) totaled 1,007. Before the final eastward leg, and in light of the refreshed interest, Eckener, Lehmann, and von Meister hosted nearly four dozen chairmen, directors, presidents, and managers from the worlds of finance, transportation, oil, rubber, chemicals, and steel on a guest flight over six seaboard states. Photographers and press and news service people were on hand as businessmen intermingled with guest politicians and U.S. officials, the German ambassador (with two attaches), and six U.S. naval officers: Adm. William Standley, acting secretary of the navy; Adm. William S. Pye, assistant to the CNO; and Adm. Arthur B. Cook, chief of BuAer, plus Fulton, Rosendahl, and Watson.[111] Next day, 10 October, the States saw the last of *Hindenburg* for the year. Before the leave-taking (delayed due to fog), Eckener expressed his hope that U.S. venture capital would charter a German-built rigid for a North Atlantic airship service.

As *Hindenburg* pushed northeastward, Lakehurst's communications unit received a signal from the fast-retreating zeppelin:

> Commander Charles E. Rosendahl U.S. Naval Air Station Lakehurst, N.J.—Our officers and crew join us in an expression of thanks and appreciation to you the officers, crew, and civilian personnel of the Naval Air Station for the splendid cooperation, excellent work, and hospitality during the 1936 demonstration flights. *Auf wiedersehen.* [signed] Eckener/Lehman.

rigid airships were not dead in the minds of lighter-than-air personnel and that, someday in the not-too-distant future, she would be joined by newer and perhaps bigger airships."[110] The comings and goings of *Hindenburg* and its transatlantic sister *Graf Zeppelin* (then in its ninth year) underscored this faith and reawakened ardent interest in commercial possibilities.

As the 1936 commercial season closed out, airship men were triumphant. The *Reederei* had shown that a service to Central Europe was practical, with flight time averaging two and one-half days—half the

Hindenburg and Los Angeles. *The 1936 transatlantic season was intended to win support for a joint German-American consortium, building upon the record of* Graf Zeppelin. *Aeronautical question of the hour: Is it possible to inaugurate regular service across the North Atlantic? (Navy Lakehurst Historical Society)*

Settled onto its landing wheel, Hindenburg *is walked to the mast by ground-crew parties. Emblematic of the big-ship era, the venerable* Los Angeles *swings to her mooring. Longest moor-out during 1936: twenty-two days. (National Archives)*

CHAPTER 7

Denouement

She is a good ship. She should be used. We must carry on.
—Capt. Max Pruss (August 1937)

For *Los Angeles* and the naval rigid, the late 1930s are sequel to the story. New ideas were explored, old data reexamined, new projects developed, constructive proposals put forth. Airship policy was studied, reviewed, and re-reviewed; policies were recommended by BuAer, seemingly accepted and paid great lip service to. But maneuvering room had narrowed; very little construction was even promised. These were the dark, stranded years of rhetorical flying, of evaporating hope—a depressing time in the wilderness during which, finally, the ZR vanished altogether.

In the meantime, grounded by a wait-and-see policy stance, navy lighter-than-air diminished to an interwar ebb.[1] As 1937 opened, the experimental program exploiting *Los Angeles* was realizing further useful data. For the purposes of conducting haul-up tests, checking superheat cycles, and studying camouflage effects, *L.A.* swung to the mobile mast on 11–12 January. Thereafter, during January, March, July, and in August, *L.A.* was placed on shores with overhead suspensions to support ship's deadweight during gas-cell changes.

Meantime, the next Hull Board convened at Lakehurst had submitted its report. For the period ending 31 December, and except for a few minor units—septic tanks, certain ballast bags, and water-recovery storage tanks—the aircraft's general condition (Rosendahl's words) remained "remarkably sound." This was an officer who refused to quit. "While the present personnel allowance is not sufficiently large to permit maintenance of the LOS ANGELES in the readiness for flight condition which is desirable," the lieutenant commander added, "the ship could be made ready for flight within a few weeks if required and adequate personnel were furnished."[2]

Germany's program sounded a happier note. Two zeppelins were in transoceanic service, a third building. In Washington and on the street, the question was being asked: Why can't we build and operate as the Germans do? There was talk of an American operating company chartering a German-built airship.

Inaugurating the year's service, *Hindenburg*'s schedule included eighteen arrivals. First landing: early on Thursday, 6 May. Slated for quick turnaround, the base would be busy cleaning up, servicing, and loading the visitor for the return crossing. Rather than docking, *"129"* would ride to the mast—taking on fuel, hydrogen, food, stores, freight, express, mail, passengers.

LTA Class X qualified that spring in nonrigids, but no designations resulted. "Our status was somewhat differ-

On the gangway to his duty station, a mechanic fights the slipstream as he crosses to one of Hindenburg's starboard power cars. Note the felt slippers, to combat sparks. (Kevin Pace courtesy Navy Lakehurst Historical Society)

Angeles]. The watch required varying the trim of the ship as well as ballasting as weather conditions changed during the night and day." This operating season, the pair were slated to join *Hindenburg* for its first eastbound leg plus the return. "Needless to say, we were both thrilled to death at the prospect of our trip and by the time of that fateful day we were fully packed, we had obtained our German marks, and were in all respects ready to leave."[3]

Sunrise, 6 May, brought no zeppelin. Nor did it arrive later, due to persistent headwinds. Having loitered over New York, Lehmann gained the field shortly after 1600 but headed to the southeast, to the coast. There the diesels were idled as *Hindenburg* cruised the beach, awaiting instructions. On station, many hundreds had collected, all impatient and—thanks to local, scattered downpours—many soaked. In addition to people meeting passengers or accompanying departing fares, assorted functionaries waited. Agents from American Airlines were on hand, as were U.S. officials from the port of Philadelphia, postal representatives, police, and the gentlemen of the press and radio—plus naval and marine personnel and *their* families.

Fourteen-year-old Owen Tyler, son of Ty Tyler, was one of these. "Thanks to Dad's connections with the personnel of the Zeppelin Company," he remembered, "I'd been hired as a courier to run dispatches between the company's offices in one of the HTA hangars and the ship that night. Complete with uniform and official Zeppelin bicycle cap, I was the envy of all my friends on the station and undoubtedly felt pretty self-important." Not far off, well positioned on the sun-porch roof of their quarters, ent than in 1936," Andrews remembered, "although our [with Lieutenant Antrim] knowledge and experience in rigid airships was still lacking as before." That particular requirement stood; by law, airship men had to log big-ship time. Stand-ins therefore were exploited. The year before, the two had assisted in handling *Hindenburg*, and each had received orders to it as an observer. "From time to time," Andrews continued, more pensive than bitter, "we stood watch in the [moored-out *Los*

Moments before the fateful ignition, 6 May 1937. Captain Max Pruss is in command. (Dean Price courtesy Robert Canace)

his sister bided the arrival. "With weather conditions as 'iffy' as they were that evening, the waiting period seemed interminable but as the *Hindenburg* neared, we were elated that it was coming in so close to where we were standing. Mother and I could clearly see the smiling faces lined up at the windows of the lounge and we and the passengers gaily waved to one another. Then . . ."[4]

Seasoned eyes also were upon her. Here was von Meister, American agent for *Deutsche-Zeppelin Reederei*, on the landing field with Rosie. He noticed the tail dip several times during the approach, a movement he thought unusual. The ship was "obviously tail-heavy." At the mooring-out circle, Lieutenants Antrim and Ben May (with enlisted men) were atop the railroad mast. In charge of the group catching the car, Andrews, as assistant ground handling officer, was directly beneath *Hindenburg*, now hanging cloudlike, headway off. "The bow line was dropped," he recalled of the high landing, "and walked forward to our mooring mast and connected up. The two forward lines and quarter lines were dropped and walked out by their crews—everything was proceeding beautifully. The ship was riding on an even keel, the passengers could be seen waving from the lounge windows on the port side, and the German officers could be seen in the control car—some giving orders to the mast mooring officer and the ground handling officer. Everything was under control. A slight wind was blowing from the east but it was of no consequence."[5]

An instant later, this workmanlike scene changed, utterly. Von Meister saw a flame burst out atop the hull, just ahead of the fin. Through the cover he noted a shine, a glow that progressed downward, deeper in. Abruptly, the entire stern was afire. Louise Tyler recalls an "audible whoosh," many thought there was an explosion. The concussion threw some to the ground, among them Monty Rowe; with two sailors, the aviation metalsmith second class was on the main mooring line. He too had seen a "rosy glow." Rosie's executive officer, Reichelderfer, was in his quarters packing for "the most perfect kind of travel that I know." Then came a dull thud. Looking at his wife, "Uh oh," he murmured. "The force was terrific," Andrews later recalled. "I can remember that those of us who were more or less below the ship when the explosion occurred were somewhat doubled up by the force of the downward explosion."

Hypnotized through an endless, flaming descent, hundreds looked on—transfixed by spectacle and horror. The effect was indelible—or left a blank. "I was standing by at the hangar anxiously waiting for the ship to land so that I could start my duties," Owen Tyler continues, "when the sudden small burst of flame appeared on top of the ship, near the stern. As it became totally engulfed in fire, the shock and indescribable horror of the scene numbed my mind to the extent that I'm unable to remember *anything* of what I did or where I went for the next 4 or 5 hours."

Near the field, mother and sister waited in an agony of concern. "Knowing that Dad was in charge of the landing crew and that Bud was heaven-knew-where as a messenger boy that night, we were completely distraught."[6]

A ghastly night ensued. As it continued to burn and smoke, the hulk was attacked by stunned naval personnel and city fire departments. Responding to the call, ambulances, and doctors from area hospitals had rushed stationward. The Medical Department, with the assistance of civilian doctors, nurses, and volunteer workers, rendered emergency treatment. People requiring hospitaliza-

Hindenburg is consumed. The view is northwest, from the area of Hangar No. 1. The fire proved an irrevocable blow for naval lighter-than-air as well as the transoceanic passenger airship. (H. J. Applegate Collection)

On the landing field, dazed passengers are led to safety. (San Diego Aerospace Museum)

tion were transferred via ambulances to civilian centers. In Lakewood, Ernst Lehmann died the following day. This loss in particular "was widely mourned, along with all the other tragic fatalities," Owen Tyler recalls.

At the circle, a watch was established—to keep the unauthorized clear as fire and rescue workers hunted the remains for the injured and dead. Meantime, a temporary mortuary was set up in the big hangar, until identification could be established. To help manage the consequences of the disaster, two emergency details arrived in reply to Rosendahl's telephone orders: a lieutenant and 125 enlisted men from the U.S. Coast Guard Station, Cape May, and seventy men from the navy yard at Philadelphia. Meanwhile, American Airlines completed its transport of prospective zeppelin fares back to New York. Allen Hagaman, a civilian member of the ground crew, died of third-degree burns at the dispensary.

The station log for the midwatch of 7 May sums up a morning after:

> U.S.S. LOS ANGELES docked on north side of hangar #1. Non-rigid airships *K-1, G-1* and *ZMC-2*, docked on south side of hangar #1. Non-rigid airship *J-4*, deflated in hangar #2. Landplanes T4M-1, O3U-1 and XJW-1 se-

cured in H.T.A. hangar. HINDENBURG wreckage under military guard at number one mooring out circle.[7]

A black mood prevailed. "It was not a night of rest for anyone," Andrews said. "The day after full realization set in as we looked at the still-smoldering hulk lying so dead and inert on the landing field." If no airships were flying, airplane traffic did carry on, some of which was accident related. Late on the 7th, a transport departed for Anacostia bearing the German ambassador and a pair of German military attaches, accompanied by Fulton.

Inquiries were launched. "The Germans assembled a group of their experts [including Eckener] who were assisted as necessary by American naval officers, enlisted personnel and others who might be able to contribute to the investigation. Most of us who assisted in the landing operation were brought before the board and questioned as to where we were at the time of the explosion, what we saw and what we did immediately before and after." When the investigation had concluded, civilian contractors made short work of the remains. "In a remarkably short time," Andrews added, "[the contractor] had cleared all evidence of same from the landing field although there was evidence of the disaster from the blackened earth which clearly outlined the shape of the stricken ship."[8]

Just why the fire ignited is unknowable today. One fact, though, was unassailable: the *Reederei* dared not continue as before. Germany (Berlin announced) would not to fly on hydrogen again. In seconds, the era of airships as commercial transport had come to a shattering end. Losses are an inevitable concomitant of aerial enterprise. Yet this melancholy event has come to symbolize the consequences of lighter-than-air transport. Imprinted on the popular mind, the accident represents a massive public-relations loss; in 1937, it was a near knockout blow for the large airship.

Ever hopeful, airship men persisted. All was not (yet)

DENOUEMENT

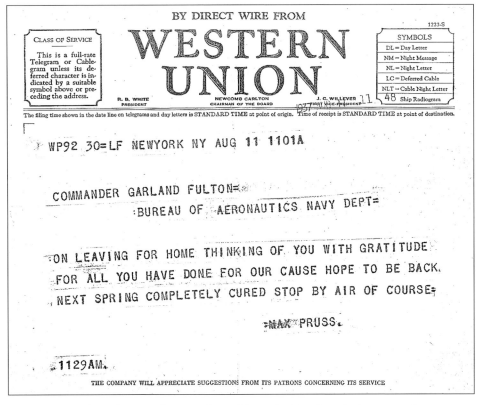

Telegram to Cdr. Fulton on the occasion of Captain Max Pruss's departure for Germany, by sea, following three months' recovery in a New York hospital. Pruss was the most badly hurt. Note the last words, "by air of course." (Langdon H. Fulton)

lost. Commercial hopes rested now upon German access to U.S. helium.[9] Sentiment proved strong for making the gas available for the successor ship, adding weight to moves under way to ease export restrictions and permit foreign sales "in quantities of no military importance." Shocked by the accident, the Military Affairs Committees of both houses invited Dr. Eckener to present his views (von Meister interpreting). The White House seemed kind. Moving to consensus, Congress approved the Helium Act in September—four months after the fire. Airships flying to and from the United States, it specified, must use helium. Despite rising international tension, helium, it seemed, would reach Nazi Germany.

A "revocable permit" for the use of Lakehurst was drawn up, and export of up to twelve million cubic feet—enough to fill LZ-130, the *Graf Zeppelin*—was authorized. This project was carried forward to the point where initial shipments were ready to be ship-loaded. Events elsewhere progressed swiftly. Germany invaded Austria; support vaporized. Mail to congressmen poured in, urging—demanding—that Washington withhold the promised gas. These sentiments reached the secretary of the interior, Harold Ickes. Invoking the "military value" argument, he killed the deal "in a fanfare of publicity and anti-Nazi propaganda," including a public verbal brawl with Rosendahl.[10] Pleas from Eckener, who hoped for continued traffic in a collaborative venture, from Rosendahl, and from others proved futile.

The helium imbroglio drifted toward 1939.

The zeppelin's destruction had piled agony upon desperation; the political nadir was at hand. Navy Department policy toward airships, stalled since loss of *Akron*, had stayed so. Now *Hindenburg* framed the commercial debate—and, for most observers, decided the matter.[11] The DC-3 was flying, transoceanic service by flying boat an operational fact. The airship had no status in law as a common carrier. Further, the naval attitude was discouraging an airship merchant marine; with no expectation as to federal support for commercial construction, private capital remained reluctant. "It is becoming increasingly evident," Fulton growled two months following the fire, "that the lack of a clear definition of naval policy, or attitude, toward airships is not only throttling airship progress within the navy, but is blocking commercial airship progress as well." Other agencies, and Congress, awaited their cue. "Only a few days ago, the Post Office Department indicated reluctance to include airships in its foreign air mail plans because the Navy Department seemed to consider airships as having but little National Defense value."[12] Other interested observers felt similarly:

> The Naval attitude is against airships and is so unfriendly that what little activity is allowed must be carried out virtually without friendly cooperation. The present lighter-than-air naval effort, measured in dollars, or in men, is just one per cent of the total naval aeronautical program. The most that any airship enthusiast asks is to build up this to something like seven or eight per cent—and to use it to reinforce airplane activities.[13]

In January 1938, Ernest J. King was promoted to vice admiral. As Commander Aircraft, Battle Force, he was responsible for all aircraft carriers of the U.S. Fleet. Invited to join King's staff, Fulton declined, telegramming his former boss that "my background and interests in airships together with present airship situation indicate I should stick with it until airships rise or fall." The rudderless drift persisted; as months added, rational planning sat aground, stuck fast. Worse, political support was fast evaporating. "I have a very definite feeling," Adm. William D. Leahy, the CNO, informed Rosendahl, "that with the present insistence on economy in the expenditure of government funds it would be difficult, if not impossible, to induce Congress to appropriate additional funds for the construction of a rigid lighter-than-air ship, even if the navy should make a favorable recommendation." Jacob Klein (now a prospering broker) weighed in. "We have been hasty. Too hasty. The cost has been appalling. Germany has taken things characteristically slowly. We get too big with our ships. The Germans built one craft, increase its size gradually. Never, have they built a bigger Zeppelin by jumping its size too rapidly."[14]

As for a moderate-sized ZR, the proposal was viable—that is, if any of the type were operating. None were. A gleam in the bureau's eye since *Akron*'s loss, the prospective trainer (in various forms) had made repeated rhetorical reappearances. Recommended by a joint congressional committee in mid-1933, it had been endorsed by the bureau in 1933, 1934, and 1936; by the General Board in 1934 and subsequently approved by the secretary; endorsed by the Federal Aviation Commission in 1935; included in the Durand Committee's recommendations in 1936; and recommended by the Business Advisory Council that same year. Yet when funds toward construction were included in early drafts of 1938 estimates, they were eliminated—the secretary referring all aspects of the airship situation to the General Board for consideration.

After conducting hearings on airship policy early in 1937, the General Board submitted recommendations. While acknowledging the naval uses of large airships, especially those carrying airplanes, it did not feel that any should be built "at this time." Nor did the board speculate when any might be built. Instead, it recommended procurement of the long-discussed rigid "for training and development purposes," specifying that it be about three million cubic feet and carry "at least" two airplanes. This was the controversial ZRN of 1938–40. Cost: about three million dollars. Endorsing the feasibility of commercial airships, the General Board considered it in the interest of preparedness, and a duty, to cooperate in the establishment of commercial services, thereby addressing the navy's own problems.

In 1939, the General Board confirmed this view; it resubmitted its report to the secretary of the navy, recommending approval.[15] Nonetheless, the absence of a clear, robust policy topside was both strangling the naval program and blocking commercial progress. The situation continued to decay, beyond control, beyond even understanding. "The attitude of the 'powers that be' towards naval aviation in general, and lighter-than-air in particular," King deplored, "continues to be beyond my comprehension."[16]

All operations had retreated to Lakehurst. In the fall of 1937, Class XI reported for the regular course. Among its half-dozen students: Ensigns Alfred L. Cope (No. 6752), M. Henry Eppes (No. 6753), and George E. Pierce (No. 6748).[17] With *Hindenburg* mere scrap and Washington refusing helium, the opportunity for ZR time had vanished. *Los Angeles* now held secondary status, even for training. "We were just exposed to it as the only example of a rigid airship that the Navy had at that time," Eppes told the writer. "It wasn't an integral part of our training course." As then–training officer Howard Coulter remarked of the retailored curriculum, "The primary aim [in 1937–38] was the qualification of non-rigid airship pilots."

Under the immediate supervision of the flight officer, tutelage aloft consisted (as before) of familiarization hops in the old kite and free balloons, and in blimps. Flight instruction was emphasized in the early fall and spring, ground school during the winter months—or whenever inclement conditions obtained. "We had the benefit," Coulter added, "of several very knowledgeable instructors with the ability to impart their knowledge to their students, e.g., Cal Bolster, George Whittle, Connie Knox and others, including several experienced Chief Petty Officers in the practical areas."[18]

In the barn, students were taken through the timeworn *Los Angeles*, to acquaint them firsthand with the type. When *L.A.* was moored out, mast-watch duty was logged.

While in training, Eppes accumulated 32.9 hours in free balloons during nine sorties—the longest a fourteen-hour overnight solo. (The kite requirement had been dropped, the type granted mere mention.) As for powered craft, the ensign's totals reached sixty-eight hours with *J-4,* fifty-five in *ZMC-2,* sixty-two hours with *K-1,* seventy-one in that army hand-me-down *TC-14,* and

DENOUEMENT

The ex-Army TC-14 on its first trial following re-erection, 11 February 1938. Given weak support from Congress and its own high command, the Army had abandoned lighter-than-air. Dockside, L.A. is secured in the south berth. (U.S. Navy Photograph courtesy Lieut. Cdr. Leonard E. Schellberg)

twenty-nine hours in *L-1*. *J-4*'s open car proved especially memorable.[19] The last of her kind, *J-4* was generally conceded by pilots to have had the best flying characteristics of any nonrigid then in inventory.

As for naval-aviator qualification, the navy had no ZRs in flying commission or abuilding. For its part, the *Reederei* was lofting no passengers. Yet the 150-hour requirement remained, with the result, already noted, that officers completing the course could not secure designation. Detached in November 1938, Eppes (for one) held orders to the surface fleet, to a nonflying billet.[20]

Tests on *Los Angeles* continued. The California Institute of Technology and Ohio's Guggenheim Airship Institute held navy contracts for studies of

Model of ZRS-4 fitted with Mark II fins suspended in the vertical wind tunnel of the Daniel Guggenheim Airship Institute in Akron, Ohio. Using its 20-foot wind tunnel at Langley, NACA tests in 1936 had confirmed earlier (1932) results: the presence of "very large pressures" near the leading edge of airship fins. (University of Akron Archives courtesy Eric Brothers)

Model of USS Akron *under test in the water channel at the Guggenheim Airship Institute, late nineteen-thirties. Note the Goodyear-Zeppelin type girders in the model carriage. Established in the salad days of the large airship, the Akron Institute was to close in 1949. (University of Akron Archives courtesy Eric Brothers)*

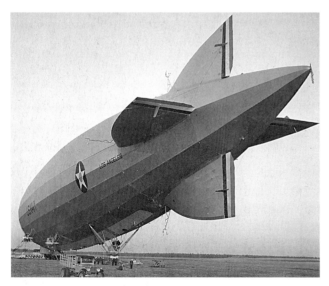

L.A. on the field for "Guggenheim gust experiments," 8 November 1937. Obtaining load measurements on a full-scale airship in free-flight had proven difficult. And though many wind-tunnel tests of scale models had been made, the results obtained were in some cases "of questionable value." Note the "sand truck," used to shunt watch personnel and ballast to the mast site. (Warren M. Anderson)

gust structure. In this connection, BuAer thought it desirable to include measurements of the gust forces on a full-scale airship, with *ZR-3* moored out. The contract was therefore amended to cover such work. Ohio would furnish advisory assistance and special equipment as well as prepare the required reports; the station skipper would install certain instruments and devices and "cooperate fully and to make the LOS ANGELES available as required."[21]

Among the essential equipment for the "Guggenheim gust experiments" were four outriggers mounted at frames 23 and 25. Located about fifteen feet out from the hull, one set was installed forward of the fin's leading edge. These supported recording wind direction and velocity instruments, wired to a central recording station in the keel, amidships, at frame 96.25. (All framing was numbered, recall, according to metric distance forward or aft of the rudder post, which was frame 5.) A tall, movable "gust tower" was incorporated into the tests, along with a special instrument for recording the airship's (momentary) position—the latter mounted on the stern ride-out car. All sensors were wired to an oscillograph set in the keelway.

The tests were slated to commence late in October, but their number and duration were as yet uncertain, dependent on weather.[22] Five to fifteen test runs—with a total of ten to thirty hours of records—had been suggested. Desired conditions: gusty winds and wind shifts. ("No records of value can be obtained with smooth air conditions, so in so far as this study is concerned the LOS ANGELES need not be moored out for such periods.") As for test procedure, the gust tower would be relocated by the watch for each run, its placement depending on the anticipated range of wind directions during any run—such information to be furnished by Aerology.

This particular project proved but fleeting. On the morning of 18 November 1937, *ZR-3* was moored out for gust-structure measurements but docked late in the afternoon that same day. The ship never again left her hangar berth.

As it had since the world war, the Navy Department—the custodian of big-ship development—held the key to both commercial and naval applications. Washington, however, was doing very little about it. By the early months of 1938, President Roosevelt, a nonbeliever in the rigid type anyway, could write, "The Navy Department does not recommend at the present time the construction of rigid airships. This decision is based on past experience with these airships, and on the fact that, in the opinion of the Navy Department, their scouting missions can be better accomplished at much less cost with long range flying boats." In other words, it was too late. Despite genuine progress, the strategic value of ZRs as fleet scouts, or aerial carriers, was unproved. As for commercial prospects, "The Navy Department feels further that the rigid airship is better adapted for commercial than for military uses, and for this reason considers their future construction and development to be proper functions of commercial interests."[23]

The bright aeronautical dawn beheld in 1924 had, in the span of fourteen years, faded to dusk.

It should not be imagined that during 1938–39 all airship men were reflexively supportive of the ZRN. With its limited performance, a new trainer seemed unlikely to convince higher naval authority of the value of big airships for war.

> We are building a training ship without having anything to train for; we are building an experimental ship which can only elaborate on material experiments and operational technique already proved. . . . Therefore, if administratively and diplomatically possible, I recommend that the Navy hold out for not less than two large airplane carrying scouting or carrier rigid airships, or else abandon rigid airships, and use all available funds for a comprehensive non-rigid program.[24]

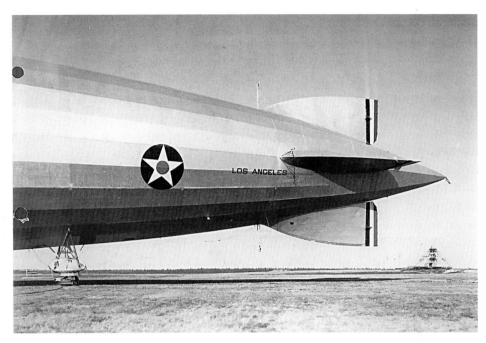

Stern of ZR-3 *rigged for gust-structure experiments. The outriggers support wind-direction and velocity instruments wired to a central station in the keel, amidships. Docked that afternoon,* ZR-3 *never again left her hangar berth. (National Archives)*

Nose-view of Los Angeles *at the rail mast. Note the boom and anemometer secured from ship's bow. (Mrs. Charles M. Ruth)*

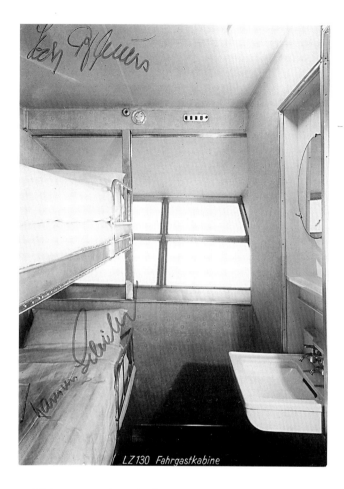

This same spring, the House Committee on Naval Affairs acted on the General Board's recommendations by authorizing the trainer in the first Vinson Naval Expansion Act. A contract, it was anticipated, would be placed, and actual work could begin prior to December. Construction time: approximately twenty-four months. While BuAer was preparing specifications, however, President Roosevelt took a sudden interest in its size and stipulated that it could be no longer than 325 feet. This was exactly half the length the bureau wanted. A 325-foot hull could not enclose a volume of much more than one million cubic feet; nor could it carry even one airplane. The ensuing controversy over size (as Smith writes) "would smother the ZRN in its cradle and carry the rigid airship to its political vanishing point in the Navy."

Meanwhile, however, Congress reluctantly appropriated initial funds, and BuAer revised its specifications to the dwarf size. In October 1938, the department called for designs and bids. Only the Metalclad Airship Corporation (ever active) and Goodyear-Zeppelin bid. The decision was made for Goodyear and its alternate bid—a fabric ship, the proven type of construction. For its part, the Navy Department did not especially want an airship of *any* size. Nonetheless, Assistant Secretary of the Navy Charles Edison sought the president's consent for the larger design. The additional expenditure, Edison argued, held a number of technical advantages—speed, military load, range.[25] Annoyed by what he thought was an attempt to evade his decision for a million-cubic-foot ship, Franklin Roosevelt directed that all bids be rejected. "Thereupon the Senate Appropriation Committee (largely Senator [James F.] Byrnes' influence) threw out the funds that might have been used for re-advertising for new bids after the President had been straightened out." In July, Roosevelt—reversing himself—agreed to reopen the matter of bids, and the item of three hundred thousand dollars was accepted "all along the line until it ran into Senator Byrnes who knifed it and refused to consent to its going back in conference. It all happened in the last minute rush."[26]

Net result: a year or more lost. For a complex of reasons, then, no contract was let before the appropriation expired, although the authorization remained in force.

In Germany, meantime, the *LZ-130* stood complete. Politically, the international weather had decidedly soured. "The new ship," Rosendahl reported from abroad, "shows many weight-saving details over previous ships and some marked improvements. It seems a shame to delay the airship art so needlessly by keeping this ship laid up."[27] The "130" would fly—with hydrogen. On 14 September, Eckener commanding, the first trial was logged. Equipped with four Daimler-Benz diesel engines, *LZ-130* ship could

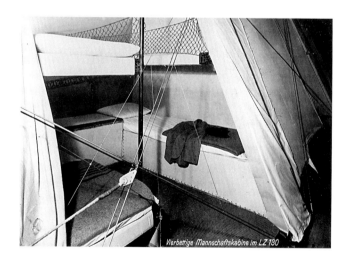

Passenger cabin (top) and a crew space off the keel of LZ-130, sister ship of Hindenburg. Its final months beset by war, the "130" never lofted a paying passenger. (Luftschiffbau-Zeppelin)

make 82.6 miles per hour. Inside the hull, nineteen double-berth cabins for thirty-eight passengers graced its helium version. (Hydrogen inflation permitted up to seventy.) The German water-recovery equipment was to prove as troublesome as the U.S. Navy's own apparatus. "The main difficulty on long journeys was that the coolers would get clogged up with soot. We would notice this by an increase in exhaust compression pressure which reduced engine performance and led to increased fuel consumption."[28]

This last-ever zeppelin never lofted a paying fare. In Franklin Roosevelt's America, anti-Nazi sentiment had all but obliterated any fresh attempt to supply helium. "As for any purely American plan of commercial airships," von Meister told Fulton, "I am not very hopeful. The job of organizing the necessary political support to get the required financial aid to start these airships will be difficult to obtain unless the group that does it has plenty of money to spend on propaganda and education. The rigid airship is so closely allied with Germany, in the public mind, that anything German at the present time seems to be taboo with the politicians."[29]

Operating from the new Frankfurt-Main terminal, *LZ-130* logged thirty "circus" missions in twelve months. At least nine, in which the ship was fitted with electronic sensing gear, were for the Luftwaffe Department of Signals. Undertaken at the request of the Reich Air Ministry, these spy flights were for radio interception, then for probing British (and Russian) defensive radar networks.

Too soon, the Axis powers were to cover the world with horror.

Back in New Jersey, the results of hull inspections for *ZR-3* were dutifully routed to Washington. Despite general surface corrosion on wiring, cables, and parts, and dirt and dust throughout, the basic hull structure was still sound. The cutting of samples for examination and testing had continued. These results, the opinions of independent experts, and other evidence confirm that *Los Angeles* remained in remarkable condition. It deserves note that the burden of maintenance and upkeep—six years idle—had fallen to the tiny staff at Lakehurst, an outpost short of funds even in more freewheeling times. Frustrated by an unsympathetic Navy Department and a miserly Congress, the aircraft's general soundness represented a near-heroic achievement.

Nonetheless, the situation was "very dark" for big airships; the greatest moments were astern. In a sigh of acceptance to his friend King, Fulton wrote of "renewed talk of washing out the whole business.... Of course, if this is what considered naval policy demands, then we had better amend war plans and get out of the airship business."[30]

A week later, *Los Angeles*'s hull board report for the period ending 30 September was submitted. It was the last ever prepared. Through years of wear and tear, equipment *had* now deteriorated, and certain replacements were in order, such as the emergency and ordinary ballast bags.[31] As for the gas cells (save for number thirteen, retailored *Akron-Macon* spares had been fitted), all were in fair to poor condition. Numerous lattices had been broken during work and inspections. Further, evidence of surface corrosion was evident throughout the hull structure. Nonetheless, insofar as the board could determine, subject to its inability to examine certain portions covered by ship's cells, *ZR-3* was basically sound. By this late date, probably more than a thousand pieces of structure had been cut free and tested in Washington; the last specimens had been forwarded in April 1937. Upon its analysis, BuStandards had had to conclude in part: "The conclusion to be drawn from these tests is: Whatever increase in corrosion or change in strength may have taken place in the last five years is not great enough to be shown by the tests."[32]

Since decommissioning, Congress had expressed considerable interest in *ZR-3*; more than a few inquiries had reached the Navy Department as to the ship's condition and general airworthiness. In particular, the changed situation brought about by the loss of *Macon* generated pressure upon the department to reactivate. All were politely dismissed. The 1933 material inspection had shown *Los Angeles* unfit for flying service, and that was good enough for the secretary. Nor had the CNO been willing to accept the responsibility attached to operating an aircraft of such advanced age for its type.

Department responses from this period have a decidedly perfunctory quality; one senses an unstated yet powerful desire to be rid of the ZR platform as an instrument of naval warfare. If it was not evident in that long-ago, it is clear today, in hindsight, that an unpleasant matter was being tolerated until at last it could be disposed of.

In March 1937, for example, Representative Carl Vinson, chairman of the Committee on Naval Affairs, proposed to authorize the immediate procurement of a rigid of moderate size to replace *L.A.*, for use in training personnel. Swanson's reply is instructive with respect to his contempt toward large airships. Swanson quashed the notion, explaining that "as its future policy regarding rigid airships is now under consideration, the Navy Department prefers

Christmas 1938, NAS Lakehurst. (courtesy David Collier)

not to make any recommendations on the bills HR 5529 and HR 5530 at this time." Nor did he speculate when it would be propitious to do so. Unfortunately, naval policy on the type had been awaiting formulation for some years; it would continue to languish until, at last, time ran out for *Los Angeles* and the rigid airship in the U.S. Navy.

By the final weeks of 1938, the test program having trailed off, *Los Angeles* had lain in berth for months. She had aged gracefully, but her condition now was largely immaterial. Struggling for relevance, the ZR program was in a desperate state—adrift, starved for support, moribund, shuttered up at Lakehurst. War breezes were blowing. As 1939 dawned, few persons could imagine any military value in the rigid airship in general, let alone an obsolete, hangar-bound design conceived in another era. Commercial operations had terminated—a consequence of both *Hindenburg's* fatal landing and reaction to the Berlin regime. In sum, an aeronautical era was nearing its end. On 17 December, the fourth material inspection of *ZR-3* was ordered by the CNO; for the next two weeks, the bureau assembled information bearing on material condition for the Board of Inspection and Survey.

The board convened on 6 January 1939.[33]

A detailed visual inspection was made, after which a preliminary report was forwarded to the CNO late in January. To assess conditions more conclusively, however, airframe testing and structural sampling were ordered. Since these required preliminary calculations as well as special equipment, it was decided to postpone the taking of samples until after the test program. On 18 January, two thousand dollars was authorized for this work. In the meantime, pending laboratory results, the board formed the opinion that *Los Angeles* would require extensive overhaul for flight. Upon completion of the full program of tests, the board would summarize her material condition, estimate the costs of reconditioning, then render its recommendations as to disposition.

Garland Fulton had worked actively to help realize acceptance of large airships in both the navy and in commerce. This same month, in remarks before the Institute for Aeronautical Science, Fulton addressed himself to the overall situation. His frustration is plain.

> I know there is a tendency in some quarters to say that the airship was a nice old fellow but he long ago reached the limit of his development and has been outmoded by the airplane. Anyone who has carefully examined the situation knows this is far from the truth. Definite improvements are possible in airships through application of aerodynamics; utilization of improved materials; more refined structural analysis; and in the installation of very much lighter and more efficient power plants.... Many of the fundamentals of aeronautical science were developed around airships and I feel it would be a retrograde step, much to be deplored, if the lighter-than-air branch is allowed to become the "forgotten man" of aeronautics.

As for the commercial sphere, the *LZ-130* (Fulton continued) was flying with hydrogen "and is now awaiting a solution to the problem of obtaining helium or a decision to resume commercial operations with hydrogen. Construction of another airship *LZ-131*, somewhat larger, is being proceeded with."[34]

Nursing a very faded hope, Rosendahl pleaded for a stay of execution. "As to the disposition of the LA," he wrote Fulton, "for heaven's sake don't scrap her yet. Even if she never flies, take out engines, etc., and put her at the mast with enough attention merely to keep her ballasted down heavy, and let her ride for a year if she will. Recording instruments should be aboard to measure loads, etc., but let's don't break up a full scale model in the shed—yet, anyhow."[35]

On 10 February, the first of four "shear tests" was applied to the hull. The premise here: that structural soundness was mainly dependent upon the condition of girders and wires, especially wire tensions. Accordingly, the tests consisted essentially of applying stress on two main frames, to note how tensions on the most severely stressed shear wires adjusted themselves. Why the tests? Irregular slack had been found in the wiring, raising concern as to possible existence of irregular shear forces upon the hull—and consequent excessive stress concentrations. *L.A.* therefore was put into equilibrium and horizontal trim, and cradles were wheeled beneath the wing cars. Tethering lines to the deck were secured amidships, surge lines installed fore and aft. So held, *ZR-3* stood supported along the center one-third of its hull length; forward of frame 130 and aft of frame 70, the structure would not be subject to any external forces save those intentionally imposed. For access to girder work and the recording shear-wire tensions, outer-cover panels near cells three and twelve were removed.

The test commenced.

Maximum downward loads of five thousand pounds were applied at the tops of frames 25 and 175, counterbalanced by paired upward forces of 2,500 pounds applied at frames 40 and 55, and at frames 145 and 160. The test loads were imposed in 25 percent increments, and the corresponding wire tensions recorded. Tensions were again measured during the unloading cycle under both the 50 percent and no-load conditions. (All the recorded wire tensions were later graphed for the final report.) Five days following the initial test, a second vertical shear test was conducted. This time the loads were applied so as to change the sign of the shear. A third test was held on 28 February, the last on 6 March—in both cases the loads being applied horizontally to the hull.

Results? The shear wiring system was sound. The tensions produced during the initial and final loads proved to be "remarkably near alike," demonstrating to bureau engineers that the hull structure had behaved elastically and had acquired no permanent "set." Nevertheless, if *Los Angeles* were to be reconditioned, complete replacement by a modern system of tight wiring was recommended, in order to increase general strength.

On 13 March, the bulkhead wires were tested. The fullness in cell seven was changed in progressive steps from 100 percent down to 10 percent, then back to complete fullness. (During this procedure, the adjacent cells were inflated to about 90 percent.) As the test proceeded, the behavior of the bulkhead wires was checked by measuring tensions in the wires at frames 85 and 100—the bulkheads that, between them, formed the bay for number seven. This particular trial was a severe one,

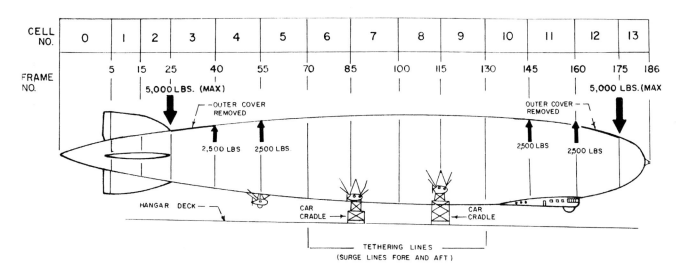

In No. 1 hangar, five tests were conducted as part of Los Angeles's last material inspection. The first, a vertical shear test, was made on 10 February 1939. Incremental loads up to 5,000 lbs. were imposed then removed while wire tensions were measured near Cell Nos. 3 and 12.

Civilian riggers take shear-wire tensions, 15 February 1939. Outer cover panels have been peeled from Frames 25 to 40 and longitudinals 2 to 4 (port side), to expose the structure. Note Cell No. 3 and the gas-cell netting. The white strip is a calibrated fullness chart—read by patrolling riggers at the point where the cell bottom "left" the tape. (National Archives)

and it broke a wire. Nonetheless, the result was deemed satisfactory.

Removal of the panels from frame 25 to 40 and from 160 to 175 had afforded an opportunity to examine that area for the first time in years. Not surprisingly, dust and dirt had accumulated in some of the girder channels and joints; when brushed away, the surface beneath was found to be slightly corroded. Nonetheless, the aluminum alloy appeared to be in substantially the same condition as the lower part of the hull, or somewhat better. The shear wires were in fair condition, with slight corrosion at some of the terminals. No shear or gas-cell netting wire failures were observed. However, the wires were rusted, causing pronounced rust streaks on the gas-cell fabric.

In-dock tests complete, structural sampling commenced. During April, more than twenty-one pounds of specimens were removed, tagged, and forwarded to Bu-Standards, each carefully selected at random from the hull structure as well as from high-interest locations, such as corroded spots.[36]

That May, the examination and testing were completed and written up. Though the aluminum-alloy tubing from the main water-ballast line was found to be badly corroded, the steel wire, brass tubing, and channels and lattices were in satisfactory condition. Indeed, as far as BuAer could determine, its data failed to reveal *any* important reduction in strength due to corrosion since the airship's 1932 decommissioning. Further, a review of test results since 1926 showed that "so far as deterioration of the material, whatever may have taken place has not been sufficient to lessen appreciably the strength of the girders in the ship."

Late in May, the final report was submitted. Based on its inspection, tests, and deliberations, the board pronounced

Taking wire tensions during tests Nos. 3 and 4, topside, 7 March. The hull structure has been exposed from Frames 160 to 175. (National Archives)

ZR-3 to be in substantially sound condition. Rigid airships are large and complex structures, however; although viable, the aircraft was hardly flight ready. For an estimated cost of from $420,000 to $742,000, the board found, *Los Angeles* could be restored to a condition approximating its original strength and airworthiness, and at least equal to its condition when removed from service. Such bandaging-up was beyond the capacity of Lakehurst, so a short, fair-weather flight to "a contractor's plant" (in Akron) was proposed, for modernization and recommissioning.[37] Though not a modern design, *ZR-3* had value for training and experimentation, as was again pointed out: if operated conservatively after major overhaul (with special regard to bad weather), she possessed a real, if limited, usefulness for naval service.

Having thus assessed condition and possible utility, the report went to the heart of *Los Angeles*' predicament in the navy in the waning years of the 1930s: "The desirability or necessity for recommissioning the LOS ANGE-LES must be predicated upon the Navy policy or program with reference to acquiring or operating, at some future time, a rigid airship of like size or larger." As we have seen, a "policy or program" awaited definition. In its recommendations, the board concluded that a major overhaul would put *ZR-3* in condition for limited service, primarily as a training airship—and went on:

> Whether or not the overhaul is warranted depends not alone on material questions, but also on matters of policy with reference to the future of airships of this class. In view of the age of the LOS ANGELES and the short life to be expected from a reconditioned airship of obsolescent design and in view of the high costs involved, amounting to perhaps 20% of the cost of a new airship, the Board considers such an extensive overhaul as herein outlined should not be undertaken.
>
> If the LOS ANGELES is not to be recommissioned, it is recommended that she be made available for such tests not involving flight as may be prescribed by the Bureau of

Station visitors pose alongside Los Angeles, *19 July 1939. The end now was certain: there would be no reconditioning. In these final months, the public was permitted entry onto her keel corridor. (U.S. Navy Photograph courtesy Mrs. Charles M. Ruth)*

Aeronautics and that upon completion of such tests, the airship and usable parts thereof be scrapped.[38]

Rear Adm. John H. Towers had just become chief of the bureau (and received appointment to the NACA). One of the earliest naval fliers, a recent commander of carrier *Saratoga* and no friend of Rosendahl, Towers coldly dismissed the issue, informing Fulton early in June that there would be "definitely no reconditioning." The admiral *was* open to suggestions for destructive tests on the airframe. The end had become a dead certainty.

Meanwhile, off Britain's east coast, *LZ-130*, fitted with special gear, listened for radar transmissions. "The attempt failed," Winston Churchill records, "but had her listening equipment been working properly, the *Graf Zeppelin* ought certainly to have been able to carry back to Germany the information that we had radar, for our radar stations were not only operating at the time, but also detected her movements and divined her intention."[39] On 20 August, the *130* sortied briefly, overflying Essen and Mulheim with mail—the last flight by any zeppelin, the last anywhere by rigid airship. The Soviet-German nonaggression pact was signed four days later; abruptly, a flight scheduled for the 26th was canceled. Thought to have no military utility, the airship was grounded. On 1 September, German troops poured across the Polish frontier.

That portentous September, the CNO forwarded the board's report to the secretary of the navy with the endorsement that it be approved. The secretary did so, that very day. On 3 October, via memorandum, the CNO in turn approved the recommendation of the Board of Inspection and Survey not to overhaul. Instead, *Los Angeles* would be made available for tests, then scrapped.

Three weeks later, on 23 October, Towers so advised Lakehurst's commanding officer. Elaborate structural tests would not be made, with the possible exception of "one or two simple air pressure tests" conducted without delaying the scrapping operation. In the meantime, Washington was investigating methods of disassembly and disposal. BuAer requested estimates of time and costs as to what parts of the dismantling process could be handled by the base without interfering with other on-board activities. As an alternative, a contract with an outside wrecking firm was suggested, with Lakehurst supervising the work. Pending a decision as to method, however, preliminary dismantling could begin at the discretion of the air-station commanding officer.[40]

On 24 October 1939, USS *Los Angeles* was stricken from the Navy List.

Disassembly began immediately. The air station's skipper, Cdr. Jesse L. Kenworthy, Jr., was anxious to complete the project—and with no fanfare. (Tests concluded, pressmen were kept uninformed till early December.) The "approaching need" for additional hangar space, he told Washington, and the impending cold weather made it desirable "to proceed with dispatch."[41] By way of preparation, *L.A.* had been supported in the north berth, under the overhead cranes, bow to the west. For the last time, her wooden cradles were wedged into place beneath the main car and power cars. The main rings also were shored from the deck, using cedar poles along each side of the keelway

DENOUEMENT

for the length of the ship and beneath the keel apex girder. Tethering lines were secured amidships by civilian riggers, who also attached overhead suspensions to each main and intermediate frame. Each cable was then individually tensioned.

So held, *ZR-3* stood rigidly supported, as a unit, from above and from below.

Dismantling commenced with the out-fittings. In less than seven days the tanks, ballast bags, lines, and pumps from the fuel and water-ballast systems were unshipped, or very nearly. So too were crew quarters, cargo racks, the major and minor controls, oil tanks, and flight instruments.[42] Of the fifty-two aluminum fuel tanks, eleven were German originals. Several of these were cut into strips and shaped into picture frames—for private distribution as souvenirs. Personal initiative was operating here; though not officially sanctioned, a sizeable number of girders, annunciator bells, and miscellaneous items were retained by former crew and other personnel attending the cut-down. A framed photograph of Grace Coolidge, which had hung in the control car since November 1924, also was saved.[43]

Within four days about 40 percent of the outer cover was stripped from the hull. It was in poor condition, and little time was wasted in removing the sealing strips between the individual panels of fabric or in loosening the flax cord lacing them on. Instead, cord and fabric were unceremoniously cut from the framework and thrown to the deck. All fourteen cells were in place, each deflated to 50 percent or less. By the 27th, a mere four days following notification of the decision to scrap, *ZR-3* was an eviscerated hulk—dismantled to the point where final disassembly of the hull structure itself could commence.

As Kenworthy saw it, the station command was fully

The end. Stricken on 24 October, ZR-3 awaits final disassembly, November 1939. The outfittings and outer cover have been stripped from the hull, the cars partially dismantled, all but two gas cells removed. The disassembly—unique in U.S. aeronautics—marked the United States Navy's explicit rejection of the rigid-airship scout. (Official U.S. Navy Photograph, Karl Arnstein Collection/University of Akron Archives courtesy Eric Brothers)

capable of concluding the project on its own; Washington was so advised on 27 October. Bringing in a contractor would involve preparing proposals and advertising for bids, by which time the station itself could conclude the work. Further, Lakehurst's civilian personnel were familiar with the hangar equipment (cranes, trolleys, and so on). The station, he estimated, could complete the disassembly and remove all scrap in less than two months. Cost for labor and equipment was put at six thousand dollars. These estimates made no allowances for the destruction tests recommended by the board and mentioned by the bureau. At this point, the base was uncertain as to the nature of these tests and the work involved in their execution. Anyway, the additional time and expense were deemed undesirable.

Dismantling work near the control car. Here a sailor attacks the bow hatch with cutters; behind lies ship's climbing shaft. (U.S. Navy Photograph, Karl Arnstein Collection/University of Akron Archives courtesy Eric Brothers)

On the 27th, Lakehurst requested a project order for final tests, for six thousand dollars—and asked that they be made as simple as possible.[44]

In early November BuAer approved. A project order for nine thousand was issued on the 9th, to cover the estimated costs, including air-pressure destruction tests (outlined later). It was Washington's desire that the disassembly work proceed without delay.

So large a structure was going to produce a great deal of scrap and, in turn, present special disposal problems. On the 4th, Lakehurst had been advised to contact firms that might be interested in bidding on the metal, thus ensuring its preparation "in a manner acceptable to the trade." The concern that had bid successfully on the remains of *Hindenburg* two and a half years before was suggested as a likely candidate. Further, BuAer wanted steps taken to ensure the complete disposal of all scrap materials, without its diversion to "unauthorized purposes." Apparently, this embraced historical purposes as well—very little was saved or found its way into national museums.[45]

As November ebbed, the destruction tests were finalized and preparations effected for a final service to the navy. In the meantime, the hulk was deflated, and dismantling of the control car and engine cars was begun. Within the hull, the gel latex cells were purged into the hangar's exhaust line and thence to the Helium Plant, for repurification and storage. Then, one by one, thirteen cells were disconnected at the handling patches and dropped to the keel, where each was pulled out for discarding. The remaining cover was stripped off. The helium exhaust trunks and hoods topside as well as the climbing shaft at frame 145 also were removed at this time.

Los Angeles once again infiltrated the national press. "LOS ANGELES DISMANTLED; TESTS MAY SETTLE DIRIGIBLES' FUTURE," one New York sheet headlined that December. "In the great hangar," the *Tribune* announced, "the fabric covering already has been removed from the sturdy old airship. . . . Workmen are beginning to take apart the duralumin framework and within a few days the dirigible will exist only as piles of neatly stacked girders." In public remarks, Kenworthy explained his view. "The *L.A.* was a grand ship, but today she is as much out of date as a 1923 automobile would be on the highways." He added, "She always has been an experimental ship and she is an experiment today—an experiment to destruction." The scrapping did not mean the navy had lost faith in rigids, he was sure. "On the contrary, according to high officers here," one writer told readers, "it is proof that every effort is being made to amass the most accurate and complete information in the world about them as a background for future activity in the dirigible field. . . . Although lighter-than-air officers are not counting their dirigibles before they are constructed, it is no secret that they are hopeful of soon getting at least one new one of advanced design. They want an airship of 10,000,000 cubic feet capacity; this would be capable of serving as a carrier for eight or ten bombing planes and could cruise 10,000 miles without refueling."[46]

On the 4th, the Bureau of Supplies and Accounts advised regarding the prospective scrap metal. The station was to segregate the remains into lots that, in the CO's judgment, would realize the best prices when finally offered. A survey covering all scrap materials resulting from the project was requested.

On 4 and 5 December, the final strength tests were held—a last performance. On the 4th, the tail's cruciform girders were assessed. With no shores aft of frame 40, the test load—two tons of weights on the horizontal fins, at frame 15—was applied uniformly in 25 percent increments, the overhead suspensions being taken up by equivalent amounts at each step. Under full load, cell one (fin area) was then inflated with air under pressure until, at 1.6 inches of water, a major failure occurred—buckling of the lower horizontal cruciform girder in frame 15. Thirty or more "distinct sounds of failure" of the gas-cell netting and netting wires were audible before this buckling. (No shear wires failed, and the cell remained intact.) Next day, amidships, the 297,357-cubic-foot number-seven cell—between frames 85 and 100—was overinflated with air under pressure. The test continued until, at one inch of water, the bulkhead wire in frame 100 failed, followed almost immediately by girder failures. Three days later, a report of the tests—all carefully recorded—was dispatched to Washington.[47]

By mid-month, the control car (stripped and partially dismantled) stood detached from the hull. The mooring fittings, winches, and related gear at its bow also lay on the deck. One by one, the twelve-cylinder Maybachs had been unshipped. The sealing strips were peeled from the fabric panels of each car and the lacing undone, after which the sides were lifted off. Unbolted from their beds, the power plants were lifted onto a wheeled engine stand and shunted to the shops.[48] There each received preservation attention. Draped with moisture-proof paper, the engines were crated for shipment to facilities at Norfolk. David Patzig, Aviation Chief Machinist's Mate, supervised this work, after which he and his crew were rotated

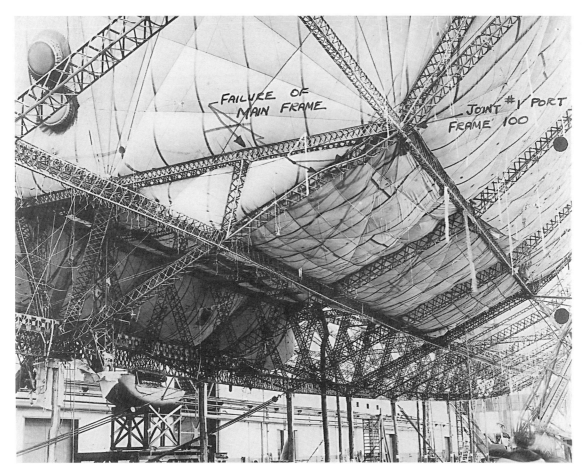

Destruction tests, 5 December. Cell No. 7 was deliberately over-inflated with air until a major failure—in this instance, a main-ring failure at Frame 100. Note the pole shoring beneath the keel girder. (National Archives)

Ship's framework shell, 15 December 1939. (Acme Photo courtesy Lieut. Cdr. Leonard E. Schellberg)

Dismantling of the hull structure. Note the work platforms from the overhead crane (top); working from movable staging, the dismantling crew had full access to the hull circumference. Cutting began topside, at the apex girder, then advanced down and around near the main rings. Here workmen sever girders near the keel. (Lieut. Cdr. Leonard E. Schellberg)

to the deck area, to assist in the ongoing disassembly. The five power cars were cut loose from the airframe; cell seven was cut from its bay and withdrawn. By the third week of December, *Los Angeles*—now but an empty framework shell—was ready for final dismemberment.

Considerable thought had been given to this final phase. The disassembling was without precedent—the requirements were novel, even to experienced hands—so various ways and means were investigated. The details were worked through—and approved. Docked from below on shores and cradles, and from the overhead, the full structure was supported as unit. Yet each section of the great hull—from main ring to main ring—was independently held as well. So supported, the full framework could be progressively cut away without affecting the stability of the whole. Once severed, each hull section could then be lowered to the deck by crane for final disassembly. Though not regularly used, the two platform cranes would be exploited as staging for the men attacking the hull.

Chief Boatswain's Mate (CBM) Bill Buckley was assigned responsibility for the concluding work. Under his direction, a dismantling crew was mustered from trainees at the Airship Training School, augmented by such seasoned hands as Bill Baker, Patzig, and Ducky Ward. Civilian employees rounded out the cut-down force. In all, a team of about two dozen was directly involved, organized into groups.

Some men rotated between the deck and the work topside—that is, on the crane platforms.

For the actual cutting, three-foot bolt cutters were chosen. Baker was a wizard in any workshop; his talents challenged, he set to modifying perhaps a half-dozen tools for snipping aluminum girders. His solution: cutters fitted with unusually long blades. So modified, the jaws proved formidable, easily capable of severing the thin alloy. Although axes, wire cutters, and ordinary bolt cutters also saw use, the modified cutters were employed throughout and very much accelerated this phase of the work.

Around midmonth the cutting-down began. Commencing at the stern, the control surfaces were detached and lowered away. The men then set to work on the framework hull. Topside, the two cranes were moved into position, one over each flank of the open circumference. Hanging beneath the operator, two or three men worked from movable staging hung from the work platform of each crane. Monorails along the undersides provided lateral mobility, while, simultaneously, the entire assembly could be repositioned both vertically as well as horizontally, anywhere along the 658-foot cylinder. Working from this staging and from beneath it, the men had ready access to virtually the entire framework.

Cutting began at longitudinal six—the centerline girder at the apex of the hull circumference. Severing number six near a main ring, the men top-

Advancing from aft forward, the hull-structure surrendered in about fourteen working days. (Lieut. Cdr. Leonard E. Schellberg)

A view from the deck during cut-down. Here two men work from the keel walkway. (Wide World Photo courtesy Lieut. Cdr. Leonard E. Schellberg)

A cutting crew at work, high off the hangar deck. (University of Akron Archives courtesy Eric Brothers)

Final cutting was accomplished on the deck. Here two intermediate rings and the connecting longitudinals are sliced into more or less uniform lengths. (Lieut. Cdr. Leonard E. Schellberg)

side followed that ring down, cutting through each longitudinal in turn, advancing down and around both flanks of the circumference toward the keel, at the bottom centerline of the cross-section. When the keel girder work and all twenty-six longitudinals had been sliced, the now-free section was lowered to the deck, where a team was waiting to receive it. The gas-cell netting and wire bracing were readily cut away as work parties reached them. Aloft, via crane, the men were shunted forward to the next main frame, where the process was repeated and that section of framework detached and dropped. On the deck, meantime, the girder work was disassembled and the longitudinal and transverse girders cut into roughly equal lengths, after which the material was stacked. As hull sections were dispatched and the work progressed forward, their particular suspensions were unshipped, the shoring stowed, and deck space cleared of all gear.

The hull shape surrendered. Courtesy of planning and Baker's handiwork, this final phase advanced smoothly. Disassembly work schedule: eight hours each day, five days a week. Within about fourteen working days the job was done, and *ZR-3* was gone.

By the third day of the new decade, the hull structure, tail surfaces, power cars, and all appurtenances had been completely disassembled and the hangar itself cleared of all scrap. What had been *Los Angeles* had become unsightly piles near the Power House, segregated into lots, awaiting final disposal. By February (if not earlier), the formal survey and preparation of manuscript had been completed and, as ordered, forwarded to the Bureau of Supplies and Accounts.

As predicted, the dismantling had experienced scant disruption: the work had progressed easily, briskly, without accident. Emotions varied. To the school trainee-inductees, the cut-down had been an uncommon assignment, nothing more. To professional airship men, however, the matter was altogether different, on another level. Many officers had been lieutenants junior grade or midshipmen when *ZR-3* had first reached Lakehurst from Central Europe. Among the enlisted ranks, scores of careers had set sail with her. *L.A.* had served long and well. In a life work of memorable moments (and, yes, brushes with tragedy), *Los Angeles* had proven a lucky ship—and was well loved for it. For the airmen who knew her, then, scrapping *ZR-3* transcended mere personal loss: it represented a fatal blow to the program for which they had labored. It had to have rankled. Mrs. Settle may have spoken for most: "I hated to have her dismantled."[49]

On 12 January 1940, personnel who had assisted the dismantling and cut-down project were recognized by the air station's new commanding officer. Into each of their service records was typed a single-paragraph entry:

> Commended for skillful and efficient workmanship under the leadership of Chief Boatswain Buckley U.S.N. in dismantling the rigid airship USS *Los Angeles* in a better manner and in less time than had been specified to the Chief of the Bureau of Aeronautics.
>
> [signed] G. H. Mills

Girder scrap prior to storage and sale. The viscera of aluminum and steel wire realized less than $4,000 for the Navy Department. (Acme Photo courtesy Lieut. Cdr. Leonard E. Schellberg)

Not long after the order to destruct, news of the dismantling stimulated at least four requests for specimens. Lakehurst routed these on to the bureau for comment. Requests for personal souvenirs were politely but firmly refused. To inquiries of a more official nature, BuAer returned a "no objection." Thus it was that during December–February pieces were shipped to a textile school in North Carolina, a New York museum, and to Rensselaer Polytechnic Institute. (The latter had requested typical girders and fittings for its Aeronautical Structures Laboratory.) In March, a selection of eleven or so typical joints and structural fittings were forwarded to Goodyear Aircraft.

As per bureau request, also, channels and lattices were set aside at Lakehurst—the equivalent of two lots of specimens like those selected for BuStandards testing. "It is proposed to arrange for tests of these specimens in 1944 and 1949 by way of rounding out and adding to

The big-ship era done, Navy lighter-than-air in 1939–40 was in orphan status. Now the blimp held center stage. Each airman named here had served aboard Los Angeles. *(Lieut. Cdr. Leonard E. Schellberg courtesy Navy Lakehurst Historical Society)*

the history of light alloy materials already built up around the LOS ANGELES."⁵⁰

Bids on the scrap were opened. The award, to Reynolds Metals Company of New York City, realized $3,667.80.⁵¹ On 7 March, Lakehurst informed the chief of the bureau that the project to dismantle the rigid airship USS *Los Angeles* had been closed out as of 29 February.

For the ZR concept, the tide had fully ebbed. The 1939 cut-down was not simply the forfeiture of one aircraft but the abandonment by the United States of an entire technology. The heroine of this history (some might argue) had achieved no great things; operated to no special acclaim, her career had been solid and effective but largely unspectacular. Instead, *L.A.* had simply endured—perhaps her greatest achievement. Regardless of viewpoint, the last ZR had vanished. *Los Angeles* was finally, irretrievably gone.

An era of naval aeronautics ended with her.

Epilogue

In the interwar environment, the U.S. Navy's ZR project failed to prove its value to naval warfare. Objects of special scorn, the big ships were through.

The rigid airships' probation had been protracted, troubled, and ever controversial. Factional wrangling, botched initiatives, fiscal strangulation, and operational failures—all had drained away potential. Following loss of *Macon*, a sustained policy vacuum had ensued; the drift persisted until the experiment itself was liquidated—or nearly so. In mid-1940, Secretary of the Navy Charles Edison tried to reinstate the ZRN appropriation. In his view, the airship had been a victim of its high cost and overreaction to crashes; the Navy had neither built nor operated a sufficient number of units for a fair assessment. On 5 June, Towers argued the ZRN's case before the House Subcommittee on Appropriations and asked that funds be made available for the three-million-cubic-foot trainer—a final, faltering gasp.[1] At Lakehurst, airship men held to their vision as, in Washington, salesman Rosendahl (with King's endorsement) mapped out programs for nonrigid construction. "the neutrality patrol is an ideal case for airships," C. P. Burgess observed. Though a rigid offered greater value, "The very fact that there is a war on is bound to make the airship situation more hopeful."[2]

In the two decades available to the naval rigid, the department had endured an uneasy, indifferent stewardship. The surface fleet had seldom worked collaboratively with airships. Unfortunately for lighter-than-air, the pace of current events had outrun its promise; by 1940, the crying naval need was for ships and airplanes, not rigid airships—let alone a marginal ZR trainer. Heavier-than-air partisans had forever insisted that funds could be better applied to flying boats or to carrier aircraft—more units at comparable cost. In the end, Congress decided that funds wanted for the ZRN would be better spent on long-range patrol planes and in resuscitating the rigid's smaller, unglamorous cousin—fabric blimps.

So the airplane retained its (wide) developmental edge. Within two generations, the rigid type—incomprehensibly large, costly, a craft from a different era for a different world—was allowed to pass into oblivion. Rosendahl may have helped explain the demise: "To the layman there is still something almost fantastic, something all but incredible about airships. The very size of the craft, involving an enormous expenditure of time and effort and money, has made its progress slower and magnified the mishaps and failures that accompany every pioneering effort. The dirigible has had to battle its way through seas of doubt and skepticism."[3]

With war declared in September 1939, proponents were left to patch together a program. Proposals for a revival notwithstanding, the decision against the rigid type proved final.

"I think the [rigid] airships had about had their day," Admiral Harold B. Miller later remarked, looking back. "In my opinion it was the PBYs which brought their usefulness to an end. For the $4,000,000 the ships cost one could provide a squadron of 12 PBYs which were certainly less susceptible to destruction and could cover a wider and faster scouting search curve than the airships."[4]

As America prepared to scrap its last ZR, a *Flight* editorial had endorsed the airplane-carrying rigid as a naval scout. One, possibly two, German pocket battleships were at large in the Atlantic and, perhaps, the Indian Ocean.[5] As the editorialist tersely noted, "The first and most difficult problem is to find them." He recommended airplane-carrying rigid airships for the purpose. Operating from bases along Britain's maritime trade routes, the big ships could patrol the empire's strategic threads at relatively high speeds and at a cost less than that of cruisers, the traditional scout for the fleet. In short, with airships, the vital sea search could be "much simpler."

At sea in two oceans, the United States faced its own dilemmas. The first line of defense might be thousands of miles from American shores. "Use of airships as plane carriers," the *Herald Tribune* noted months into the Euro-

Power cars from LZ-130 and LZ-127 (r.) at Frankfurt-am-Main, March or April 1940. The Air Ministry had ordered the dismantling of both Graf Zeppelins. *The figure is Anton Wittemann, one of Dr. Eckener's cadre of zeppelin captains. This photograph was taken covertly. (Friedrich Moch courtesy Douglas H. Robinson)*

pean war, "would greatly increase the mobility of a surface fleet's scouting arm, the airship experts say. If an emergency should arise on the Pacific Coast calling for the presence of planes to help the fleet scout and attack an enemy, they assert, a carrier airship could be there in from forty-eight to sixty hours after leaving the East Coast, while a surface carrier going through the Panama Canal would require from ten to twelve days."[6]

This had been the promise since 1918—an instrument of very-long-range reconnaissance over vast desert seas into which an enemy force might "disappear." For a variety of reasons, however, the type had failed to deliver. As the 1940s opened, the assumptions underlying a naval rationale—technological, military, political—had been undercut by progress in heavier-than-air aeronautics. They were also the victims of international forces; given the urgent pressures of 1939–41, few could imagine the rigid airship having genuine military significance. In too many minds, time simply had run out on promises.

In Central Europe, the old *Graf Zeppelin* was suspended in its Frankfurt shed, then placed on exhibit. In the terminal's new (1938) Hangar No. 2 hung *LZ-130*—the world's last, most modern zeppelin. Deeming them militarily useless, Reich Air Marshal Hermann Goering wished to be rid of both, and their sheds. Veteran airmen like Max Pruss and Hans von Schiller bitterly opposed their destruction, which they viewed as wanton. On the last day of February 1940, the two zeppelins were ordered dismantled. Little was saved—and that surreptitiously. On the third anniversary of the *Hindenburg* fire, 6 May, the sheds themselves were demolished (using dynamite) at Goering's direct order.

The commercial transoceanic field was on hold—for the duration. Germany, certainly, no longer carried on. For

EPILOGUE

Aerial of air station Lakehurst about the time of the Oahu raid. To help meet the national-security emergency, taxiways have been graded to circles 1 through 4; here three blimps ride to the mast. The big ships but a memory, the abandoned circles for Los Angeles, Akron, and Hindenburg are still evident. (National Archives)

America, only the nonrigid type was flying. Yet the enveloping national emergency brought a renaissance. The revived U-boat threat shifted antisubmarine warfare to center stage. Replicating its 1914–18 experience, the lowly blimp would again prove a superb antisubmarine asset.

At the Naval Air Station Lakehurst, the mission had been reconceived; now the blimp predominated. "[Lakehurst] is and will remain the training center for all officer and enlisted crews for airships," Cdr. George H. Mills, the station's skipper, wrote days before Pearl Harbor. "It will also be the main overhaul base for Atlantic coast airships and will serve as a base for some of the airship patrol squadrons."[7] "With the advent of preparations for war," Louise Tyler Rixey said recently, "of course things changed dramatically on the station. . . . The tremendous influx of personnel arriving for paratrooper and flight training turned NASL into a veritable beehive of activity. I can only imagine the stress involved for the officers and others in charge, but Dad's hours at work certainly became longer."[8]

Garland Fulton was a force in U.S. lighter-than-air aeronautics; as one historian notes, during 1925–40 Fulton exerted a greater influence on the direction of the navy's LTA program than any other individual not in a policy-making position. In December 1940, Captain Fulton voluntarily retired, terminating a career of twenty-eight years, at least eighteen of which had been devoted to lighter-than-air aeronautics in general, and to the rigid airship in particular. If there was hope for the ZR, it vanished when Fulton cleared his desk at the bureau.[9]

The possible contributions of big naval airships to World War II will always be a matter of debate. There is something to be said for the antisubmarine and convoy-escort potential of the rigid type; in 1941–42, such missions held utmost urgency for fleet commanders. Operating over midocean wastes far to seaward of the coastal blimps, ZRs could have augmented overextended destroyers, aircraft carriers, and patrol planes. For the

Lieut. George H. Mills. A graduate of Class VIII, Mills trained on board ZR-3, was billeted to Macon, and, in January 1940, read his orders as commanding officer, NAS Lakehurst. He received command of the war's first Airship Group in 1942, and, in 1943, was promoted to Commander, Fleet Airships, Atlantic. (U.S. Navy Photograph courtesy Navy Lakehurst Historical Society)

Battle of the Atlantic, the reconnaissance "eyes" of *Los Angeles* or a replacement represent a concept that is hardly far-fetched. But it was the role of lighter-than-air "carrier" that held the maximum potential. Steaming from U.S. coasts, and, yes, the Hawaiian Islands, these aerial strategic scouts could have patrolled far to sea and, perhaps . . .

To millennial eyes, the rigid airship seems an obvious anachronism, a machine rendered obsolete by its own technology and a waning utility—in short, prey to the modern tide of fast, long-range airplanes. Yet, interestingly, during the 1980s the U.S. Navy assessed the surveillance airship again. Generous Reagan-era defense budgets brought the naval airship to the brink of renewal. And there it stalled. The challenges of a new century appear to favor long-range, heavy-lift aerial transport. The military's strategy and structure are under review. Reluctantly, the Pentagon is shifting its geographical priorities toward Asia. Each of the armed services will be compelled to move quickly over long distances, and, without ample overseas bases, haul its own weapons, fuel, and supplies.

The ZR represents the first multimillion-dollar weapons system of the modern era, born of twentieth-century technology, that was terminated without ever being tested in combat.

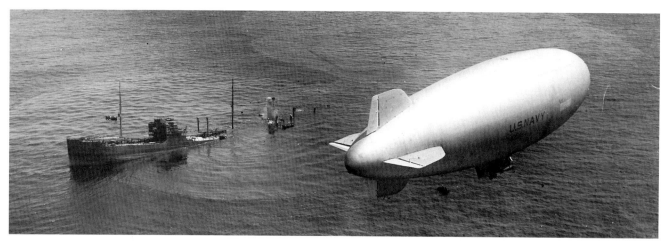

K-4 circles the torpedoed S.S. Perisphone, 25 May 1942. The tanker had taken two hits from U-593—probably the only vessel torpedoed while under blimp escort in two world wars. Long treated like a stepchild, the nonrigid patrol airship (or ZP) became the workhorse of the wartime program. Soldiering into the cold-war sixties, naval lighter-than-air operations were ended on 31 August 1962. (U.S. Navy Photograph)

APPENDIX 1

Characteristics and Performance: LZ-126/ USS Los Angeles (ZR-3)

General Dimensions:

Total volume of gas cells, 100 % inflation, cu. ft. 2,599,110
Total volume of gas cells, 100 % inflation, cu. m 70,000
Air displacement of hull, cu. ft. 2,764,461
Overall length, ft. (m) 656.2 (200)
Maximum diameter of hull, ft (including pneumatic bumper) . . . 90.68 maximum Height 101.71

Hull Structure:

Materials of structure . . . . Duralumin
Composition . . . . Variable: Al (92% min.), Cu (4%),
                                                                                            Mn (0.75%), Mg (0.50%)
Tensile strength, lbs. per sq. in. 50,000
Normal spacing of main frames, ft. (m) 49.2 (15)
Normal spacing of intermediate frames, ft. (m) 16.4 (5)
Number of sides to polygons of framing 24
Load-wire strength, lbs. (0.071 to 0.118-in dia.) . . . . 1,040 to 2,670

Fuel, Oil, Ballast:

Number of 648 lb. gasoline tanks (American condition) 49
Number of 672 lb. fuel tanks 10
Number of 300 lb. half-tanks 2
Fuel capacity, lbs 40,000
Number of oil storage tanks 3

Number of alcohol tanks 2
Number of kerosene tanks 1
Number of service ballast bags, 2,200 lb. capacity 14
Number of emergency ballast bags, 550 lb. capacity 8

Gas Cells:

Number of cells 14
Volume of largest cell (No. 8), cu. ft. 99,829
Diameter of largest cell, ft. 91.54
Weight of rubberized fabric, ozs. per sq. yd. 5.5
Surface area of No. 8, sq. ft. 27,864
Total weight of gas cells (German), lbs. 9,158
Number of maneuvering valves, No. 8 1
Number of automatic valves, No. 8 2
Diameter of automatic (pressure-relief) valve, in. 19
Gas-cell netting . . . . one-eighth-inch Ramie cord

Outer Cover:

Material . . . . grade BB cotton, panels varied in thickness according to strength required at specific locations
Treatment . . . . four coats of dope, two clear and two mixed with aluminum powder
Unit weight of material, lbs. per sq. ft. 0.045
Tension, lbs. per ft. 40–80
Breaking strength, lbs. per in. 45

Powerplant:

Engine . . . . reversible Maybach Type VL-1, 12-cylinder water cooled, 60-degree V-type, 400 hp at 1,400 rpm; later Type VL-2
Dryweight one engine (VL-1), including air starter, lbs. 2,437
Speeds, rpm
    Full 1,450
    Cruising 1,375
    Half 1,050
    Slow 600
Number of engines 5
Total horsepower . . . . 2,000 (VL-1), 2,750 at 1,500 rpm (VL-2)
Maximum propeller rpm 1,450
Diameter of propeller arc, ft. . . . . 12 (two-bladed, wood)
Generators . . . . (2) 5 KW in Nos. 4 and 5 power cars 24 volts DC

Complement and Crew:

Complement (1931–32) . . . . 82, 13 officers and 69 enlisted
Flight crew (1931–32) . . . . 43, 10 officers and 33 enlisted

CHARACTERISTICS AND PERFORMANCE

Performance:

Total lift, nominal gas volume and helium at standard value, lifting 0.062 lbs. per cu. ft., lbs.	153,500
Deadweight of structure, lbs. (late American condition)	90,860
Useful lift, including crew, fuel, oil, ballast, lbs.	62,640
Designed maximum speed, mph	79
Cruising speed, knots	50
Endurance at 50 knots, hrs.	78.5
Range at 50 knots, naut. mi. (statute mi.)	3,925

APPENDIX 2

Roster of Graduates, Lighter-Than-Air Pilot Training, NAS Lakehurst (1923–1940)

CLASS I (1923–24)

*McCRARY, F. R.
*LANSDOWNE, Z.
*NORFLEET, J. P.
*TYLER, R. F.
*LAWRENCE, J. B.
*HOUGHTON, A. R.
*BAUCH, C. E.
*NULL, T. B.
*REICHELDERFER, F. W.
*BUCKLEY, W. A.
*HUNSAKER, J. C.
*ROTH, L. J.
*GOWAN, J. H.
*THURMAN, E. C.
STEELE, G. W.
KLEIN, J.
ROSENDAHL, C. E.
DEEM, J. M.
PIERCE, M. R.
WEYERBACHER, R. D.
MAYER, R. G.

HANCOCK, L.
SHEPPARD, E. W.
ARNOLD, J. C.
MILLER, R. J.
ANDERSON, J. B.
WHITTLE, G. V.

*LTA—Experience previous to convening of Class I.

CLASS II (1924–25)

SETTLE, T. G. W.
HENDLEY, T. B.
THORNTON, J. M.
RICHARDSON, J. G.
CARTER, J. B.
WILEY, H. V.

CLASS III (1926–27)

WALKER, M. J.
CLARKE, V. A.
McCORD, F. C.

CLASS IV (1927–28)

RODGERS, B. J.
STEVENS, J. H.
*PECK, S. E.
DENNETT, R. R.
WATSON, G. F.
JACKSON, E. S. (Observer)
LANGE, K. (USNR)

CLASS V (1928–29)

BRADLEY, M. M.
BAILEY, S. M.
ZIMMERMAN, W. E.
MacINTYRE, A.
MAY, B.
DUGAN, H. J.
BUSHNELL, W.

CLASS VI (1929–30)

SHOEMAKER, H. E.
BERRY, F. T.
DRESEL, A. H.
SEVERYES, J. H.
ROLAND, C. W.
MILLER, C. F.
PICKENS, H. H.
REDFIELD, M.
WHELAN, T. M.
GREENWALD, J. A.
ROUNDS, C. S.

CLASS VII (1930–31)

SEYMOUR, P.
CLAY, A. T.
MAGUIRE, C. J.
DANIS, A. L.
SEWELL, A. T. (USNR)
REPPY, J. D.
CALLAWAY, G. H.
SAYRE, R. E.
SHAFER, D. W.
CLENDENING, C. T.
CAMPBELL, G. W.
CLAPP, A. B.
WALSH, GEO.
CALNAN, G. C.

CLASS VIII (1931–32)

JENKINS, B. P.
KENWORTHY, J. L., Jr.
COCHRANE, E. F.
WORDEN, F. L.
MILLS, G. H.
PHILLIPS, W. K.
FISHER, J. L.
MacLELLAN, H. E.
CUMMINS, D. E.
MACKEY, D. M.
CROSS, R. F.
COULTER, H. N.
HARRIGAN, D. W.
BOLSTER, C. N.
SULLIVAN, E. J.
YOUNG, H. L.
WESCOAT, H. M.
BOETTNER, J. A. (USNR)
KENDALL, G. H.
COCKELL, W. A.
ZOOK, H. M.
VAN SWEARINGEN, E. K.

CLASS IX (1934–35)

KNOX, C. V. S.
GILMER, F. H.
SACHSE, F. C.
ORVILLE, H. T.
ZURMUEHLEN, G. D.
FLAHERTY, M. F. D.
WEINTRAUB, D. J.
WILSON, R. D. (USNR)
FURCOLOW, L. P. (USNR)
McKEE, L.
MORAN, H. G.

CLASS X (1936–37)

ANTRIM, R. M.
ANDREWS, R. S.

CLASS XI (1937–38)

WERTS, C. L.
COPE, A. L.

PIERCE, G. E.
McCORMICK, J. J.
EPPES, M. H.
RICE, T. A

CLASS XII (1938–39)

PERLER, W. R.
BLOUNT, C. E.
HANGER, W. M.
WILLIAMS, R. J.

ENTLER, D. M.
KIRKPATRICK, H. G.

CLASS XIII (1939–40)

BLY, R. E.
BOYD, W. W.
KEEN, W. H., Jr.
GOODLOE, G. C.
BARKLEY, R. L.
BURFEIND, H. F.

APPENDIX 3

Flight Log, USS Los Angeles, 1924–1932

Flight Number	Take-off Date	Take-off Time	Landing Date	Landing Time	Duration (h.m.)	Total Flight Time (h.m.)	Remarks
1924							
1	27 August				2.10	2.10	Builder's trial—Germany
2	6 September				8.46	10.56	Trial—Germany
3	11 September				7.59	18.55	Trial—Germany
4	13 September				4.42	23.37	Trial—Germany
5	25–26 September				33.26	57.03	Trial—Germany
6	12–15 October				81.32	138.35	Delivery: To NAS Lakehurst
7	25 November						To Anacostia NAS

Commissioned 25 November 1924 at Anacostia NAS, Washington, DC
Cdr. Jacob H. Klein, USN, Commanding

Flight Number	Take-off Date	Take-off Time	Landing Date	Landing Time	Duration (h.m.)	Total Flight Time (h.m.)	Remarks
8	11-25	1704	11-25	2100	3.56	3.56	Return Lakehurst
9	12-22	1229	12-22	1410	1.41	5.37	
10	12-30	0907	12-30	1307	4.00	9.37	
1925							
11	1-9	1014	1-9	1608	5.54	15.31	From high mast
12	1-9	1626	1-9	1813	1.47	17.18	Flying moor
13	1-15	0939	1-15	1615	6.36	23.54	Moored to *Patoka*, Baltimore
14	1-15	1705	1-15	2042	3.37	27.31	

Capt. George W. Steele, USN, assumed command on 19 January 1925

(Continued)

Flight Number	Take-off Date	Take-off Time	Landing Date	Landing Time	Duration (h.m.)	Total Flight Time (h.m.)	Remarks
15	1-24	0520	1-24	1540	10.20	37.51	Photographing eclipse, off Block Island
16	2-20	1520	2-22	0036	33.16	71.07	To Bermuda and return (nonstop)
17	4-16	0910	4-16	1600	6.50	77.57	
18	4-21	1435	4-22	0636	16.01	93.58	Moored *Patoka*, Bermuda
19	4-23	0857	4-24	0534	20.37	114.35	
20	5-3	0958	5-4	2215	36.17	150.52	Moored *Patoka*, Puerto Rico
21	5-6	1019	5-6	2035	10.16	161.08	-do-
22	5-8	1433	5-10	0431	37.58	199.06	
23	5-15	1030	5-15	1738	7.08	206.14	33 military & civilian passengers
24	5-26	1145	5-26	2155	10.10	216.24	Calibrating radio compass (rc), NAS Lakehurst
25	5-27	1144	5-27	1800	6.16	222.40	
26	6-2	0904	6-2	1346	4.42	227.22	Moored *Patoka*, Annapolis
27	6-2	1551	6-2	2012	4.21	231.43	-do-
28	6-3	1014	6-3	1728	7.14	238.57	Return Lakehurst, deceleration tests
29	6-7	0130	6-8	1342	36.12	275.09	Minneapolis flight (aborted)

1926

Flight Number	Take-off Date	Take-off Time	Landing Date	Landing Time	Duration (h.m.)	Total Flight Time (h.m.)	Remarks
30	4-13	1701	4-13	1853	1.52	277.01	Post-repair trials
31	4-14	1411	4-14	1630	2.19	279.20	-do-
32	4-27	0943	4-27	1740	7.57	287.17	NACA tests
33	4-30	0943	4-30	1623	6.40	293.57	-do-
34	5-7	1650	5-7	2005	3.15	297.12	-do-

Lt. Cdr. C. E. Rosendahl, USN, assumed command on 10 May 1926

Flight Number	Take-off Date	Take-off Time	Landing Date	Landing Time	Duration (h.m.)	Total Flight Time (h.m.)	Remarks
35	5-13	0940	5-13	1942	10.02	307.14	-do-
36	7-26	0828	7-26	2005	11.37	318.51	Calibrating rc station, NJ
37	7-27	1353	7-27	1930	5.37	324.28	-do- (NY)
38	8-3	1400	8-3	1845	4.45	329.13	
39	8-4	0939	8-4	1848	9.13	338.26	Calibrating rc station, RI; moored *Patoka*, Newport
40	8-5	0905	8-5	2318	14.13	352.39	-do-
41	8-6	0320	8-6	0710	3.50	356.29	
42	8-9	1025	8-9	1950	9.25	365.54	Calibrating rc station, Long Island (LI)
43	8-10	1012	8-10	1900	8.48	374.42	-do- Lakehurst
44	8-23	1338	8-23	2043	7.05	381.47	-do- Cape May, NJ
45	8-26	1353	8-26	2053	7.00	388.47	-do- (DE)
46	8-30	1027	8-31	0401	17.34	406.21	-do- (VA)
47	9-1	0930	9-1	1600	6.30	412.51	-do- moored *Patoka*, Cape Charles, VA
48	9-1	1652	9-1	2330	6.38	419.29	-do- (NC)
49	9-7	1523	9-8	2342	8.19	427.48	Moored *Patoka*, Cape Charles
50	9-8	0908	9-8	2030	11.22	439.10	Calibrating rc station (NC)
51	9-10	1212	9-10	1425	2.13	441.23	Philadelphia Air Races

Flight Number	Take-off Date	Time	Landing Date	Time	Duration (h.m.)	Total Flight Time (h.m.)	Remarks
52	9-10	1512	9-10	1941	4.29	445.52	Return Lakehurst
53	9-15	0924	9-15	1810	8.46	454.38	
54	9-20	1027	9-20	1832	8.05	462.43	Moored *Patoka*, Narragansett Bay, RI
55	9-21	1025	9-21	2355	13.30	476.13	
56	10-4	1154	10-4	1747	5.53	482.06	
57	10-7	1115	10-7	1649	5.34	487.40	Moored *Patoka*, RI
58	10-8	1027	10-8	1742	7.15	494.55	
59	10-12	1227	10-12	1720	4.53	499.48	
60	10-13	1150	10-13	0028	12.38	512.26	
61	10-14	1107	10-15	0538	18.31	530.57	Moored Ford mast, Dearborn, MI
62	10-15	1542	10-16	0700	15.18	546.15	Return Lakehurst
63	11-1	1057	11-1	1644	5.47	552.02	
64	11-3	1132	11-3	1718	5.46	557.48	
65	11-12	1055	11-12	1830	7.35	565.23	
66	11-21	1056	11-21	1830	7.34	572.57	Moored *Patoka*, Chesapeake Bay, VA
67	11-22	0951	11-22	1702	7.11	580.08	Return Lakehurst high mast
68	11-23	1115	11-23	1645	5.30	585.38	
69	12-3	1055	12-3	1635	5.40	591.18	

<u>1927</u>

Flight Number	Take-off Date	Time	Landing Date	Time	Duration (h.m.)	Total Flight Time (h.m.)	Remarks
70	4-12	1155	4-13	0445	16.50	608.08	
71	4-15	0902	4-15	1554	6.52	615.00	Moored *Patoka*, Chesapeake Bay, VA
72	4-15	1650	4-15	2235	5.45	620.45	
73	4-16	1025	4-16	1842	8.17	629.02	
74	4-23	2038	4-24	1936	22.58	652.00	Moored *Patoka*, Pensacola, FL
75	4-25	1614	4-27	0122	33.08	685.08	Return Lakehurst
76	5-2	0506	5-2	2032	15.26	700.34	
77	5-7	0415	5-7	1932	15.17	715.51	
78	5-13	0407	5-13	1837	14.30	730.21	Sea-search missing transatlantic fliers
79	5-21	1115	5-21	1905	7.50	738.11	
80	6-11	0430	6-11	2013	15.43	753.54	Air exploration tests with UO-1 airplane
81	9-2	1021	9-2	1948	9.27	763.21	Speed, deceleration tests
82	9-6	1840	9-7	0550	11.10	774.31	-do-
83	9-7	0957	9-7	1827	8.30	783.01	
84	9-8	1545	9-8	2147	6.02	789.03	Moored *Patoka*, Cape Charles, VA
85	9-9	1623	9-9	2256	6.33	795.36	Return Lakehurst
86	9-10	0854	9-10	1732	8.38	804.14	Escorted *SS Leviathan*, NY Harbor; Philadelphia
87	9-16	0847	9-16	1939	10.52	815.06	Moored *Patoka*, Plantation Flats, VA
88	9-17	0904	9-17	1737	8.33	823.39	Return Lakehurst

(Continued)

Flight Number	Take-off Date	Take-off Time	Landing Date	Landing Time	Duration (h.m.)	Total Flight Time (h.m.)	Remarks
89	9-21	1017	9-22	1755	32.04	855.43	Moored *Patoka*, Plantation Flats, VA
90	9-23	0925	9-23	1803	8.38	864.21	Calibrating rc station, NJ
91	9-27	1009	9-27	1820	8.11	872.32	
92	10-5	1037	10-5	1753	7.16	879.48	Experimental Stub Mast
93	10-13	1713	10-13	2248	5.35	885.23	Moored *Patoka*, VA
94	10-14	1011	10-14	2216	12.05	897.28	Return Lakehurst
95	10-15	0952	10-15	1707	7.15	904.43	
96	10-24	1534	10-25	0641	15.07	919.50	Moored *Patoka*, Newport
97	10-25	0958	10-25	2255	12.57	932.47	Took on 21 ground crew as ballast for landing
98	10-26	1517	10-27	1926	28.09	960.56	
99	11-10	1100	11-10	1624	5.24	966.20	
100	11-14	1013	11-14	1653	6.40	973.00	
101	11-15	1121	11-15	2216	10.55	983.55	Moored *Patoka*, VA
102	11-16	0949	11-16	2240	12.51	996.46	Return Lakehurst
103	12-22	2036	12-23	0720	10.44	1007.30	
104	12-23	1243	12-23	1706	4.23	1011.53	
105	12-26	1700	12-28	0040	31.40	1043.33	Search for lost Grayson's plane "DAWN"

1928

Flight Number	Take-off Date	Take-off Time	Landing Date	Landing Time	Duration (h.m.)	Total Flight Time (h.m.)	Remarks
106	1-27	0354	1-27	1535	11.41	1055.14	Landed USS *Saratoga*, off Newport
107	1-27	1542	1-27	1910	3.28	1058.42	Return Lakehurst
108	2-11	1127	2-11	1742	6.15	1064.57	Moored *Patoka*, VA
109	2-11	1823	2-12	0335	9.12	1074.09	Return Lakehurst
110	2-26	0656	2-27	2247	39.51	1114.00	France Field, Canal Zone
111	2-28	1012	2-29	1303	26.51	1140.51	Moored *Patoka*, Cuba
112	3-1	1040	3-3	0358	41.18	1182.09	
113	3-3	0425	3-3	0643	2.18	1184.27	Emergency takeoff, return Lakehurst
114	5-2	1938	5-3	0413	8.35	1193.02	
115	5-30	0159	5-30	1905	17.06	1210.08	
116	5-31	1017	5-31	1947	9.30	1219.38	
117	6-15	2012	6-16	0451	8.39	1228.17	Moored *Patoka*, Newport
118	6-16	1000	6-16	1906	9.06	1237.23	Return Lakehurst
119	7-2	0850	7-2	1950	11.00	1248.23	

Lt. Cdr. H. V. Wiley, USN, assumed temporary command

Flight Number	Take-off Date	Take-off Time	Landing Date	Landing Time	Duration (h.m.)	Total Flight Time (h.m.)	Remarks
120	7-16	1928	7-17	0430	9.02	1257.25	
121	7-17	0957	7-17	1956	9.59	1267.24	Flying moor, high mast
122	7-24	0827	7-24	1911	10.44	1278.08	Flying moor, high mast
123	7-25	0206	7-26	0035	22.29	1300.37	
124	7-30	1854	7-31	0431	9.37	1310.14	
125	8-27	1405	8-27	1838	4.33	1314.47	
126	8-28	1000	8-28	1938	9.38	1324.25	
127	8-29	0958	8-29	2325	13.27	1337.52	

FLIGHT LOG, USS *LOS ANGELES*, 1924–1932   233

Flight Number	Take-off Date	Take-off Time	Landing Date	Landing Time	Duration (h.m.)	Total Flight Time (h.m.)	Remarks
128	8-31	1850	9-1	0519	10.29	1348.21	Near-collision airmail plane
129	9-1	0833	9-1	1748	9.15	1357.36	
130	9-4	0837	9-4	1728	8.51	1366.27	
131	9-5	1407	9-6	1952	29.45	1396.12	
132	9-10	0638	9-10	1801	11.23	1407.35	
133	10-6	1825	10-8	1837	48.12	1455.47	To Fort Worth, TX mast (helium plant)
134	10-9	0858	10-10	2135	36.37	1492.24	To Chicago, then Lakehurst

Lt. Cdr. C. E. Rosendahl, USN, resumed command

Flight Number	Take-off Date	Take-off Time	Landing Date	Landing Time	Duration (h.m.)	Total Flight Time (h.m.)	Remarks
135	10-30	1927	10-31	0540	10.13	1502.37	Test radio
136	12-4	1622	12-5	0022	8.00	1510.37	Speed tests dummy water recovery
137	12-5	1652	12-6	1608	23.16	1533.53	

<u>1929</u>

Flight Number	Take-off Date	Take-off Time	Landing Date	Landing Time	Duration (h.m.)	Total Flight Time (h.m.)	Remarks
138	1-8	2040	1-10	1246	40.06	1573.59	Moored *Patoka*, FL
139	1-10	1659	1-11	0804	15.05	1589.04	-do-
140	1-12	1639	1-14	0810	39.31	1628.35	-do-
141	1-14	1128	1-15	0118	13.50	1642.25	Return Lakehurst
142	3-1	1818	3-2	0712	12.54	1655.19	
143	3-4	0714	3-4	1759	10.45	1666.04	Inaugural parade, Washington, DC
144	3-27	1655	3-29	0317	34.22	1700.26	Search missing Sikorsky plane
145	4-23	1900	4-24	0021	5.21	1705.47	
146	4-24	0040	4-24	0515	4.35	1710.22	
147	5-8	1907	5-9	0442	9.35	1719.57	

Lt. Cdr. H. V. Wiley, USN, assumed command on 9 May 1929

Flight Number	Take-off Date	Take-off Time	Landing Date	Landing Time	Duration (h.m.)	Total Flight Time (h.m.)	Remarks
148	5-9	1526	5-10	1945	28.19	1748.16	Niagara Falls flight
149	5-22	0053	5-22	0634	5.41	1753.57	
150	5-22	0819	5-22	2214	13.55	1767.52	To Washington; search for schooner, LI/CT
151	6-3	1305	6-3	1930	6.25	1774.17	
152	6-3	1940	6-4	0430	8.50	1783.07	
153	6-4	0800	6-4	2233	14.33	1797.40	
154	6-17	1943	6-18	0414	8.31	1806.11	
155	7-3	0022	7-3	0526	5.04	1811.15	Drag experiments, dummy water rec. panel; UO-1 trapeze test
156	8-15	1028	8-15	1708	6.40	1817.55	Hook-on tests
157	8-15	1839	8-16	2009	25.30	1843.25	Concord, NH flight
158	8-20	1822	8-21	0520	10.58	1854.23	UO-1 trapeze tests
159	8-21	0939	8-21	2201	12.22	1866.45	UO-1 trapeze tests
160	8-27	1630	8-28	1817	25.47	1892.32	Cleveland air races; UO-1 hook-on/transferred passenger
161	8-29	1306	8-30	1610	27.04	1919.36	Return Lakehurst

Lt. Cdr. C. E. Rosendahl assumed temporary command on 31 August 1929

(Continued)

Flight Number	Take-off Date	Take-off Time	Landing Date	Landing Time	Duration (h.m.)	Total Flight Time (h.m.)	Remarks
162	9-11	1857	9-12	0548	10.51	1930.27	Moored *Patoka*, Newport
163	9-12	1021	9-12	1759	7.38	1938.05	Return Lakehurst
164	9-20	1847	9-21	1814	23.27	1961.32	
165	9-21	1920	9-22	0053	5.33	1967.05	Flying moor stub mast

Lt. Cdr. H. V. Wiley, USN, resumed command on 24 September 1929

Flight Number	Take-off Date	Take-off Time	Landing Date	Landing Time	Duration (h.m.)	Total Flight Time (h.m.)	Remarks
166	9-25	1637	9-25	2315	6.38	1973.43	Undocked/unmoored via mobile mast; maneuvering w/ J ship
167	10-7	2024	10-8	0614	9.50	1983.33	
168	10-8	0833	10-8	1734	9.01	1992.34	Hook-on tests/last moor high mast
169	10-8	1945	10-9	0618	10.33	2003.07	
170	10-9	0836	10-9	1735	8.59	2012.06	UO-1 trapeze tests
171	10-9	1945	10-10	0556	10.11	2022.17	
172	10-10	1648	10-10	1816	1.28	2023.45	UO-1 trapeze tests; Moffett, Asst SecNav on board
173	10-14	0911	10-15	0601	20.50	2044.35	UO-1 trapeze tests
174	10-15	0920	10-16	0020	15.00	2059.35	UO-1 trapeze tests
175	11-5	1715	11-6	0016	7.01	2066.36	Fuel consumption tests
176	11-7	0403	11-8	0125	21.22	2087.58	Akron, Cleveland flight
177	11-12	1319	11-12	1854	5.35	2093.33	
178	11-20	0933	11-20	1710	7.37	2101.10	UO-1 trapeze tests
179	11-20	1719	11-21	0632	13.13	2114.23	
180	11-21	0934	11-21	1626	6.52	2121.15	Flying moor to stub mast
181	11-21	1727	11-22	0648	13.21	2134.36	Washington, DC flight
182	12-9	1135	12-9	1712	5.37	2140.13	
183	12-9	1731	12-10	0010	6.39	2146.52	

<u>1930</u>

Flight Number	Take-off Date	Take-off Time	Landing Date	Landing Time	Duration (h.m.)	Total Flight Time (h.m.)	Remarks
184	1-16	0728	1-17	1328	30.00	2176.52	
185	1-20	0714	1-20	1628	9.14	2186.06	Washington, DC flight
186	1-23	1755	1-24	1739	23.44	2209.50	To Parris Island, SC mast
187	1-25	1017	1-26	0715	20.58	2230.48	Calibrating rc station, St. Augustine, FL
188	1-26	1017	1-26	2225	12.23	2243.11	Return Lakehurst
189	1-31	1048	1-31	1240	1.52	2245.03	Glider (Barnaby) launch
190	2-15	0705	2-15	1517	8.12	2253.15	
191	3-28	1934	3-29	0010	4.36	2257.51	

Lt. Cdr. V. A. Clarke, USN, assumed command on 31 March 1930

Flight Number	Take-off Date	Take-off Time	Landing Date	Landing Time	Duration (h.m.)	Total Flight Time (h.m.)	Remarks
192	3-31	0807	3-31	1809	10.02	2267.53	In-flight data for BIS
193	4-4	2045	4-5	0528	8.43	2276.36	
194	4-9	2024	4-10	0522	8.58	2285.34	Scranton, PA flight
195	4-23	0543	4-23	1916	13.33	2299.07	Flying moor mobile mast
196	4-23	2105	4-24	0127	4.22	2303.29	
197	4-25	0136	4-25	1914	17.38	2321.07	Scranton, PA flight
198	4-28	0515	4-29	0453	23.38	2344.45	Intercepted USS *Leviathan* off NYC

FLIGHT LOG, USS *LOS ANGELES*, 1924–1932 235

Flight Number	Take-off Date	Take-off Time	Landing Date	Landing Time	Duration (h.m.)	Total Flight Time (h.m.)	Remarks
199	5-8	1940	5-9	0439	8.59	2353.44	
200	5-9	0719	5-9	1932	12.13	2365.57	
201	5-9	2114	5-10	0440	7.26	2373.23	
202	5-15	1325	5-15	1515	1.50	2375.13	UO-1 hook-on
203	5-20	0510	5-20	2023	15.13	2390.26	Presidential fleet review/UO-1 hook-on
204	5-20	2240	5-21	0442	6.02	2396.28	
205	5-26	0510	5-26	2223	17.13	2413.41	Calibrating rc station, Nantucket Island
206	5-31	0118	5-31	2015	18.57	2432.38	Glider (Settle) launch over Washington, DC; UO-1 hook-on
207	6-9	1020	6-9	1320	3.00	2435.38	
208	6-11	0840	6-11	1914	10.34	2446.12	
209	6-11	2143	6-12	0447	7.04	2453.16	
210	6-19	0649	6-19	1852	12.03	2465.19	Admiral Byrd welcome, NYC
211	6-20	0455	6-20	1810	13.15	2478.34	
212	6-23	1938	6-24	0422	8.44	2487.18	Moffett on board
213	6-26	0449	6-27	1120	30.31	2517.49	Floyd Bennett airport dedication
214	7-3	1135	7-3	1535	4.00	2521.49	
215	7-8	0723	7-8	1916	11.53	2533.42	Water recovery tests
216	7-8	2033	7-9	0437	8.04	2541.46	
217	7-15	0650	7-15	1835	11.45	2553.31	Moored *Patoka*, Newport
218	7-15	2205	7-16	0446	6.41	2560.12	Return Lakehurst
219	7-16	2018	7-17	0445	8.27	2568.39	Moffett, Asst SecNav on board
220	7-30	1854	7-31	0500	10.06	2578.45	
221	7-31	0824	7-31	1905	10.41	2589.26	Moored *Patoka*, Newport (twice)
222	8-1	0754	8-1	1540	7.46	2597.12	Return Lakehurst
223	8-6	0136	8-6	1715	15.39	2612.51	Testing radio facsimile
224	8-11	0740	8-11	1934	11.54	2624.45	Moored *Patoka*, Newport
225	8-12	0757	8-12	1832	10.35	2635.20	-do-
226	8-13	0754	8-13	1812	10.18	2645.38	-do-
227	8-14	0755	8-14	1715	9.20	2654.58	Return Lakehurst
228	8-19	0305	8-19	1610	3.05	2658.03	
229	8-19	1351	8-19	1847	4.56	2662.59	
230	8-20	0832	8-20	1820	9.48	2672.47	Airplane hook-on, "making pictures"
231	8-25	0820	8-25	1728	9.08	2681.55	Parachute tests, hook-on W/J-3 "for pictures"
232	8-26	0816	8-26	1334	5.18	2687.13	
233	9-4	2333	9-5	1717	17.44	2704.57	Moffett on board
234	9-19	1251	9-19	1749	4.58	2709.55	
235	9-22	0902	9-22	1759	8.57	2718.52	
236	9-23	0830	9-23	2336	15.06	2733.58	
237	9-29	0851	9-29	1718	8.27	2742.25	
238	9-29	1816	9-30	0605	11.49	2754.14	
239	10-1	0840	10-1	1740	9.00	2763.14	Moffett on board
240	10-1	1925	10-2	0556	10.31	2773.45	
241	10-2	0846	10-3	0600	21.14	2794.59	

(Continued)

Flight Number	Take-off Date	Time	Landing Date	Time	Duration (h.m.)	Total Flight Time (h.m.)	Remarks
242	10-7	0800	10-7	1750	9.50	2804.49	Boston flight
243	10-10	0920	10-10	1729	8.09	2812.58	
244	10-11	0812	10-11	1613	8.01	2820.59	
245	10-21	0916	10-21	1904	9.48	2830.47	
246	10-27	0830	10-27	1726	8.56	2839.43	
247	11-3	0900	11-3	1652	7.52	2847.35	
248	11-10	1740	11-11	0706	13.26	2861.01	Moffett on board
249	11-14	1136	11-14	1338	2.02	2863.03	

1931

Flight Number	Take-off Date	Time	Landing Date	Time	Duration (h.m.)	Total Flight Time (h.m.)	Remarks
250	1-14	0909	1-14	1605	6.56	2869.59	Post repair trials, tests
251	1-22	0912	1-22	1645	7.33	2877.32	-do-
252	1-23	0911	1-23	1715	8.04	2885.36	-do-
253	1-30	0728	1-30	1122	3.54	2889.30	-do-
254	2-4	1726	2-5	1733	24.07	2913.37	To Exped. mast, Guantanamo Bay, Fleet Prob. XII
255	2-6	0820	2-7	0824	24.04	2937.41	Moored *Patoka*, Costa Rica
256	2-8	0808	2-9	2034	36.26	2974.07	Moored *Patoka*, Panama Bay
257	2-12	0845	2-12	1851	10.06	2984.13	Moored *Patoka*, Panama Bay
258	2-15	1629	2-16	0650	14.21	2998.34	Moored *Patoka*, Panama Bay, Asst SecNav on board
259	2-16	1635	2-18	0700	38.25	3036.59	Fleet Prob., moored *Patoka*, Gulf of Dulce
260	2-18	1835	2-20	0643	36.08	3073.07	Fleet Prob., Asst SecNav on board, moored *Patoka*, Panama Bay
261	2-20	0840	2-20	1844	10.04	3083.11	Moored *Patoka*, Panama Bay
262	2-22	0822	2-23	2125	37.03	3120.14	To Exped. mast, Guantanamo Bay
263	2-27	1000	2-28	0843	22.43	3142.57	To Exped. mast, Parris Island
264	3-1	2310	3-2	1758	18.48	3161.45	Return Lakehurst
265	3-18	0841	3-18	1801	9.20	3171.05	
266	3-31	0103	3-31	1740	16.37	3187.42	
267	4-9	0823	4-9	1813	9.50	3197.32	
268	4-14	1915	4-15	0538	10.23	3207.55	
269	4-15	0706	4-15	2000	12.54	3220.49	Calibrating rc station (L.I.)

Cdr. A. H. Dresel, USN, assumed command on 21 April 1931

Flight Number	Take-off Date	Time	Landing Date	Time	Duration (h.m.)	Total Flight Time (h.m.)	Remarks
270	4-30	2038	5-1	0631	9.53	3230.42	Calibrating radio compass
271	5-4	2050	5-5	2020	23.50	3254.12	Cape Hatteras flight
272	5-18	0951	5-18	2025	10.34	3264.46	
273	5-26	2000	5-27	0546	9.46	3274.32	
274	5-27	0810	5-28	0621	22.11	3296.43	Hook-on practice approaches; to Norfolk
275	6-2	0813	6-3	0621	22.08	3318.51	
276	6-9	1825	6-10	0612	11.47	3330.38	Moffett, Asst SecNav on board; hook-on
277	6-17	0845	6-18	0535	20.50	3351.28	Hook-on; intercepted SS *France* (NYC)

FLIGHT LOG, USS *LOS ANGELES*, 1924–1932

Flight Number	Take-off Date	Take-off Time	Landing Date	Landing Time	Duration (h.m.)	Total Flight Time (h.m.)	Remarks
278	6-18	0817	6-18	1951	11.34	3363.02	Hook-on tests
279	6-25	1704	6-26	0550	12.36	3375.38	Wilkes-Barre Convention
280	7-11	2001	7-21	0610	10.09	3385.47	Moored *Patoka*, Long Island
281	7-15	0915	7-15	1631	7.16	3393.03	Return Lakehurst
282	7-27	1030	7-27	1745	7.15	3400.18	F2Y hook-on; King & Queen Siam
283	8-4	1942	8-5	0135	5.53	3406.11	Moffett on board; test Mark VI condenser
284	8-14	1000	8-14	1751	7.51	3414.02	Hook-on tests
285	8-17	2000	8-18	0621	10.21	3424.23	
286	8-18	0910	8-18	1915	10.05	3434.28	Hook-on tests (by 2 planes)
287	8-31	0820	8-31	1945	11.25	3445.53	Moored *Patoka*, Newport
288	8-31	2137	9-1	0713	9.36	3455.29	
289	9-1	0945	9-1	1932	9.47	3465.16	
290	9-8	0842	9-8	1926	10.44	3476.00	Hook-on exercises (64 "landings"); moored *Patoka*, RI
291	9-8	2111	9-9	0823	11.12	3487.12	Return Lakehurst
292	9-9	1048	9-9	1848	8.00	3495.12	Hook-on practice
293	9-18	0853	9-18	1852	9.59	3505.11	Hook-on practice; New England flight
294	9-29	0824	9-29	1723	8.59	3514.10	Dozen passengers Baltimore-Washington
295	9-29	1830	9-30	0735	13.05	3527.15	Night hook-ons
296	9-30	0952	9-30	1832	8.40	3535.55	Taking pictures over NYC
297	10-13	0821	10-13	1801	9.40	3545.35	Hook-on practice
298	10-19	0820	10-20	0606	21.46	3567.21	Up Hudson River
299	10-23	0837	10-23	1650	8.13	3575.34	XF9C hook-on
300	10-26	0830	10-27	1840	34.10	3609.44	XF9C hook-on; to Parris Island mast
301	10-28	1232	10-28	2125	8.53	3618.37	Return Lakehurst
302	11-2	0543	11-3	1727	35.44	3654.21	In formation with USS *Akron*
303	11-11	0849	11-12	1621	31.32	3685.53	Armistice Day flight
304	11-23	1715	11-24	0523	12.08	3698.01	Sonic altimeter tests
305	12-1	1734	12-2	0703	13.29	3711.30	RCA radio facsimile tests
306	12-2	0835	12-2	1649	8.14	3719.44	RCA radio facsimile tests
307	12-8	1624	12-8	2120	4.56	3724.40	
308	12-15	1622	12-16	0713	14.51	3739.31	
309	12-16	0829	12-16	1654	8.25	3747.56	
<u>1932</u>							
310	1-7	1624	1-8	1005	17.41	3765.37	
311	1-14	1624	1-15	0655	14.31	3780.08	
312	1-19	1712	1-20	1818	25.06	3805.14	Boston flight; Calibration rc station, Nantucket
313	1-25	1836	1-26	1441	20.05	3825.19	Officer into cell No. 9

Cdr. F. T. Berry, USN, assumed command on 1 February 1932

(Continued)

APPENDIX 3

Flight Number	Take-off Date	Take-off Time	Landing Date	Landing Time	Duration (h.m.)	Total Flight Time (h.m.)	Remarks
314	2-9	1700	2-10	0916	16.16	3841.35	Off NJ coast & up Hudson River
315	2-16	1624	2-17	0142	9.18	3850.53	
316	2-19	0714	2-19	1747	10.33	3861.26	
317	2-24	0714	2-24	1746	10.32	3871.58	Washington/Annapolis flight
318	3-3	1756	3-4	0620	12.24	3884.22	
319	3-18	1820	3-19	0604	11.44	3896.06	Offshore Long Island & NJ
320	3-24	1104	3-24	1945	8.41	3904.47	
321	3-24	2034	3-25	0630	9.56	3914.43	Moffett on board; cruised Jersey coast
322	4-5	0701	4-5	1814	11.13	3925.56	
323	4-5	1922	4-6	0520	9.58	3935.54	
324	4-18	1850	4-19	1839	23.49	3959.43	Augusta, ME flight
325	4-19	2108	4-20	0523	8.15	3967.58	Night flight Philadelphia/Wilmington, DE
326	5-18	2023	5-19	0540	9.17	3977.15	
327	5-19	0810	5-20	0534	21.24	3998.39	GE "narrow-casting" test (NY); Newport
328	5-23	0811	5-23	2027	12.16	4010.55	To Washington, DC following coast
329	5-23	2051	5-24	0537	8.46	4019.41	Moffett on board; offshore NY, CT, NJ
330	5-24	0807	5-24	2118	13.11	4032.52	To Cape Charles Light
331	5-31	0800	6-1	0531	21.31	4054.23	To VA-NC border, following coast
332	6-8	0607	6-9	0515	23.08	4077.31	Along coast to Washington/Annapolis
333	6-9	0810	6-9	2105	12.55	4090.26	
334	6-23	0618	6-23	2205	15.47	4106.13	Following NJ coastline
335	6-24	0753	6-24	2037	12.44	4119.00	-do-
336	6-24	2135	6-25	0511	7.36	4126.36	To Philadelphia, NYC, offshore NJ

USS *Los Angeles* decommissioned 0815, 30 June 1932

APPENDIX 4

Ship's Crew, USS Los Angeles, 1931–1932

Commander A. H. Dresel, USN Commanding

Lt. F. T. Berry
Lt. R. F. Tyler
Lt. G. C. Calnan (CC)
Lt. C. J. Maguire
Lt. G. F. Watson
Lt. (jg) M. M. Bradley
Lt. (jg) A. MacIntyre
Lt. (jg) B. May, 2d
Lt. (jg) T. M. Whelan
Lt. (jg) C. S. Rounds
Machinist A. B. Clapp
Machinist B. C. Hesser

STUDENT OFFICERS

Lt. Cdr. B. F. Jenkins
Lt. Cdr. J. L. Kenworthy, Jr.
Lt. Cdr. E. F. Cochrane
Lt. Cdr. V. A. Clarke, Jr.
Lt. Cdr. F. L. Worden
Lt. Cdr. D. E. Cummins
Lt. Cdr. J. L. Fisher
Lt. Cdr. H. E. MacLellen
Lt. Cdr. W. K. Phillips
Lt. V. Bailey

Lt. G. H. Mills
Lt. D. M. Mackey
Lt. R. F. Cross
Lt. H. N. Coulter
Lt. (jg) H. M. Wescoat
Lt. (jg) E. K. Swearingen
Lt. (jg) E. J. Sullivan
Lt. (jg) W. A. Cockell
Ens. H. M. Zock
Ens. G. H. Kendall

CREW

Adams, Charles E.	Cox
Baldwin, Harry M.	Y.1c
Barnes, William M.	AMM3c
Bartens, Charles F.	AMM3c
Bernard, Martin J.	Sea2c
Beyer, William R.	AMM3c
Brook, William	BM1c
Buffaloe, Vernon R.	Cox
Carroll, Ansel F.	AMM3c
Cavadini, Cesare P.	CRM
Chenoweth, William	BM2c
Coleman, Lester K.	ACMM
Conover, Wilmer M.	AM2c

Cooper, Fred	AMM2c	Meredith, Lee I.	AMM2c
Danley, Charles W.	QM1c	Miller, Martin O.	ACMM
Echipare, Balbino	Matt1c	Moser, George W.	CBM
Edwards, Newton E.	Sea1c	Muelken, Walter W.	BM2c
Erwin, Moody E.	AM3c	Oliver, Arthur M.	Sea1c
Estridge, Clarence D.	AMM2c	Olsen, Carol G.	AMM2c
Fickel, Wesley S.	AMM3c	Peak, Richard S.	SC1c
Gavigan, Eugene M.	AMM3c	Peckham, Frank L.	CBM
Goode, Peter A.	SF1c	Rowe, Herbert R.	Sea1c
Haskell, Kenneth E.	Sea1c	Rutan, Lucius W.	ACMM
Hingsberger, Andrew J.	AM1c	Ryan, James "D"	Sea1c
Humphrey, Frank L.	Sea1c	Safford, David R.	AMM2c
Hill, Malvin M.	Cox	Santiago, Pacifico	Matt2c
Jandick, Paul A.	AMM1c	Selden, Gerald	AMM1c
Jennings, James C.	ACMM	Sprague, Xensel A.	AMM2c
Johnson, Rufus B.	BM2c	Stevens, Orville L.	AMM3c
Kerson, Ivan J.	AMM2c	Styles, Wade	Cox
Lane, Carl W.	Cox	Taylor, Milan R.	ACMM
Lanozo, Moises	Matt2c	Terry, Arthur C.	AMM1c
Leahy, Philip M.	AMM3c	Tobin, Frederick J.	CBM
Leonard, John F.	AMM2c	Walton, Arthur H.	SK1c
Leonard, John J.	CBM	Ward, Reginald H.	CBM
Leverone, Arthur A.	CY	Westphal, William C.	AMM2c
Liles, Leon D.	C Aerog.	Webber, Robert C.	Sea2c
Long, Glenn L.	AMM3c	Woicinski, Ed	AMM1c
McCracken, John C. Jr.	AMM3c	Wolf, Carl Jr.	AMM3c
Magnuson, Fridolf R.	AMM3c	Woodard, Harvey J.	Sea1c
Manley, Harley E.	CRM	Youngblood, Joe W.	BM1c

APPENDIX 5

Navigational Equipment, USS Los Angeles

(a) Instruments:
Chronometer (with correction and rate)
Standard sextant
Two-bubble sextant
Pioneer drift and speed indicator
Gortz speed and drift indicator
Two azimuth circles
Two stopwatches
One comparing watch
Navigraph (broken)
Altimeter
Airspeed meter
Aerograph
Clock
Two compasses
Variometers

(b) Books:
OOD notebook
Navigator's workbook
Navigator's reference book
Logs
Bowditch (2)
Almanacs current (2)
Azimuth tables, red and blue
Star identification tables
Coast pilot
Light list
Aviation Pilot and Airway Bulletin
HO 208 and *20*

(c) Equipment:
Log sheets, scratch pads
Cross-section paper
Barograph charts
Release forms for passengers
Two parallel rulers
Two flashlights
Two binoculars
12 pencils
2 erasers
Star finder
Small protractor
Arm protector (3)
Triangles
Navigators case
Extra dividers
Compass spider
Batteries for sextant
Flares (20)

— 241 —

Very pistol and ammunition
Parachute message
Breakdown shapes

Navigraph charts
Ply board
Charts

Notes

Preface

1. Theodore von Karman with Lee Edson, *The Wind and Beyond: Theodore von Karman, Pioneer in Aviation and Pathfinder in Space* (Boston: Little, Brown, 1967), 160; Norman J. Meyer, audio interview by author, Alexandria, Va., 30 January 2001. Trained as an aeronautical engineer, in 1941 Mayer was hired by Goodyear Aircraft; he joined the Navy's Bureau of Aeronautics in 1951 as one of its senior advisors.

Introduction

1. Sixth Annual Report of the National Advisory Committee for Aeronautics, 1920, 55.

2. Harry Vissering, *Zeppelin, The Story of a Great Achievement* (Chicago: Privately printed, 1922), 49. Vissering was a Chicago industrialist who had flown on *Bodensee*, a postwar commercial Zeppelin. He returned home an ardent admirer of the venture and an agent of *Luftschiffbau-Zeppelin* in the United States. This lovely volume has been called "persuasive propaganda" by historian Douglas H. Robinson.

3. Hunsaker, "Some Lessons of History," Second Wings Club "Sight" Lecture, 26 May 1965. On his return, Hunsaker built the first U.S. wind tunnel capable of testing all-scale models—an immense leap for aerodynamics. His 1914 paper, "The Present Status of Airships in Europe," *Journal of the Franklin Institute* 16, no. 6 (June 1914) "marked the beginning of serious thinking about Airships in the United States." Note, Fulton to Charles L. Keller, 10 February 1960, courtesy Keller.

4. On duty as an Academy instructor, King also served as Discipline Officer on Fulton and Byrd's dormitory floor. "So excited was King about aviation that he would frequently break all the rules of 'discipline' by encouraging groups of midshipmen to sit up all night with him in lively discussion of aviation and its possible future meaning to the Navy. [My father] would later reflect that most of the early aviation, particular Bancroft Hall floor, and thus had been heavily influenced in later pursuit of aviation duty by Lieut. King's dynamic enthusiasm." Langdon H. Fulton, letter to author, 9 December 1996.

5. Notes by Cdr. J. C. Hunsaker, USN (CC), *The Development of Naval Aircraft; Non-Rigid Airships for the Navy*, 1923, V. 2, Box 150, History of Aviation Collection University of Texas at Dallas [hereafter HOAC].

6. Capt. Garland Fulton, USN (Ret.), "The General Climate for Technological Developments in Naval Aeronautics on the Eve of World War I," in Doyce B. Nunis, Jr (ed.), "Recollections of the Early History of Naval Aviation: A Session in Oral History," *Technology and Culture*, 4, no. 2 (Spring 1961), 161–62.

7. Already the lines of division were deep. This endemic fear and loathing—making the harshest possible view—was to persist, holding LTA proponents in a chronic defensive crouch and yearning for status.

8. Archibald D. Turnball and Clifford L. Lord, *History of Naval Aviation* (New Haven: Yale University Press, 1949), 73–74. "A zeppelin," commented Lt. John Towers, USN, an early naval aviator destined to rise to admiral, "can do the work of a light cruiser."

9. *Rochester (NY) Union & Advertiser*, 5 August 1916.

10. The intelligence process involves analysis as well as information. On the whole, much of this information was valueless.

11. Notes by Cdr. J. C. Hunsaker, USN (CC), *The Development of Naval Aircraft; Rigid Airships*, 1923, VI.1, Box 150, HOAC. "My experiences up to that time," Hunsaker continued, "were limited to a knowledge of the scientific literature regarding rigid airships which was very meager; and I had been allowed to inspect the outside of the passenger Zeppelin *Viktoria Luise* and to take a short flight in her over Berlin in 1913."

12. No serious production efforts were attempted until 1919. Development of a suitable aluminum-copper-manganese-magnesium alloy for girder elements "at first

baffled our best efforts," Hunsaker was to remark. "The development of the art of making and fabricating duralumin was taken up as a cooperative effort between Dr. Blough of the Aluminum Company, Dr. Burgess of the Bureau of Standards, and the Bureau of C&R. Eventually our American duralumin had become both cheaper and more uniform than the foreign metal and of slightly higher strength. ZR-1 was built entirely of American made duralumin." Humsaker, *Rigid Airships,* op. cit., p. VI.1–4.

13. At that point, Army interest in rigids was at least equal to that of the Navy. "I was a little surprised," Fulton was to recall, "to find that a larger number of Army students had been sent to [Hunsaker's] course at M.I.T. than Navy students. There were at least a half dozen Army fliers there . . . The aeronautical course included work on dirigibles (airships) and airplanes." Fulton, op. cit., p. 163.

14. The British Admiralty, no less impressed by zeppelins, appreciated their vulnerability to hydrogen. It therefore began "quietly investigating" for sources of helium in Canada. And, soon, the U.S. was being pressed to expand its production and supply England. The Admiralty even sent "observers" to keep in touch with the progress.

15. Paolo E. Coletta, *A Survey of U.S. Naval Affairs, 1865–1917* (Lanham, Maryland: University Press of America, 1987), p. 200.

16. Hunsaker, *Non-Rigid Airships,* op. cit., p. V.2.

17. "In my book," Fulton opined, "Taylor was among the early giants in giving support to aviation . . ." Letter, Fulton to Keller, 5 March 1961.

18. During the war, Goodyear's Akron facilities were to manufacture more than half of the LTA craft used by the government—free balloons, kite balloons, blimps. As for flight training, "We didn't know a thing about flying those things [blimps]," Peck, an Akron graduate, recalled—"had to teach ourselves. But we did!" Rear Admiral Scott E. Peck, USN (Ret), interview by Michael C. Miller, Chula Vista, CA, 9 February 1980.

19. Hunsaker, *Non-Rigid Airships,* op. cit., p. V.4.

20. "War Activities of the United States Navy," address to the Harvard Alumni Association, Cambridge, Massachusetts, 1918.

21. J.A. Sinclair, *Airships in Peace and War* (London: Rich & Cowan LTD., 1934), p. 60.

22. It was soon realized that, for convoy duties especially, high speed was not essential. More useful was the airship's ability to loiter at sea for long periods.

23. As Fulton was to recall, "Soon after we entered the War the word came along that aviation activities were growing by leaps and bounds and needed additional staffing. They were looking for young constructors who had some smattering of aeronautical knowledge. I volunteered and was ordered to Washington in May 1918 and became Hunsaker's executive assistant." Fulton, op. cit., p. 164.

Chapter 1. *LZ-126:* Birthplace Friedrichshafen

1. "The years 1919–22 were pretty lean ones for aeronautical activities everywhere. War surpluses had to be used up. Retrenchment was the word." Letter, Fulton to Keller, 17 June 1960.

2. Late in 1917, *L-59* demonstrated the practicability of long-distance air transport. Aloft for ninety-five hours and with fuel still on board on arrival, the zeppelin flew 4,230 miles—Germany to Berlin's Africa colonies and the return. These air miles exceeded the distance between Berlin and Chicago.

3. Aboard as working members of ship's crew were Lt. Cdr. Zachary M. Lansdowne, USN (westbound) and, on the return, Col. William N. Hensley, Jr., of the Army Air Service. By virtue of his being detailed to *R-34,* Landsdowne, Naval Aviator (Airship) No. 100, became the first American to cross the Atlantic by air. As assistant naval attache at Berlin, he would assist the negotiations for construction of *LZ-126/ZR-3.*

4. Why? "For one thing," Robinson points out, "the Germans planned it that way, and sullenly refused all operating and technical information to their French and British opposite numbers. For another, after two years of neglect, the ships were in bad condition, particularly their gas cells. Lastly, the Zeppelins in use at the end of the war were highly efficient war machines, designed expressly for bombing from great altitudes, and so lightly built that only highly trained crews could fly them safely." Douglas H. Robinson, *The Zeppelin in Combat,* rev. ed. (Sun Valley, Calif.: John W. Caler, 1966), 344–45.

5. Cdr. J. C. Hunsaker, (CC), "The Development of Naval Aircraft—Rigid Airships (Notes by), 1923, VI.11–12, box 150, HOAC. In 1920, Hunsaker made a thorough study of British practice and visited Friedrichshafen. "I was permitted to inspect superficially the airships *Bodensee* and *Nordstern* which are hung up in the sheds there awaiting final disposition. The principal interest in these ships lies in the fact that they were designed after the Armistice and incorporate features which the builders believed to be advantageous from their war experience." Hunsaker to Director Naval Intelligence,

8 September 1920, Hunsaker Collection, box 22, folder 9, Smithsonian.

6. Letters, Fulton to Charles L. Keller, 8 November 1958, 14 April 1959, courtesy Keller.

7. Britain's first passenger airship was the *R-36*, launched in 1921. "In view of the interest which is being taken at the moment in the use of airships for commercial purposes, it is worthy of note that an airship of the 'R. 38' class, adapted for transport, could carry 40 passengers and 2 tons of freight in a nonstop flight to Egypt in about 48 hours." *Flight* 8, no. 23 (9 June 1921), 389.

8. *ZR-3* files, Garland Fulton Collection, National Air and Space Museum (NASM), Smithsonian Institution; Richard K. Smith, *The Airships Akron and Macon: Flying Aircraft Carriers of the United States Navy* (Annapolis, Md.: U.S. Naval Institute, 1965), 13. Hensley was no maverick; an allotment of $360,000 was transferred to German marks in Paris, then placed in Hensley's credit—funds for purchase. At the October 1919 exchange rate, this was equivalent to eight million marks. While "getting the option and obtaining signatures on the contract," the colonel flew sixteen passenger trips on *Bodensee*. Fulton, ibid. Despite the fiasco, the army remained anxious for one or more rigids.

9. Zeno Wicks, "A Short History of Naval Aviation, Particularly of Lighter-than-Air, as Experienced by Z. W. Wicks," undated carbon copy, 3–4, courtesy Langdon H. Fulton.

10. Garland Fulton, "Helium through W.W.'s I and II," undated paper, Fulton Collection, NASM, Smithsonian. The NACA deemed helium "imperative" for airships; in 1920 it called upon the secretaries of war, the navy, and the interior to encourage production so as to increase demand, hence supply, thus lessening the cost of production. *Sixth Annual Report of the National Advisory Committee for Aeronautics* (Washington, D.C.: U.S. Government Printing Office, 1920), 19.

11. The army and navy bore the costs on a fifty-fifty basis, with each to receive half the helium produced. Though still high, the cost per unit volume was on a steadily decreasing curve.

12. "Up until the mid-twenties N.A.C.A. devoted very appreciable time and thought to airship programs. After mid-twenties, N.A.C.A. veered away from airship work—or rather they allowed HTA work to have priority. Their interest in *materials*, especially duralumin, continued however largely, I suspect, because HTA craft stood to benefit therefrom." Letter, Fulton to Keller, 19 October 1958.

13. On the executive committee also in 1924: Rear Adm. David W. Taylor (since 1917), Maj. Gen. Mason M. Patrick, U.S. Army, and Orville Wright. Hunsaker's name is another that appears early. Over the years, various standing committees, with subcommittees, were established. In 1927, Taylor was chairing the Committee on Aerodynamics; its seven-member Subcommittee on Airships, organized that September, included Fulton, Dr. Karl Arnstein, the army's Major William E. Kepner, and Ralph H. Upson.

14. *Sixth Annual Report of the National Advisory Committee for Aeronautics*, 1920, 19.

15. The Americans were but minor players. "We had not ratified the Peace Treaty," Fulton noted, "and were definitely at the low end of the table." Letter, Fulton to Keller, 20 April 1959.

16. See Robinson and Keller, *Up Ship! U.S. Navy Rigid Airships, 1919–1935* (Annapolis, Md.: Naval Institute Press, 1982), 116–23, for an in-depth account.

17. Luftschiffbau Zeppelin to Foulois, 13 June 1921, Fulton Papers, courtesy Keller.

18. Despite disaster, the NACA, in a special resolution to the president and secretaries of war and navy, urged the continuance of airship development. Its September statement, among other points, called for "the development and checking of the theories used in the general design of airships."

19. "I wish to impress upon you first, the embarrassing position in which the loss of *ZR-2* has placed the Navy Department and second, and consequently, the urgent necessity for pushing this negotiation for a German Zeppelin as replacement for *ZR-2* in our program through and, in spite of Allied opposition, to a successful issue." Letter, Secretary of the Navy [hereafter SecNav] to Secretary of State [hereafter SecState], 19 September 1921, Fulton Collection, courtesy Keller.

20. Ibid.

21. Cable, SecState to U.S. Ambassador, Paris, 7 November 1921, Fulton Collection, courtesy Keller; quoted also in Chairman (Moffett) to Aeronautical Board, 22 October 1923.

22. Memorandum, Capt. Frank B. Upham to SecNav (Bureau of Aeronautics [hereafter BuAer]) via Director Naval Intelligence, 9 March 1922, NASM Archives, file A3U-721000-01, Smithsonian.

23. Kurt F. Bauch, interview by author, audio cassette, Millville, N.J., 29 June 1977. Bauch, trained as a mechanical engineer, was hired by Luftschiffbau-Zeppelin in 1922. Following a two-month apprenticeship, he was assigned to the *"126"* project.

24. Letter, Moffett to Beehler, 10 May 1922, box 4016, unmarked file, Fulton Collection, Smithsonian. In commanding tones, Moffett insisted upon complete authority: "I need not inform you that the Department can tolerate no interference with the construction of this ship from any Commission or Control. The second difficulty promises to be with our own good friends the Army. As you may know there is an element in the Army Air Service, which is sincerely and zealously advocating an independent air service, and use anything they can lay their hands on as material to advance their arguments. The reparations ship is a NAVY project with the Army having, upon completion, a joint interest in the ship." Ibid.

25. Letters, Kennedy to Geiger, 25 May and 5 July 1923, box 4016, Fulton Collection, NASM, Smithsonian; Chief, Air Service, Confidential Report, 28 September 1925, NASM Archives, file A3U-721400-01, Smithsonian.

26. "I do not agree and the lawyers would not agree," Fulton wrote, "but am quite willing to try to pry the stuff loose with as little actual friction as possible, even though this means following the German style of long wearing arguments over trivial technicalities. If we use a big stick we prejudice the *quality* and very possibly the *quantity* of data" [emphasis in original]. In an earlier letter: "They simply want to go off in a corner and develop their design." Fulton to Hunsaker, 18 January 1923 and 22 July 1922, Hunsaker Collection, box 2, folder 3, NASM, Smithsonian.

27. Dissatisfaction with respect to cooperation is plain from the record. "If shop work is underway," Fulton rebuked, "there should be progress reports, inspection records, tests of materials and similar data available. None have been furnished." Aso, "Practically no information has been supplied on the 400 HP Maybach engine." Memorandum, 31 October 1922, box 4016, Fulton Collection, NASM, Smithsonian. Yet the INA could be gentle, as one note shows: "Dear Dr. Arnstein, if you have the time, I would appreciate your straightening me out." Ibid.

28. Hunsaker to Fulton, 16 January 1923, Hunsaker Collection, box 2, folder 3, Smithsonian.

29. For a recounting of the interallied negotiations that produced the *ZR-3* contracts as well as Washington's protracted headaches with the Luftschiffbau, see Robinson and Keller, *Up Ship!*, chap. 9.

30. Luftschiffbau's original stress calculations covered two extreme conditions: light load (gross lift 114,000 pounds) and full load (gross lift 189,000 pounds). The aircraft's gross lift in normal flight condition was, in 1929, about 140,000 pounds. C. P. Burgess, "Static Bending Moments and Stresses in U.S.S. LOS ANGELES in Present Normal Flight Condition," *Design Memorandum No. 92,* December 1929, 1.

31. C. P. Burgess, "The Application of the Principle of Least Work to the Primary Stress Calculations of Space Frameworks," *Aeronautical Engineering* 1, no. 3 (July–September 1929), 131. A civilian official in BuAer and a peerless engineer, Burgess wrote the only textbook published in the United States (1927) concerning rigid-airship design.

32. Today, finite-element analysis allows designers to depict a structure and apply forces to it; by using the characteristics of its individual elements, they can determine how the arrangement itself will respond. Norman J. Mayer, audio interview by author, Alexandria, Va., 30 January 2001.

33. Bauch. The contract stipulated three copies of all plans as well as "complete and indexed" set of drawings within sixty days of U.S. acceptance. "By the way," Bauch added, "this was also the time when we did not have yet all the calculators and the calculating machines. So the entire stress analyses was carried out on two fifty-inch slide rules. I still have mine (laughing) in Akron." A great deal of local strengthening was incorporated, to accommodate stresses imposed during ground handling and at mooring masts.

34. Letter, Fulton to Moffett, 26 February 1923, box 4016, Fulton Collection, NASM, Smithsonian. The army had revived the question of ultimate allocation and wanted a decision. "Personally I can't see any legitimate Army function for large airships, unless in future wars the Army and Navy are not going to co-operate with each other at all. The Navy, through its earlier work with rigid airships, is the only service competent to acquire *ZR-3* and get maximum benefit from such acquirement." Ibid.

35. Letter, Hunsaker to Fulton, 10 April 1923, Fulton Collection, courtesy Keller. A copy resides also in the Hunsaker Collection, box 2, folder 3, NASM, Smithsonian.

36. "The apprenticeship at Maybach was then regarded as one of the best in Germany," recalled Eugen Bentele, a mechanic assigned to *Graf Zeppelin* and *Hindenburg*. "But it was hard training." Eugen Bentele, *The Story of a Zeppelin Mechanic; My Flights 1931–1938*, trans. Simon Dixon (Friedrichshafen, Ger.: Wolfgang Meighorner-Schardt, 1992), 17.

37. Letter, Fulton to Upham, 2 May 1923, Fulton Collection, courtesy Keller.

38. *Aviation,* 24 December 1923, 774.

39. In view of the possibility of his also commanding *ZR-3*, Klein was to make the crossing if at all possible. "As a matter of personal interest to you, the Admiral took up with General Patrick the idea of substituting Klein for Kennedy. I thought the Admiral had a good deal of nerve to even suggest it, but I was strong for the idea. Naturally the request was delivered with thanks!" Memorandum, Land to Fulton, 19 September 1924, Fulton Collection, courtesy Keller.

40. Memorandum, Pennoyer to Halsey, 13 September 1923, ibid.

41. Memorandum, Chairman (Moffett) to Aeronautical Board, 22 October 1923, NASM Archives, file A3U-721400–01, Smithsonian. In Moffett's view, national interests would best be served by continuing established policy until both ships were thoroughly tested in naval service—*ZR-1* "to prove her practicability with the Fleet; *ZR-3* to prove her usefulness by experiment in determining the feasibility of the commercial use of rigid airships." Ibid.

42. Letter, Hunsaker to Fulton, 20 September 1923, Fulton Collection, courtesy Keller.

43. Letter, Moffett (Land) to Fulton, 19 May 1924, ibid.

44. Ibid.

45. Memorandum, unknown to Land and Lt. Mitscher, 22 January 1924, box 4008, Fulton Collection, NASM, Smithsonian.

46. Starting months before the preparations and trials, Goodyear-Zeppelin had worked to get its own story "injected" into the news pertaining to *ZR-3*. Objective: favorable public opinion in America toward lighter-than-air craft.

47. "Trial Trip Mishap to ZR 3," *Star* (London), 28 August; "All Five *ZR-3* Engines to Be Strengthened," *New York Times*, 29 August 1924. The trial had revealed other, minor disappointments. Mounted outside near the bridge, the windmill providing current for the gyro compass made an annoying hum. "We must have that noise stopped," Eckener remarked, "or we will all go mad!" The dynamo furnishing current to the electric kitchen also proved noisy. But, as Steele wrote, "It was surprising to note how little noise was made by the engines, even at full power. There was no vibration in the passenger car and very slight rolling of the ship, even in the disturbed air in which we flew." "Life on an Airship," *Aero Digest* 5, no. 6 (December 1924).

48. Bauch.

49. The airship had been insured for nearly $750,000 for the crossing, the policy underwritten by English, German, Dutch, and Danish companies. *Aviation*, 22 September 1924, 1027.

50. Letter, Fulton to Keller, 23 May 1961.

51. "Third *ZR-3* Test Trip a Success," *Washington Star*, 12 September 1924.

52. *Aviation*, 29 September 1923, 1065.

53. This was to determine drag, from which was calculated the airship's drag coefficient and the ratio of horsepower required to total shaft horsepower. This furnished a basis for comparison of different craft and was of use for future design.

54. Record Group 72, Bureau of Aeronautics General Correspondence, 1925–1942 [hereafter RG 72, BuAer GenCor], box 313, National Archives.

55. As a chart makes plain, Newfoundland and Labrador were crucial to great-circle traffic. As yet, Canada was not preparing North Atlantic forecasts. Hence, while Washington had full access to its data, the Canadian Meteorological Service played no part in preparing these transmissions; Dr. Morley K. Thomas, unpublished manuscript. In 1930, in support of *R-100*'s leap to Montreal, the service was active in transatlantic forecasting.

56. Most changeable and violent forms of weather occur at or near the boundaries between opposing air masses. These zones of interplay are termed "fronts," so named by the Norwegian meteorologists whose studies yielded the modern concept of the structure of storms. Principal discover: physicist Vilhelm Bjerknes.

57. Within a decade, a complex system of airways weather reporting was in place. Based on regular observations and reports of actual conditions, its dispatches gave pilots definite information of unfavorable weather along their routes. One outcome: fewer forced landings due to *unexpected* bad weather. This close-network type of organization, serving fixed routes, had limited application to transoceanic flying.

58. F. W. Reichelderfer, "Aviation Weather Service, Particularly for Ocean Route, 1934 Aviation Commission Hearing (report to)," 9, courtesy Navy Lakehurst Historical Society. The department, recognizing advances made by the "Frontal School," dispatched officers to Europe to study the latest techniques. "The results of these studies in each case have been of considerable value to our naval meteorologists in their daily analysis of the ocean weather map for use of the Fleet. The results have verified the early belief that the frontal method holds the greatest promise for improved weather service at sea." Ibid., 10–11.

59. "Deadhead passengers and ambitious correspondents who had hoped to stow away on the ship," the

Times reported, on the 13th, "found their footsteps dogged by a squad of plains clothesmen, which the Zeppelin company engaged for the purpose of guarding the hangar and shooing away suspicious loiterers." Some were making life miserable for Eckener. In the end, company sleuths saw to it that all remained behind. Among the disappointed: the Spanish military attache.

60. Cold grants extra lift; accordingly, the intent was to get away before the sun and rising air temperatures.

61. Besides Eckener, these were Ernst Lehmann, Hans Flemming, Hans von Schiller and Anton Wittemann.

62. Why not Fulton? Closing out the INA office entailed numerous loose ends—and he wanted "to make a finished job of it." "Somebody responsible has got to stick around—particularly regarding data which is just as important as ship." "My own personal idea is that you should lay off as I don't see that there is any advantage to it," Hunsaker had advised. "The risk is unjustified. You have experience that can't be replaced." Fulton to Hunsaker, 28 February 1923, Hunsaker Collection, box 2, folder 3, NASM, Smithsonian; Hunsaker to Fulton, 10 April 1923, Fulton Collection, courtesy Keller.

63. Hans von Schiller, "Log of the *ZR-3*," *Aero Digest* 5, no. 5 (November 1924), 266–68, 300. Subsequent remarks are from this work.

64. Bauch.

65. Peter Fritzsche, *A Nation of Fliers: German Aviation and the Popular Imagination* (Cambridge, Mass.: Harvard University Press, 1992), 138. According to the *Vossische Zeitung*, ZR-3 "will prove to Americans that the shackles which now fetter German airship construction must fall if true progress is to be made in the direction of resuscitating a sick world." As quoted in *New York Times*, 13 October 1924.

66. To ensure an ample reserve, extra tanks had been mounted. In "American condition," seventy-five would be carried. Of that number, fifteen were "service" tanks for the power cars' direct use. The remaining sixty were considered "reserve." The latter was pumped into emptied service tanks as required.

67. Allotment for the Americans: one suitcase and three blankets each. Stashed in their belongings were playing cards, cartons of chewing gum. and chewing tobacco. *Public Ledger,* 10 October 1924.

68. Hugo Eckener, *My Zeppelins*, trans. by Douglas H. Robinson (London: Putnam, 1958), 21. *LZ-126* was not to be the last dirigible; "it turned out better than we had expected." Bauch.

69. Eckener, ibid., 23.

70. Seeking to minimize interference, the U.S. Navy had requested the cooperation of radio groups and individuals. "On the other hand, the Department asks for the heartiest cooperation of the radio public in the matter of listening in for messages transmitted by *ZR-3*, as such assistance may be extremely valuable in maintaining communication with the ship." *Aviation,* 22 September 1924, 1027.

71. It should not be assumed that Eckener was naive with respect to North Atlantic conditions. Eckener himself had an extraordinary "weather eye"—one reason for his consistent success in the air. Further, his organization had conducted simulated flights based on weather maps.

72. Capt. Ernst A. Lehmann with Leonard Adelt, *Zeppelin: The Story of L-T-A Craft,* trans. Jay Dratler (London: Longmans, Green, 1937), 236.

73. Eckener, 23–24.

74. Nevil Shute, *Slide Rule: The Autobiography of an Engineer* (London: William Heinemann, 1954), 111. Shute was chief calculator for Vickers during its *R-100* construction project.

75. Surface-ship deployment dates from 1919, when several were stationed along the route of *R-34*. For *ZR-3*, the Navy Department was coordinating aerological and radio-relay services from cruisers. Via dispatch from Steele, and in the interest of "surer communication," Eckener had requested that *Milwaukee* shift southeasterly, but keeping within radio range of *Detroit* and shore stations. *New York Times,* 13 October 1924. The inauguration of a system of weather reports to *ZR-3* "might be considered as a forerunner of a service regularly maintained for commercial airships engaged in trans-Atlantic commerce." Navy Department press release, 7 September 1924, NASM Archives, file A3U-721800-01, Smithsonian.

76. Eckener, 27; von Schiller. In connection with the arrival, Boston Airport, Mitchell and Langley Fields, and the Marine base at Paris Island, South Carolina, had been designated as emergency landing fields.

77. Eckener, 27.

78. *New York Times,* 16 October 1924.

79. Ibid.; von Schiller. "All in all," von Schiller observed, concluding his on-the-scene-report, "the trip was marvelous. Surely it has proved the entire feasibility of using airships for fast mail traffic."

80. Eckener, 27.

Chapter 2. *ZR-3*: Homeport Lakehurst

1. Melvin Cranmer, an air-station electrician, was an ardent radio amateur. Hearing of the contact, he rushed home. "I listened to it coming in. And that was

quite a thrill." Interview by author, audio cassette, Lakehurst, N.J., 12 February 1978.

2. In 1924, Reichelderfer was one of *two* aerological officers in the Navy, all others having resigned their commissions or switched duties. With Lt. J. B. "Bruce" Anderson away on *Shenandoah* to the West Coast and back, "Reich" had taken his place. Capt. F. W. Reichelderfer, USN (Ret.), interview by author, audio cassette, Washington, D.C., 18 June 1977.

3. Reichelderfer, ibid. Interestingly, the wartime zeppelins had operated with scant meteorological support. The ambiguities of stress-analysis theory notwithstanding, no major structural failures were recorded.

4. "Drastic Safety Measures Taken to Prevent Explosion of *ZR-3*," *Washington Star,* 14 October 1924. Pierce's posted orders required that all hands "use only air-lock doors for communication between shops and hangar floor. No smoking by anybody on landing field while *ZR-3* is on the field. No smoking anywhere within hangar, shops or offices opening directly into hangar." Ibid.

5. Capt. Ernst A. Lehmann with Leonard Adelt, *Zeppelin: The Story of L-T-A Craft,* trans. Jay Dratler (New York: Longmans, Green, 1937), 240; Lt. Cdr. Leonard E. Schellberg, USN (Ret.), interview by author, audio cassette, Toms River, N.J., 6 February 1977. That November, Schellberg and other men were transferred to the station for further transfer to *ZR-3*.

6. Cargo included mail as well as a thousand or so tiny toy figure souvenirs—the first-ever merchandise by air from Europe to America. These were brought to the New York and Philadelphia stores of John Wanamaker. The first set went to President Calvin Coolidge; the others were used to raise Christmas funds for children.

7. Custody of *ZR-3* passed from its German crew when the Navy landing party took it in hand. Until Steele read his orders and assumed command, Klein, as station skipper, would be responsible for the ship.

8. Commanding Officer [hereafter CO], Naval Air Station [hereafter NAS] Lakehurst to Chief, BuAer, 31 October 1924, box 4010, Fulton Collection and (also) NASM Archives, file A3U-723400-01, Smithsonian.

9. Though it is fifty miles away, Lakehurst lay within the Philadelphia Customs District. Accordingly, all foreign arriving aircraft had to be cleared by federal officials from the port city.

10. Hugo Eckener, *My Zeppelins,* trans. by Douglas H. Robinson (London: Putnam, 1958), 28.

11. "I remember," Reichelderfer told the author, "what an impressive and splendid man [Eckener] was.

He was very competent, large, dignified. He was always very businesslike. You knew he was a man of some rank." Reichelderfer.

12. F. W. von Meister, interview by author, audio cassette, Peapack, N.J., 7 February 1977. "The arrival in the United States of this strictly modern Zeppelin," one publicist predicted, "will no doubt create a wonderful interest as the American people have never seen a real Zeppelin and it will give a great impetus to airship activities throughout the world." Harry Vissering, *Zeppelin: The Story of a Great Achievement* (Chicago: privately published, 1922), foreword.

13. "Remarks of Dr. Hugo Eckener before Annual Meeting of National Advisory Committee for Aeronautics, Washington, D.C., 16 October 1924," extract from minutes. When Eckener remarked on the great comfort of airships, Steele interrupted, "I can bear that out." Fulton Collection, NASM, Smithsonian.

14. "We congratulate you upon the successful completion of your flight across the Atlantic and upon the magnificent airship you bring. You have brought to us this splendid product of German skill and of scientific ability. We wish this ship to be a symbol of peace and of friendship between the two nations here represented. We wish you to feel that when the German flag is replaced by the American flag, that the ship you have built will still fly a flag friendly to the builders, the flag of a people honestly desiring the prosperity and happiness of all the German people." File A3U-721800-01, Smithsonian; "Washington Honors ZR-3 Officers," *New York Times,* 17 October 1924. The public had suggested a dozen or so names. Wilbur hailed from Los Angeles. When the name for *ZR-3* was announced, one San Francisco newspaper pronounced it fitting for the largest gas bag in existence.

15. Hugo Eckener, *Im Zeppelin über Lander und Meere* (Flensburg: Verlagshaus Christian Wolf, 1949), 65–66; quoted in Peter Fritzsche, *A Nation of Fliers: German Aviation and the Popular Imagination* (Cambridge, Mass.: Harvard University Press, 1992), 139.

16. *Literary Digest,* 25 October 1924, 12–14. *Shenandoah*'s nineteen-day transcontinental sortie together with the delivery of *ZR-3* had, according to *Aero Digest,* "proved conclusively what may be looked for from dirigibles." *Literary Digest* 6, no. 1 (January 1925), 7.

17. In Friedrichshafen, deep uncertainty prevailed until construction revived. "In fact, the *Graf Zeppelin* was built there before we started *Akron* and *Macon*. We *envied* the people there. We were sitting here [in Akron, doing initial design work for a large passenger ship] and they

were building another airship." Kurt Bauch, interview by author, audio cassette, Millville, N.J., 29 June 1977.

18. For major refits and layups, deadweight was supported about equally between cable suspensions and cradles under all cars, together with shores—poles under longitudinal zero cut to length. The tension of the cables, which were equipped with dynamometers, was adjustable, to allow for redisposition of weights (and the work to be done) on board, and for roof movements. Care was taken to avoid undue concentration of load at any point, so as to minimize shear and bending moments.

19. Rear Adm. Calvin M. Bolster, USN (CC) (Ret.), audio cassette prepared for author, 21 April 1976; G. V. Whittle, NAS Lakehurst memorandum, 22 October 1924, NASM Archives, file A3U-721000-01, Smithsonian. Damaged during delivery of *ZR-3*, number-eight cell would await permanent repair before reinflation. *ZR-1* when completed was the first rigid airship to fly with helium.

20. The 1924 board consisted also of Captains J. G. Tawresey (CC) and I. E. Bass, and Lieutenant Commanders R. A. Burg and J. H. S. Dessez.

21. *Report of Board of Inspection and Survey*, November 1924, RG 38, Office of the Chief of Naval Operations [hereafter CNO], Board of Inspection and Survey [hereafter BIS] Reports of Acceptance Trials of Naval Aircraft, 1919–1932, box 50, National Archives. Tons of equipment and spare materials—in huge packing cases—had arrived in advance of the delivery. Contents included appurtenances for the (stripped) crew quarters and stateroom furniture, as well as such miscellany as window drapes, duralumin cooking gear, and spare lattices, channels, and rivets. *Washington Star,* 14 October 1924.

22. Lt. David F. Patzig, USN (Ret.), interview by author, audio cassette, Spring Lake, N.J., 25 March 1978.

23. The gassing main from the Helium Plant fed on each side of the hangar to outlets that in turn were connected by flexible tubing to the airships' inflation manifolds. A deflation line with exhauster carried off gas for repurification.

24. No protective coating had been applied in Friedrichshsfen. Why? The builders estimated the useful life of their ships at five years and, rightly, saw no advantage in adding protection for that duration. Lt. R. G. Mayer (CC) to CO, USS *Los Angeles,* 6 March 1929, Fulton Collection, NASM, Smithsonian. In 1926, a lengthy (sewn-together) padded canvas "walkway cover" replaced the boards. To surface sailors, a five-year surface life seems almost ludicrously short. Rigid airships were huge yet fragile structures of (new) aluminum alloy, wires, gas cells, and fabric. The science of aerodynamics was young, and builder/operators had scant precedent to guide them in appraising material condition and remaining useful life.

25. Following the fiery loss of its semirigid *Roma (RS-1),* the Army Air Service had decided upon helium. From 1920 to about 1936 the army and navy juggled the available gas between them, sharing costs and equipment, and coordinating planning through a governing body comprising army, navy, and Bureau of Mines (BuMines) representatives. BuAer acted for the navy as operating agent for the Helium Board at the Fort Worth plant.

26. In accordance with an act of Congress, in July 1925 the naval officer in charge turned over the plant to a representative of BuMines.

27. Obstacles awaited. The resolution adopted by the Council of Ambassadors in 1921 approving the acquisition of *ZR-3* had reprieved the large building shed—temporarily. To ensure complete execution of the treaty, the zeppelin works were to be destroyed, its personnel dispersed. Further, Eckener faced an empty cash box, as well as an empty shed.

28. *Vorwarts,* no. 487, 15 October 1924, evening edition, quoted in Fritzsche, 141.

29. *LZ-126* was built to accommodate passengers, freight, and mail. To counterbalance passengers' weight, freight was to be stored aft. For noncommercial deployment, the navy would compensate by trimming with fuel and water and by storing spares aft.

30. "The basic concept and philosophy of Dr. Eckener was a very clear one," Von Meister observed, "and that was that the only real practical use—commercially—of the airship was intercontinental, transatlantic/transpacific, air transport. That was *not* possible unless other nations would participate. So his desire and his aim was to get enough of the commercial and industrial and financial people in the United States interested in the airship." Von Meister. As it happened, the Navy's acquisition of a modern ship (and German know-how) was to prove of inestimable worth to the Zeppelin venture.

31. Letter, Cdr. Fred T. Berry to Mrs. Berry, 6 August 1930, courtesy Tom Berry.

32. Superheat is the temperature differential between the airship's gas and the atmosphere—usually due to the sun's heat. Since the density of the gas in the cells decreased as it expanded with heating, its lift was more (positive superheat). If superheat was expected upon clearing the shed, only part of the full flight crew embarked. Why?

If fully loaded in berth and ballasted accordingly, reballasting would be needed on the field when the ship picked up superheat. Those not yet aboard kept near the forward car, ready to embark as ordered.

33. Bolster. Little would change. In the next war, K-class blimps returned from long flights statically light. "The landings were invariably at night, often with little or no wind. The ships of that day didn't have controllable-pitch props with reversing capability. Pilots literally had to drive the ship down to within reach of the ground crews who 'bulldogged' them the rest of the way down and frantically hung ballast on them bringing the ship under their control." Lt. Cdr. R. W. Widdicombe, USN (Ret.), memorandum to Capt. Norman L. Beal, USNR, CND Office of Naval Warfare, 1 November 1985, courtesy Lt. Cdr. Widdicombe. Widdicombe enlisted in 1937 and transferred to NAS Lakehurst in June 1940 (Class IX) for enlisted LTA training.

34. In June 1927, Tyler qualified as Naval Aviator (Airship) No. 3376.

35. By arrangement, eight Air Service officers had been ordered to Lakehurst for "observation and instruction on SHENANDOAH." By November, 1924, two were attached: Col. Chalmers G. Hall and Capt. William E. Kepner. Both accompanied *ZR-3* on this day. The more experienced, Hall, was to urge his chief to "further action" to secure *Los Angeles* for the Army. Log, NAS Lakehurst, National Archives; memorandum, Chief of Air Service, 8 September 1925, NASM Archives, file A3U-721400-01, Smithsonian.

36. USS *Los Angeles*, log, 25 November 1924. Rather than "light" landings, the Germans usually weighed off to near equilibrium. Hydrogen was relatively inexpensive and thus offered greater operational flexibility.

37. "Baffling" from Eighteenth Annual Report of the National Advisory Committee for Aeronautics, 1932, 60. A system had been wanted in a hurry, so further design and test work awaited *ZR-3*. "You liked to stay away from that water-recovery water as much as possible," a rigger recalled, "because it was *black*, full of carbon, it messed everything up—you didn't want to use it if you didn't have to. It was much nicer just usin' water that you pumped into the ship from an outside water source." Lt. Herbert R. Rowe, USN (Ret.), interview by author, audio cassette, Toms River, N.J., 4 December 1977. A 1930 graduate of the LTA Training School, "Monty" Rowe served as a rigger aboard *Los Angeles*, *Akron*, and *Macon*.

38. After a light landing on 30 December (two unsuccessful tries), a Marine private was dragged by the mooring lines. The man sustained fractures of the right radius; two others received minor injuries. The station log records the accident; the ship's log does not.

39. Draft memorandum, Chief, BuAer, to CO, NAS Lakehurst, undated (January 1925), box 4010, Fulton Collection, Smithsonian. His vigorous public defense of the airship notwithstanding, Moffett had thought he detected inertia in Lakehurst's operating organization.

40. Before undocking or docking, the engines were made ready. On the field, masted or under way, all were kept warmed to the designated temperature, by being run at intervals.

41. Standard practice was to unhouse via the lee doors. When passing through the east doors stern first, into the rising sun, the after cells acquired superheat relatively rapidly, thus disturbing trim temporarily.

42. "Some of us went to Pulham yesterday," then-Cdr. Emory Land of the naval attache's office wrote, "to see the mooring mast and observe the methods of handling *R-33* in connection with the mooring mast trials under way at Pulham. We were all very much impressed with the simplicity of this method of handling rigids. The advantages of a mooring mast are so great as to put an entirely different status on the question of airship development and use." Land to Hunsaker, 24 March 1921, RG 72, BuAer GenCor, box 3714, National Archives.

43. Capt. M. M. Bradley, interview by author, audio cassette, Springfield, Va., 22 October 1977.

44. "Taking off from a mast is generally looked upon as safer than taking off from the ground because of the initial altitude," the manual advised. "However, this is generally a delusion for the mast is an obstacle to be feared."

45. "Take into consideration," the manual cautions, "loss of power and mobility of ground crew if ground is covered with snow, or if crew is cold and tired."

46. Cold-weather flight gear comprised two-piece, heavily lined flying suits, leather helmets, goggles, moccasins (no hard-soled shoes of any kind aloft), fleece-lined gloves, plus winter underwear. and so on. Each man was responsible for the gear in his custody. "They [the suits] were real heavy, I remember that," Rowe recalled. "After you carried that around for four hours on one watch, particular[ly] if you were on a wheel—on either the rudders or elevators—and it was a-tall rough, you'd work up a sweat, even in the coldest weather." Rowe, ibid. Summer wear: dungarees and denim jacket.

47. Capt. Edwin T. Pollock, "Studying the Eclipse from the Los Angeles," *McClure's* 1, no. 2 (June 1925),

9–22; NASM Archives, file A3U-721800-01, Smithsonian.

48. CO (Klein) *Los Angeles*, to Chief, BuAer, 6 March; CO, *Los Angeles*, to Chief, BuAer, 29 April 1925, file A3U-723400-01, Smithsonian; Steele, "The Airship in War and Peace; Beginning and Early History," 1926, box 4008, Fulton Collection, Smithsonian. The alcohol arrived in drums. Formaldehyde and red dye was added prior to distribution to render it unpalatable. "Quite a number of the ingenious people around decided that formaldehyde could be distilled out," a crewman remembered. Hand-made stills—usually a fire extinguisher fitted with copper coils—restored the liquid to a drinkable state. "You could distill *all* the formaldehyde that the Navy could put in it, or the Government could put in, out in no time at all." Lt. David F. Patzig, USN (Ret.), interview by author, audio cassette, Spring Lake, N.J., 25 March 1978.

49. CO, *ZR-3* to Chief, BuAer, 20 May 1925, NASM Archives, file A3U-723400-01, Smithsonian.

50. Reichelderfer. The main communications unit (a room filled with receivers and transmitters) was the lifeblood of Aerology. "The Navy Department and the lighter-than-air organization didn't skimp on communications," Reichelderfer opined. "They got the very best equipment. They did a marvelous job of intercepting the synoptic reports, the weather reports, twice or four times daily. We had excellent service"—so good that he took the facilities quite for granted. "If something went wrong, then they heard from me." Ibid.

51. Except when relieved by the skipper or his representative. The mooring officer knew, for example, that the bridge's view of the high mast's operating platform was cut off as the ship approached. Consequently, a great deal of discretion—and responsibility—was left to this officer.

52. It was customary to start a pair of side propellers first, then an additional pair. The centerline engine was not started until well clear, as the ground suction from this prop tended to pull the stern down.

53. It should not be assumed that when the ship was away Aerology and Communications stood idle. ZRs were specialized types requiring either a mast ship or special bases. Accordingly, every scrap of information was gathered (particularly concerning conditions near to and ahead of the ship), and a summary was transmitted or relayed at least every twelve hours, usually every six—and more in between if there was a change. These data were added to the map being drawn on board. Reichelderfer.

54. "Giant Dirigible Arrives Safely But Is Unable to Make Mooring," (Hamilton, Bermuda) *Royal Gazette*, 23 February 1925.

55. Ibid.; "By Dirigible to Bermuda," *Aero Digest* 6, no. 4 (April 1925). *Patoka* departed on 7 March, steaming for Beaumont, Texas.

56. The assistant secretary had requested the trip's cost. Itemized, the expenses included 2,900 gallons of gasoline ($528.67); 110 gallons of lube oil ($54.29); miscellaneous supplies ($107.44); and the cost of provisions ($141.91). RG 72, BuAer GenCor, box 5576, National Archives.

57. Memorandum, CO, Los Angeles, to BuAer, 14 April 1925, RG 72, BuAer GenCor, box 5582, National Archives. It was found that *Shenandoah*'s cells averaged a 4 percent loss of lift per month due to purity and diffusion losses.

58. Letter, Fulton to Keller, 5 March 1961.

59. Memorandum, CO, Los Angeles, to BuAer, 9 December 1926, RG 72, BuAer GenCor, box 5584, National Archives.

60. Conversely, when word was received to cut *in* water, the Maybach was stopped, and its engineer placed a plug in the atmospheric exhaust, thereby directing all exhaust gases through his condenser unit. When ready to restart, he rang up on the telegraph and was answered.

61. Charles E. Rosendahl, *Up Ship!* (New York: Dodd, Mead, 1931), 59.

62. Accompanying was a verification certificate from the colonial postmaster and staff, Bermuda, to the postmaster and staff of New York City. "I am forwarding by the U.S. Air Ship *Los Angeles* [it read in part], the first Aerial Overseas Mail, ever dispatched from Bermuda to a foreign country, and convey to you, my felicitations, at the great event."

63. "Aboard the Airship 'Los Angeles' to Porto Rico and Back," *Literary Digest*, 4 July 1925, 60–61.

64. RG 72, BuAer GenCor, box 5585, National Archives. A note by Truscott on the cover sheet reads, "Mr. H. D. Ashton was *very* unpopular both before and after!" All of the pusher-type ship's propellers were two-bladed wooden affairs, their leading edges and tips protected by plate sheeting. Diameter: twelve feet.

65. USS *Los Angeles*, log, 1200 to 1600 watch, 7 May 1925.

66. NAS Lakehurst, log, 0000 to 0400 watch, 11 May 1925, National Archives.

67. Memorandum, CNO to *ZR-3*; *ZR-1*; Lakehurst, RG 72, BuAer GenCor, box 5564, National Archives.

Given its cruising radius, it was inadvisable for *ZR-1* to try for Honolulu.

68. "The California-Hawaiian route should prove an ideal training ground for initial [commercial] operations," Hunsaker opined with *ZRS-4* building. "It is proposed to operate with mails to Honolulu until a safe and reliable service has been demonstrated and then to attempt to build up a passenger business" in preparation for expansion to the Far East. Ohio Society meeting, 10 February 1930, box 3, folder 80, Hunsaker Collection, Massachusetts Institute of Technology, courtesy William Trimble. A line of commercial airships crossing the Pacific, with the terminals and personnel that would be involved, offered a potential national defense asset as well. "There is no form of Naval auxiliary vessel which can be incorporated into the combatant part of the Fleet so directly as these airships which are essentially sister ships of the Navy's scouts now building." "Trans-Pacific Airship Service," box 4, folder 17, LTA Society, University of Akron archives, courtesy William Trimble.

69. Commercial and military air traffic had increased to the extent that by 1926 it was necessary to develop navy radio-compass stations to accommodate aircraft as well as shipping.

70. "The airship pilot of today," the manual noted, "works and calculates in the main with the dynamic performance and factors of the ship. Of course, for all that, we must not lose sight of the fact that after all we are dealing with an airship swimming or floating statically in the air, and this fact will always be brought home to us in all its undisguised nakedness every time we attempt to effect a landing."

71. RG 72, BuAer GenCor, box 5587, National Archives. When BuAer forwarded these results to the CNO, comments resulted. "The personnel at Lakehurst have begun to make real progress in solving the problems of handling an airship at the mooring mast." On the cover sheet, Fulton wrote, "They see the light of day—should have been done long ago."

72. Rosendahl, *Up Ship!*, 60–61.

73. In berth, a half-section each of riggers and engineers constituted the duty section—approximately twelve men with two on watch (four hours) in the ship continuously. The pair made their inspections alternately, keeping one man near the telephone and guarding the gangway from unauthorized entry. A weigh-off was made daily (at 0800), and a report made giving all weights and so on. ZRs were kept heavy, with ballast adjusted as necessary when crew, workmen, and visitors boarded.

74. Letter, Fulton to Keller, 19 February 1961.

75. "We lived at the last quarters on officers' row not far from the high mast," Wiley's son Gordon recalled. "LCdr [Lieutenant Commander] Lansdowne lived next door. General ["Billy"] Mitchell would visit trying to convince Lansdowne and others to join him in a separate air force. He occasionally came to our house. Admiral Moffett would visit also (not at the same time) and tell these naval aviators to stay in the Navy and that naval aviation would lead the Navy—which it did." Capt. Gordon S. Wiley, USN (Ret.), letter to author, 26 March 1997.

76. Cooling began once under way; this caused heaviness, which had to be carried dynamically—that is, using the ship as an airfoil. If dynamic load was too great, speed was sacrificed; if the ship was too light due to superheating, a downward elevator-angle (and reduced economy) or valving was required. Inclination of the ship caused a bending moment amidships.

77. Letter, Lansdowne (in BuAer) to Fulton, 4 December 1923, Fulton Collection, NASM, Smithsonian.

78. One wife, reportedly, ruled air-station society with an iron hand.

79. Letter, Mrs. Louise Tyler Rixey to author, 21 May 1996.

80. Letters, Cdr. Fred T. Berry, USN, to Mrs. Berry, 20 July and 27 June 1930, courtesy Tom Berry. The commander was part of LTA Class VI (1929–30). "Well the final examinations are a thing of the past," Berry wrote that June, "and just between you and me I don't think any of us covered ourselves in glory. Rosendahl took his final crack at the class and it was sure a good one." Ibid.

81. Don Brandemeuhl, letter to author, 23 November 1984; Lt. Herbert R. Rowe, interview by author, audio cassette, Toms River, N.J., 23 March 1978. "Most of us joined the Marine Corp to see the world (we believed those recruiting posters about foreign ports and duty stations). Lakehurst definitely did NOT fill that bill." Ibid.

82. Frederick Lewis Allen, *Only Yesterday: An Informal History of the Nineteen-Twenties* (New York and London: Harper and Brothers, 1931), 206. "For some reason," Gordon Wiley recalled, "I was at the Duty Office when the CPO of the watch told me the ship had crashed and to tell everyone. I ran down the road to our quarters shouting, 'The ship has crashed! The ship has crashed!' as people came out of their barracks or quarters." Capt. Gordon S. Wiley, USN (Ret.), letter to author, 26 March 1997.

83. During this period major alterations were ef-

fected, such as installation of a top walkway, additional water-recovery apparatus, and pressure-test gear and strain gages.

84. *Eleventh Annual Report of the National Advisory Committee for Aeronautics,* 1925, 58. "The Army is directly concerned and the commercial development of airships in America may be said to be also at stake." Development, NACA opined, should be continued. Ibid.

85. Wicks, 7–8.

86. The navy, in other words, would lay the foundation for commercial applications, after which private interests would build on them. "Throughout the history of lighter-than-air," one officer remarked, looking back, "the heavier-than-air people objected very much to it, because it took money from their programs. And if it hadn't been for Moffett, who was a lighter-than-air enthusiast, I don't think we would have ever gotten very far with it." Cdr. Joseph P. Norfleet, USN (Ret.), audio interview by author and Michael C. Miller, Cape May Point, N.J., 6 September 1975. Norfleet had reported to Pensacola in July 1915; on 6 June 1921, a lieutenant commander, he assumed temporary command of the new air station at Lakehurst.

87. *Aerostation* 12, no. 3 (Fall 1989), 7–10.

88. It should be kept in mind that print was still the supreme medium of communication—and hype.

Chapter 3. Balloons and Billets: Lighter-than-Air Training

1. W. L. Hamlen, "The First Lighter-than-Air Class at Akron," *Naval Aviation News,* November 1967, 27; James R. Shock, "U.S. Navy Pilots, 1917 through April 15, 1942: A Research of Naval Aviators (Airship) by Qualification Numbers and Dates Including Data on Balloon Pilots and Select Enlisted Men," August 1998, 3.

2. Cdr. Joseph P. Norfleet, USN (Ret.), audio interview by the author and Michael C. Miller, Cape May Point, N.J., 1 November 1976. Recalling Maxfield's advocacy of airships in those early years, Norfleet labeled his fellow officer "the father of lighter-than-air" in the U.S. Navy. Norfleet, 6 September 1975.

3. Though preeminent in the airship arts, Germany did not exploit balloons for training. During the war, the British Royal Navy Air Service had its pilots training in free balloons.

4. Shock, 6.

5. Hunsaker to Fulton, 27 February 1923, Hunsaker Collection, box 2, folder 3, NASM, Smithsonian.

6. NAS Lakehurst, Regulations (effective 24 February 1928), box 84, HOAC; George H. Mills [hereafter Mills] Collection, NASM, Smithsonian.

7. Letter, Cdr. Fred T. Berry to Mrs. Berry, 16 June 1930, courtesy Tom Berry.

8. Total flights numbered thirty-one, distributed as follows: free balloons, six; *J-3,* eleven; *J-4,* three; *ZMC-2,* one; *ZR-3,* ten.

9. "Records about them are awfully scarce," Fulton observed, "yet at one stage [kites] played an almost vital role in convoying, mine searching, anti-submarine patrols, etc." Letter, Fulton to Keller, 10 July 1961.

10. "Rigging" was defined as the assembling, adjusting and aligning of the parts, or the attachment and adjustment of the car, rudders, valves, controls, and so on, of an airship or balloon.

11. Letter, Cdr. Fred T. Berry to Mrs. Berry, 18 July 1930, courtesy Tom Berry. Berry logged fourteen kite ascents—three hours, eight minutes altogether. Stated purpose: "testing." These represented the officer's entire flying for that month.

12. By order of the War Department, in 1924 all army and naval fliers were obliged to wear parachutes. The latter's first parachute school was started at Lakehurst; its inaugural class (fifteen students) convened on 1 September.

13. Commencing in June 1918, NAS Pensacola had begun taking "upper" air soundings using instruments carried up by kite balloon. This yielded data on wind velocity and direction—information needed for the school's navigational training flights. The technique was developed by the station's meteorological officer, Lt. William F. Reed. Free balloons were a meteorological research tool as well—to follow the trajectory of body of air, for example.

14. With *ZR-3* in overhaul and *J-3* and *J-4* grounded due to weather, Lieutenant Coulter gained his flight-pay time reluctantly. "I assure you that there is no more uncomfortable aerial experience than riding a surging kite balloon basket on a snowy, gusty winter day in New Jersey. It really wasn't worth the pay!" Capt. Howard N. Coulter, USN (Ret.), audio cassette prepared for author, Los Altos Hills, Calif., July 1975.

15. Before 1922, all Navy airships had been hydrogen inflated.

16. The caustic could give a severe burn, so rubber gloves, goggles, and boots were mandatory. Some men also applied a coat of petroleum jelly to the face, to protect it from spray. The caustic could ravage a pair of

shoes in hours. See the writer's "Balloon Training at NAS Lakehurst," *American Aviation Historical Society Journal* 28, no. 1 (Spring 1983), for a fuller account of hydrogen operations.

17. Cdr. F. W. Reichelderfer, USN (Ret.), interview by author, audio cassette, Washington, D.C., 18 June 1977.

18. Alongside its regular duties, Lakehurst's aerological office carried out studies relating to airship operation. Around 1929–30, for example, abrupt wind shifts, wind gustiness, and temperature inversions received assessments. *Sixteenth Annual Report of the National Advisory Committee for Aeronautics*, 1930, 44.

19. Tyler Collection, courtesy Navy Lakehurst Historical Society. Flights orders stipulated landings within about six hundred miles of the Lakehurst station. As training officer, in 1934 Settle revised the balloon section of the *Rigid Airship Manual*. These pages reflect the knowledge, skill, and experience of a gifted officer.

20. Interview by author, Manhasset, Long Island, N.Y., 11 November 1977.

21. Lt. Herbert R. Rowe, USN (Ret.), interview by author, audio cassette, Toms River, N.J., 17 April 1977 and 9 September 1981. A 1930 graduate (seaman first class, or Sea1c) of the LTA Training School, "Monty" Rowe served as rigger aboard *ZR-3, ZRS-4,* and *ZRS-5*. Every airman with whom the author has spoken is unequivocal—free ballooning was an unforgettable experience.

22. Reflection of the sun's energy from its upper surface increased the degree of superheat—and lift—of a balloon when approaching close to the undercast. One disadvantage above a cloud cover: the flight crew could not navigate.

23. Tape (with transcript) prepared for author, 13.

24. One unit only of the E, F, and H types were built, respectively; each was specialized in design and short-lived. Though a G ship was procured in 1935, the original G design—of four hundred thousand cubic feet—was canceled in 1919.

25. *D-6* and *H-1* had been slated for transfer from NAS Rockaway, New York. They became victims of their hydrogen; on 31 August a hangar fire destroyed both, along with *C-10*.

26. On occasion, the inventory was impressed into service investigating Prohibition-era stills. Pilots sometimes were ordered to steer clear of certain areas where armed locals might resist the intrusion.

27. Walker C. Morrow, audio cassette prepared for author, Grosse Isle, Mich., 24 July 1976. Morrow joined the Aircraft Development Corporation in 1926 as a sheet-metal mechanic. He helped build and, later at Lakehurst, repair *ZMC-2*. In conventional fabric ships, shear stresses were carried through a complex system of diagonal wiring. In the metalclad, these and (to a large extent) tensile stresses were carried by the metal covering, the plating of which also reinforced the internal frames, thereby reducing weight.

28. By virtue of her skin, *ZMC-2* gained and lost superheat relatively fast, making her very rough riding. It was said that everyone knew when spring had arrived on the first day that *ZMC-2* "fell through" on approach. Personnel tended to avoid her. One student confessed that the only time he was ever airsick was on a rough day in *ZMC-2* east of the station. Students had to be rather well along in training before attempting the controls. Its tendency to roll was vicious. "She was a brute to fly" is a more or less typical comment.

29. John A. Iannaccone, *Buoyant Flight: The Bulletin of the Lighter-than-Air Society* 28, no. 2 (January–February 1981), 9.

30. After the war, the U.S. Army had announced its intention of going into the blimp business. Chagrined over losing its fight for *Los Angeles*, it probably helped along the impression that it had been charged with nonrigid development.

31. Capt. M. M. Bradley, USN (Ret.), interview by author, audio cassette, Springfield, Va., 22 October 1977; letter, 15 August 1929, courtesy Lynn Berry.

32. Letter, Cdr. Fred T. Berry to Mrs. Berry, 16 July 1930, courtesy Tom Berry. Class VI (1929–30) contained senior officers—a stratagem to get rank into the program. The qualification board—Rosendahl, Clarke, and Bauch—provoked several of its members. "Joe [Greenwald] was finally called before the board today," Berry records, "and told that he would be recommended for his wings. Joe said they were none too cordial in their congratulations and that his reaction was to tell them to go plumb to hell. Harry Shoe[maker] says if they try the same thing on him he will walk out on them." Letter, Cdr. Berry to Mrs. Berry, 18 June 1930, courtesy Tom Berry.

33. The crew's quarters had been placed near the center of gravity so that movements to and from duty stations would cause as little trim change as possible. Wash water was stored in tanks very near the point from which it was drawn, so that there was no change in moment due to using the water.

34. Lt. Cdr. Charles E. Rosendahl, "Flying with an Airship Captain," *Popular Science Monthly*, March 1930.

35. In contrast to navy practice, however, a continuous keel watch was not maintained on *Hindenburg*. Instead, a complete inspection was made each hour by a qualified rigger.

36. Used during takeoffs and landings, the phones were comparatively quiet under way. Around 1930, the German system—effective enough—was replaced by an American-made one. The latter used the ship's framework as a conductor and operated at a much higher frequency—"a marvel of the day and greatly enhanced effectiveness." Capt. George F. Watson, USN (Ret.), letter to author, 4 February 1995.

37. The elevator wheel was mounted parallel to the long axis, the better to judge the ship's angles of inclination. The instrument board mounted before the wheel included an inclinometer, an altimeter and barometer, a gas-pressure alarm, and a variometer (rate-of-climb indicator).

38. Reichelderfer.

39. An anecdote will illustrate concern for weight and trim. As *Macon*'s new CO (July 1934), Wiley resolved to increase cruising range. This included assessing gear from each department. *Every* loose item was ordered off: spare engine parts, tools, charts, even the navigator's spare pencils. "It [unnecessary gear] added up terrifically, to the point where we carried an hour or two more fuel, if not more." Thereafter, every spare item had to be justified. Capt. Earl K. Van Swearingen, USN (Ret.), interview by author, audio cassette, Alexandria, Va., 11 November 1976. "Van" (No. 3932) was a graduate of Lakehurst Class VI (1929–30).

40. Rosendahl, "Flying with an Airship Captain."

41. In August 1928, a seaman was caught sleeping on watch. He was made a prisoner at large while awaiting a summary court-martial, which sentenced him to fifteen days' solitary confinement on bread and water in the station brig, with a full ration every third day. The offender also lost eighteen dollars pay per month for two months.

42. Clifford A. Tinker, "Riding the Clouds in an Air Pullman," *Popular Mechanics*, July 1927, 104; Bradley.

43. Capt. Howard N. Coulter, USN (Ret.), audio cassette prepared for author, Los Altos Hills, Calif., July 1975. Typically, cold-weather gear was to hand for all flight personnel.

44. A framed cartoon caricature hung here—artist unknown. It depicts *L.A.* about to moor. Two men ride the mast (one with bottle and cigarette), one officer equipped with a whistle on its base. On board ship, a lasso-equipped sailor rides the bow, while an officer, topside, holds a spy glass. The skipper hangs from the main car, megaphone in hand, as another officer pulls all the helium and water-ballast cords. Above them on ship's flank is written, "Use no hooks." Courtesy Tom Berry.

45. Tinker; Navy Relief Society, "Airships," 2nd ed., NAS Lakehurst, Spring 1929, unpaginated.

46. Coulter.

47. Conversation with ADC John C. Iannaccone, USN (Ret.), and John A. Lust, Toms River, N.J., 3 October 1995. Each had held "orders to duty involving flying for enlisted men." Iannaccone reported to Lakehurst in 1929; as an AMM3c, he held a billet aboard *ZR-3* and served with the air-station command. A 1930 graduate of the LTA Training School (Sea2c), Lust trained aboard *ZR-3* and received orders to *ZRS-4*.

48. "To permit a rational basis for stress analysis, and at the request of the Bureau of Aeronautics, an extensive study is being made of the air forces which act upon the hull and control surfaces of a nonrigid airship." *Ninth Annual Report of the National Advisory Committee for Aeronautics,* 1923, 17. The investigation sought to determine air pressure, hence the loads sustained in flight and in maneuvering. Although unable to determine aerodynamic loading satisfactorily, it was the most extensive study of its kind yet attempted; the results were valuable for stress analysis and calculations. *Tenth Annual Report,* 1924, 16.

49. Norman J. Mayer, audio interview by author, Alexandria, Va., 30 January 2001. A NASA retiree and exceptionally experienced in LTA, Mayer often has been asked to consult on engineering issues related to lighter-than-air (inflated) craft.

50. S. J. DeFrance (NACA), "Flight Tests on U.S.S. Los Angeles, I. Methods of Investigation," January 1927, 2, NASM Archives, file A3U-723200-01, Smithsonian; C. P. Burgess, "Forces on Airships in Gusts," *Design Memorandum No. 32,* 31 March 1924, 9; "Notes on Research into the Structure of Wind Gusts and Violent Air Currents for use in LTA Design and Operation," Navy Lakehurst Historical Society; interview by author, audio cassette, Washington, D.C., 18 June 1977.

51. The technical development of airships lagged considerably behind that of airplanes; more knowledge on the forces on the latter in various maneuvers was one goal of aerodynamicists in this period. Though preliminary, the 1925 tests had shown "conclusively" that the stresses due to maneuvering were less than those resulting from normal flight in even moderately rough air. C. P. Burgess, "Strain Gage Tests on U.S.S. *Los Angeles* May 25 to June 4, 1925," *Design Memorandum No. 55,* 23 July 1925, 7.

52. On 30 April (for instance), in unstable air, the airship's log records: "1301 Began test for N.A.C.A. representatives. All motors 1050 R.P.M., 8 degrees right rudder. 1310 similar test, 8 degrees left rudder. 1320 continued various tests."

53. *Thirteenth Annual Report of the National Advisory Committee for Aeronautics,* 1927, 25; S. J. De France (NACA), "Flight Tests on U.S.S. Los Angeles—III. Pressure Distribution on Tail Surfaces in Rough Air," 2, NASM Archives, file A3U-723200-01; NACA Report No. 324, "Flight Tests on U.S.S. "Los Angeles," Part I-Full Scale Pressure Distribution Investigation," 3, NASM Archives, file A3U-723200–02, Smithsonian. The largest forces and bending moments corresponded "astonishingly close[ly]" to the Zeppelin Company's design assumptions. "The large airships of the future must be designed to encounter thunderstorm conditions which in the past have been regarded as avoidable hazards, and greater strength than that of the *Los Angeles* is therefore required." Burgess, "Flight Tests on U.S.S. Los Angeles, Part II—Stress and Strength Determination," *NACA Report No. 325,* 14. By 1930, *L.A.* carried gust indicators—four Venturi tubes at the bow, mounted at ninety-degree intervals.

54. The device was mounted atop the hangar, and photos were taken as *ZR-3* maneuvered over—producing a series of images on a large sheet of photographic film. From these, flight path, velocity, and attitude with respect to the camera were determined.

55. As per airship's log, Steele's time as skipper totaled 260 hours, fifty-three minutes in nineteen flights; under Klein, the figures were twenty-seven hours and thirty-one minutes during seven sorties.

56. "He got himself married to a very charming heiress and spent several months in Paris and on the Riviera before coming to Friedrichshafen," Fulton confided. "He was spoiled by his marriage from doing any more hard Naval duty; stayed with *Los Angeles* only a short while and after *Shenandoah* crack-up Lily Steele made him quit." Perhaps wistfully, he added, "Steele was an excellent officer." Letter, Fulton to Keller, 7 February 1960.

57. Despite BuAer objections, BuNav insisted on periodic sea duty for aviation officers. Rotation away to sea severed the continuity of training and flight experience—a factor that, operationally, probably contributed to error and consequent tragedy.

58. Letter, Fulton to Keller, 19 February 1961. Unlike for HTA, there were no surface billets involving airships. Consequently, an officer in need of sea service had to separate himself for a time from LTA work, thereby disrupting continuity of experience.

59. Henry L. Boswell (CBM); Robert W. Copeland (CRM); William A. Russell (ACMM), and Tony F. Swidersky (Cox) would die with USS *Akron*.

Chapter 4. Rosendahl's Reign

1. This included the air-station skipper. Though Jackson nominally held command, Rosendahl was "the real authority," repeatedly bypassing Jackson and "handling all important matters directly with Washington." See Douglas H. Robinson and Charles L. Keller, *Up Ship! U.S. Navy Rigid Airships, 1919–1935* (Annapolis, Md.: Naval Institute Press, 1982), 200–202.

2. Experience would show that helium under usual operating conditions should be purified—that is, its air contamination purged—about four times a year. Cost of purifying: about a dollar per thousand cubic feet.

3. Maintenance was a headache, since lack of a suitable cell for one position could foreclose operations. But because cells deteriorated in storage, maintaining a duplicate set was expensive.

4. When the ship was held to the deck, at least one man per engine and the steersman remained on board. "The motors must always be kept ready to start at any time," the manual advises, "and must therefore be idled every now and then, especially in cold weather, to keep them warm."

5. Shored and wedged, the weight of number-one car was off the ship (on cradles) when Settle closed out the day's log; water was used to compensate. The morning's takeoff also had buckled the apex channel of frame 60 between longitudinals one-half and one, starboard.

6. Disembarking at Detroit, Moffett was slated to proceed to Youngstown, Ohio, to dedicate an airport in honor of Zachary Lansdowne, killed in the *ZR-1* breakup.

7. Two carrier pigeons were released—at 0720 and 0735. Lakehurst's Pigeon Loft raised approximately a hundred birds, part of their training being from balloons and airships—as a means of communication while in flight.

8. This docking (east end) courted disaster. With the wind blowing cross-hangar, civilians were augmenting the regular ground crew. But the forward taxiing car jumped its tracks at 1717 (due to wind pressure). The car was cut adrift, and a snow squall struck. Stern lines slackened, *Los Angeles* began kiting toward the hangar. Engines three and five were backed. By using the starboard engines at intervals driving ahead, and the rud-

ders (to bring the stern into the wind), *L.A.* was shifted clear of the danger zone. She entered the hangar—west end—at 1916.

9. The VL-1 could still give trouble. After letting go on 7 May, *L.A.* rose statically to a hundred feet. Wind direction was different at altitude—and she drifted shedward. Two engines would not start, the aft one was not used for takeoff, and the remaining pair could not check her in time. So 1,900 pounds were dropped to clear; as one consequence, upon landing, forty-eight thousand cubic feet of helium had to be valved. Three days were expended in investigations before the decision was made to charge the failure to the eccentricities of the gasoline engine. CO, *Los Angeles* to CNO, 16 May 1927, RG 72, BuAer GenCor, box 5572, National Archives.

10. *Thirteenth Annual Report of the National Advisory Committee for Aeronautics,* 1927, 29.

11. At night, the earth radiates heat, thus cooling the surface air layer. This produces a temperature inversion (that is, warm air over cool air, whereas normally temperature drops with altitude) and, consequently, a very stable condition. At night, also, there is little unequal heating of the surface as in sunlight hours. Hence there are few convectional currents, and these are promptly checked by the extreme stability of the atmosphere.

12. On such calm summer days conditions for landing degrade. The cool morning air, warmed, gradually gives place to a superheated stratum. "We then get," the manual warns, "a very regular but absolutely incalculable play between vertical and horizontal movements of the air as the various strata of different temperatures exchange places, movements against which a slow-sailing ship, barely answering the rudder, is practically defenseless."

13. This effect occurred under way as well. For instance, ship's log records: "1410 Encountered land-sea breeze 5 miles east of Lakehurst, wind shift with attending turbulence, all engines cruising speed." USS *Los Angeles*, log, 1200 to 1600 watch, 9 May 1930.

14. The shouts were Rosendahl's. His orders: send crewmen aft—but men were already scrambling. As the rise continued, he sent word to cast off. But the angle was such that the cone could not have lifted out. Because he was the senior man aboard, Settle's judgment obtained. "The ship was *not* tripped," Settle insisted, "nor was tripping attempted." Letter to author, 18 May 1976.

15. These paragraphs are based on a number of sources, including the six-page Lakehurst memorandum "Air and Weather Conditions at Lakehurst on Thursday 25 August 1927" (author files); Vice Adm. T. G. W. Settle, USN (Ret.), letters to author, 18 May and 9 June 1976; Lt. David F. Patzig, USN (Ret.), interview by author, audio cassette, 20 March 1976; Bolster, and letter to author, 15 December 1976; Cdr. R. H. Ward, USN (Ret.), interview by author, audio cassette, Lakehurst, N.J., 11 June 1978; USS *Los Angeles*, log, 1200 to 1600 watch, 25 August 1927.

16. "This incident," Reichelderfer recalled, "made every airship officer *intensely* conscious of the phenomenon known as the sea breeze. So the first thing they would ask the Aerology office before they put a ship out was, 'Any sea breeze today? When? How strong?'" Interview by author, audio cassette, Washington, D.C., 18 June 1977.

17. Memorandum, CO, *Los Angeles* to Chief, BuAer, 26 August 1927, Fulton Collection, Smithsonian. "Although this was the most remarkable demonstration of unsuspected strength ever staged by an airship," Rosendahl later wrote, "it can hardly be called a desirable stunt for repetition." *Aeronautical Engineering* 5, no. 1 (January–March 1933), 46. It was a month for mishap. *L.A.*'s lower rudder fouled on trees four days after. The cover was holed, and girders were broken and buckled in the lower rudder and vertical fin.

18. USS *Los Angeles*, log, 5 October 1927.

19. This was hardly true for touchdowns. On 25 October, with landing stations called, *L.A.* met a wind shift that rolled her as much as four degrees. Shifting surface winds then twice frustrated approaches. Finally, trail ropes in hand on the ground, a Jacob's ladder was lowered, and twenty-one ground crew embarked to counter excessive lightness—a temperature inversion had brought a ten-degree jump in superheat.

20. Capt. Charles W. Roland, USN (Ret.), "Handling Rigid Airships on the Ground," privately printed, 1978, 6–7, courtesy Captain Roland.

21. Upon mooring, the ballast line was coupled and signals arranged with the mast watch to turn water on and off. Ordinarily, crew stayed at their landing stations until ballast was coming aboard and satisfactory trim had been gained. Crewmen also made ready to fuel, provision, and so on. The aircraft, in other words, was made ready for release immediately upon securing. German practice was similar in this regard.

22. Not that mishaps had been banished. Early in the evening of 21 September 1929, for example, as *ZR-3* was being walked to the mast, very gusty conditions were encountered due to shed "spillover"—turbulence created by the hangar, as an obstruction. The main cable was disconnected, whereupon a side gust carried away the for-

ward bumper bag. Rosendahl took to the air, valving a ton of water to clear.

23. Bolster. This problem never vanished. The broadside sail area of the ZPG series nonrigid (1952–61) was immense, and, correspondingly, the wind forces exerted during docking/undocking were immense.

24. By 1937, one pneumatic-tired mast for nonrigids (blimps) was also in use.

25. One penalty: an increase in deadweight. Also, the use of helium had reduced total lift to the extent that the useful load was less and performance was reduced.

26. Reichelderfer.

27. "While Rosie was gathering the credit (which he surely deserved)," George Watson said, "'Froggy' Fulton was fighting in the trenches at BuAer getting funds and approval from BuAer that kept us alive." Letter to author, 14 September 1995.

28. Cdr. F. W. Reichelderfer, USN (Ret.), interview by author, audio cassette, Washington, D.C., 3 July 1977.

29. Bolster.

30. Rosendahl to CNO (confirmation of telephone statement), 28 January 1928, RG 80, Office of SecNav, GenCor 1926–1940, box 4210, National Archives. Rosendahl proposed a turntable to receive the control car, thus allowing *Los Angeles* to pivot into the wind and eliminate the need for manpower to hold her to the deck. Neither BuAer nor *Saratoga*'s CO proved enthusiastic. "A turntable installation will complicate a bad job," Plans noted. "Don't believe in it. Carrier cannot be considered as a *base* for a ZR and this experiment can suffice for fueling." Cover sheet note, RG 72, BuAer GenCor, box 5580, National Archives.

31. *New York Times*, 28 January 1928; conversation, Fulton to Richard K. Smith. *Aeronautical Engineering* was no less enthusiastic: "The *Los Angeles* has demonstrated that it can land aboard a carrier and be refueled, regassed, and reprovisioned, making possible a greatly extended field of usefulness as well as demonstrating that emergencies can be met." Vol. 1, no. 1 (January–March 1929), 6.

32. Shock, "U.S. Navy Airship Pilots," 21.

33. Barometric pressure was the usual way of measuring altitude. For maximum (albeit imperfect) accuracy, the altitude scale was rotated so that the pressure read zero at the start of a flight; accordingly, the height above the starting point was displayed. (The hangar's deck is ninety-two feet above sea level, and it was another fifty-four feet to ship's center line; thus a 146-foot correction was needed.) During landing, the reading on the field was displayed so the CO could adjust his altimeter. *L.A.* was equipped with a searchlight, arranged to point vertically when she was level. Aft was a station from which an observer could measure the angle that the light made with ship's axis, thus providing a vertical right angle with a known base.

34. "If the [landing] situation is nothing out of the ordinary," the manual counsels, "it is advisable not to interfere too much with an experienced well-trained crew; let them do their work quietly, don't meddle. Nervous, excited transmission of orders from the ship is an unnecessary evil. Nevertheless, it is necessary to watch from the ship how the ground crew is working, and to support the men by suitable means."

35. Charles E. Rosendahl, *Up Ship!* (New York: Dodd, Mead, 1931), 126.

36. *Evening Bulletin—Philadelphia*, 3 March 1928. Following quotes are from this piece as well.

37. Rosendahl, *Up Ship!*, 129–33; CO *Los Angeles* to BuNav, 26 March 1928, RG 72, BuAer GenCor, box 5577, National Archives.

38. "8, Dangling to *Los Angeles* in Midair, Saved," *Evening Public Ledger*, 3 March 1928. "The flight was a splendid training cruise," another quoted. "And the all-night battle with the elements furnished a test that showed the fine stuff the men are made of"; *Sunday Call*, 4 March 1928.

39. Airship designers themselves disagreed as to maximum bending moment from aerodynamic forces acting upon an airship in flight. The *R-100* team, for example, took careful account of gusts in its design, whereas the *R-101* group chose to dismiss gust effect. C. P. Burgess, "Comparative Strength of Airships," *Design Memorandum No. 80,* June 1927, 2–3.

40. "New York is the greatest from the air," an airman wrote. "An even better sight is New York at night when everything is lighted. Every time we make that trip we fly right over Broadway from one end of it to the other. When we fly it during the day we usually come down Lexington Avenue and fly over the Chrysler and Woolworth Buildings. It is impossible for me to describe these flights and sights from the ship." "Fitchburg Boy Enjoys Flights on *Los Angeles*," *Fitchburg (Mass.) Sentinel*, 12 September 1930, Iannaccone Papers. Joining ship's company in mid-August 1930, AMM3c John A. Iannaccone served with *ZR-3* through April 1931.

41. The Sandy Hook (N.J.) radio-compass station was calibrated and a bit of radio navigation practiced as well. USS *Los Angeles*, log, 1600 to 2000 watch, 28 August 1928.

42. On her 5–6 September jaunt, *Los Angeles* was fired upon by a shotgun. A north Jersey farmer was later indicted for assault with intent to kill Wiley and his crew, and for malicious mischief. A witness for the state, Wiley testified that holes in the cover often were found after cruises, though he could not remember any perforations after the flight in question. A jury acquitted the defendant. *Newark Evening News,* 15 and 28 May; *New York Sun,* 27 May 1929. Between 1928 and 1935, *ZMC-2* was drilled by unknown snipers in three incidents. *Evening Bulletin—Philadelphia,* 1 August 1935.

43. Wicks, 6.

44. Clifford W. Seibel, *Helium: Child of the Sun* (Lawrence: University Press of Kansas, 1968), 81. "Since that time the efficiency of the plant was steadily increased and there has been a marked reduction in operating costs." Plant capacity (1933): well over twenty-four million cubic feet of helium per year. Andrew Stewart, "About Helium," *Bureau of Mines Information Circular 6745,* September 1933, 31.

45. "If planes fly close to the dirigible as it cuts speed preparatory to mooring," Lieutenant Peck explained to pressmen, "the air stream from their propellers may cause the craft to veer from its course and be very difficult to moor. After it is moored the only danger will be from flying over the huge crowds expected." "One-Way Traffic for 'Los Angeles' Visit," *Fort Worth Record-Telegram,* 5 October 1928.

46. USS *Los Angeles,* log, 0000 to 0400 watch, 9 October 1928.

47. CO, *Los Angeles,* to CNO, 12 September 1928, RG 72, BuAer GenCor, box 5573, National Archives; Garland Fulton Collection, Smithsonian. The ship's log makes no mention of the incident. Airplane pilots had complained of the risk of collision unless airships were conspicuously marked, especially at night. There was no international agreement as to what running lights should be carried or where they should be placed.

48. To preserve night vision, the ship was kept quite dark. Floodlighting was not used until *after ZR-3* lay in hand, after which all possible light was wanted. This included many a flashlight. "It looked like fireflies from the various [handling-line] positions," one airman recalled. "It looked like lights were just flickering all around." Lt. Herbert R. Rowe, USN (Ret.), interview by author, audio cassette, Toms River, N.J., 23 March 1978.

49. "Record of Navy Dirigible and Improved Methods of Airship Construction Point to Progress in Lighter-than-Air Craft," *New York Times,* 23 September 1928.

50. As for his crew, the majority had flown *ZR-3* to America, and all watch officers had been commanding officers of wartime zeppelins.

51. J. Gordon Vaeth, *Graf Zeppelin: The Adventures of an Aerial Globetrotter* (New York: Harper and Brothers, 1958), 23.

52. Reichelderfer.

53. Walter Whiteley Hubbard, "The Transoceanic Dirigible; Graf Zeppelin Starts Transatlantic Passenger Service," *American Aviator: Airplanes and Airports* 1, no. 4 (November 1928), 9.

54. "Dirigible to Be Flying Air Base," *New York American,* 17 October 1928.

55. When, around 1800, the doors were opened to house *J-3* and *J-4,* 1,900 pounds of water were added to *ZR-3*'s disposable weights (sand and water), in order to compensate for the additional lift due to superheat. USS *Los Angeles,* log, 1600 to 2000 watch, 30 October 1928.

56. Fulton was a member of NACA's Subcommittee on Airships, Committee on Aerodynamics. The minutes of its 20 December 1928 meeting state this: "Commander Fulton reported that the Navy had recently made tests with a skin type of water recovery apparatus, in which a dummy apparatus was installed on the LOS ANGELES and pressure measurements were made at various points on the apparatus out from the hull with a view to getting an idea of the interference effect. He said that one set of measurements was made two weeks ago and that they would repeat the measurements in a different location next week" (p. 4). Courtesy Navy Lakehurst Historical Society.

57. It became routine to carry an aerological officer and, from 1930, an aerological assistant. The four daily weather maps were drawn in flight, and airway reports were intercepted hourly when conditions proved unsettled. Especially after completion of the 0800 and 2000 maps, the skipper held conferences with the aerological officer, attended usually by his exec, navigator, and sometimes other heads of departments and watch officers. The intent was to project the probable influence of conditions on proposed operations.

58. Rosendahl, *Up Ship!,* 115; Capt. M. M. Bradley, USN (Ret.), interview by author, audio cassette, Springfield, Va., 22 October 1977.

59. In 1927–28, seven VL-1s went to Germany for modernization to the VL-2 configuration. Some of those modified in the States were not given new camshafts, however, and therefore were still rated at four hundred horsepower at 1,400 rpm. The new rating, with the new camshaft: 520 hp at 1,600 rpm, or 500 hp at 1,500 rpm,

depending on rating source. RG 72, BuAer GenCor, box 5584, National Archives. The bureau felt that the 520-horsepower engines, with their higher rpms, were pushing the wooden props to their strength limits. So two-bladed metal props were fitted.

60. Bradley; Rosendahl, *Up Ship!*, 114. On this day, to refuel, the pilots of *J-3* and *J-4* chose Anacostia. Both nonrigids were so badly pounded that their envelopes were ordered ripped; the pair were returned home via rail.

61. ADC John A. Iannaccone, USN (Ret.), interview by author, audio cassette, Lakewood, N.J., 11 July 1996.

62. Frederick Lewis Allen, *Only Yesterday: An Informal History of the Nineteen-Twenties* (New York and London: Harper and Brothers, 1931), 337.

63. "New York Thrilled as Dirigible Soars over City in Moonlight," *New York Telegram*, 24 April 1929.

64. In 1930–31, he had served in the Plans Division of BuAer, after which Rosie assembled the nucleus crew for *Akron* and had charge of fitting her out. Meantime, "Rosendahl as I understand is none too happy at the prospect of being detached," Berry wrote. "However he is to go to Washington which should keep him in intimate touch." Letter, Cdr. Fred T. Berry to Mrs. Berry, 16 June 1930, courtesy Tom Berry.

65. As in any organization of competent, motivated individuals, differences in opinion arose. Rosendahl—betwixt and between—wanted to remain in dock. "Read your orders," the new skipper snapped, "and I'll take the ship out." Cdr. F. W. Reichelderfer, USN (Ret.), interview by author, audio cassette, Washington, D.C., 3 July 1977.

66. Extensive repairs and maintenance were carried out, including overhaul and replacement of two engines, installation of larger radiators in all power cars, installation of two new steel props, and cleaning and varnishing. "Practically the whole structure now has a protective coat of varnish," Wiley recorded, "and it is hoped that corrosion may be kept in check, as the ship is approaching the end of its predicted span of life." CO, *Los Angeles* to CNO, 30 September 1929, RG 72, BuAer GenCor, box 5573, National Archives.

67. "I certainly would like to go in her," Berry wrote, "but no such luck. The station is all ready to receive her. I have no particular duties other than to assist the C.O. in the reception of notable visitors." "Right now," he continued next day, Friday, "about all one hears around here is 'Graf Zeppelin.' She is due according to reports sometime Sunday morning and the station goes on zero hour at midnight Saturday. There will be a hell of a hullabaloo around here for a few days." Letters to Mrs. Berry, 30 July and 1 August 1929, courtesy Lynn Berry.

68. "Aeroplanes could be specially designed to operate at the low speeds of airships (in the neighborhood of 80 m.p.h.). Comparison would then show airships to use much less power per ton. Thus they are the more economical at low speed, and better suited to flights of long range. But at air speeds much over 100 m.p.h. their advantage over the aeroplane begins to disappear." N. A. V. Piercy, *Aerodynamics* (London: English Universities Press, 1937), 99.

69. *Fifteenth Annual Report of the National Advisory Committee for Aeronautics*, 1929, 17. In 1929 also, the Daniel Guggenheim Fund for the Promotion of Aeronautics announced the establishment of an airship research institute at the Akron Airport, in Ohio, to be operated by the California Institute of Technology.

70. James R. Hansen, *Engineer in Charge: A History of the Langley Aeronautical Laboratory, 1917–1958*, NASA History Series (NASA SP-4305) (Washington, D.C.: National Aeronautics and Space Administration, 1987), 55. A reserve of speed for airships represented a safety factor, and, if provided, would have helped maintain schedules for commercial platforms.

71. Memorandum, CO, *ZR-3* to Chief, BuAer via Commander, Rigid Airship Training and Experimental Squadron [hereafter ComRATES] and CO, NAS Lakehurst, 1 October 1929, RG 80, Office SecNav GenCor, box 4210, National Archives.

72. In a first endorsement, Rosendahl thought that the board might go even farther in its study and recommend what additional "useful missions" could be performed after active flying operations had ceased, and what the final disposition of *ZR-3* should be. "This is necessary in order to plan intelligently for the future."

73. Memorandum, Chief, BuAer to SecNav via CNO, 12 November 1929.

74. The board consisted of Rear Adm. Sumner E. W. Kittelle, president; Capt. Henry T. Wright (CC); Lieutenant Commanders Vincent A. Clarke, Jr., and William K. Harrill; and Lt. George C. Calnan (CC). On 15 May, Harrill was relieved by Lt. Thomas P. Jeter. This small committee was assisted by Lieutenants George W. Henderson (CC) and Bolster—and it also received "valuable advice and assistance" from Goodyear-Zeppelin, Luftschiffbau-Zeppelin, BuStandards, and from Alcoa. These outside organizations were consulted because of their specialized knowledge relating to airship design, materials, and construction.

75. "In connection with the annual survey to determine the condition of the U.S.S. Los Angeles," NACA reported that year, "about 230 samples of channels and lattice material were tested. Although the tests indicated a slow progress in the corrosion since the previous sampling, no sample was found in which the corrosion had progressed enough to reduce the strength of the girders below the designed values." *Sixteenth Annual Report of the National Advisory Committee for Aeronautics*, 1930, 40.

76. BIS to CNO, 3 September 1930, RG 80, Office SecNav GenCor, box 4210, National Archives. The structure was not without problems, however. The gas-cell wires passed through holes in the longitudinals that originally had been reinforced by eyelets of insufficiently hard metal. Wire movements had cut through the eyelets—particularly in that region over which the level of zero gas pressure fluctuated—and then, speedily, through the soft duralumin. This caused a serious weakening of the girders, which was repaired. "It is understood that this trouble is again occurring in some places in the ship. It is believed that this involves a far more serious weakening of the structure than any loss of strength by the present rate of corrosion. It should be watched for most carefully, and all cut channels replaced." Ibid.

77. High regard for Settle was universal. A peerless airman, he might well have altered the course of lighter-than-air aeronautics. "If [ZR] airships had lived a little longer," Rosendahl told the writer, "Settle would have commanded one sure as hell." Interview (notes), Toms River, N.J., 9 November 1975. Tex was not one to disparage his fellows. Decades later, his wife still displayed anger at her husband's having been subordinated. Everyone knew of Rosendahl, Fay Settle observed, but not of the dedicated efforts of other airship men. Tired of the phrase "senior surviving officer of *Shenandoah*," Rosendahl, she reminded the writer, had not commanded the bow section of the ZR-1—Roland Mayer had. Joint interview with author (notes), Washington, D.C., 21 April 1979. "The thing that got me about Rosendahl," Scotty Peck remarked, "was that he became Mr. Lighter-than-Air after a while. Hell, he was a Johnny-come-lately in LTA, he didn't get involved until 1919 or thereabouts, after the First World War. I was in LTA during the war." Rear Adm. Scott E. Peck, USN (Ret.), conversation with M. C. Miller, Chula Vista, Calif., 9 February 1980.

78. Settle interview. "He was inexperienced at the time [Rosendahl took command], and the success of the subsequent operations of [ZR-3] are ascribable to his outstanding ability and to his incessant work." Memorandum, Settle to CO, NAS Lakehurst, 20 April 1934, Mills Collection, box 2, folder 40, NASM, Smithsonian.

79. Bolster.

Chapter 5. Testbed for the New Ships

1. Having no hangar, Goodyear would have had to use Lakehurst to erect ZRS-4. So various schemes were explored to accommodate a contractor there. "The erection of the new ship should take precedence and the LOS ANGELES [overhaul] work made to accommodate itself to it, even if the LOS ANGELES is forced to be idle for a time." Draft memorandum, 1 March 1927, box 4008, Fulton Collection, NASM, Smithsonian.

2. At midday, *Los Angeles* overflew Akron. Two days later, the CNO berated the Commandant, Fourth Naval District, for not having had prior knowledge of the sortie. RG 72, BuAer GenCor, box 5573, National Archives.

3. His INA duties concluded, Settle decided that his future held no ZR command. Further, he was "frustrated" by chronic friction and intrigues. In mid-1934, Settle took command of a gunboat in China, returning to Lakehurst in August 1938 as executive officer of the station. Letter to author, 21 April 1979.

4. Two Navy ZRs were thought the minimum necessary to establish this on a firm basis. Two would better prove or disprove their utility, since single-unit operations might prove inconclusive; also, a competitive effort was desired.

5. It should not be assumed that such planning dated only from 1928. "From the start of thinking about rigids, we were concerned with ways and means for establishing a basis on which to build up the art in the United States." Letter, Fulton to Keller, 19 February 1961.

6. Letter to author, 16 November 1975. "The airship appears to be the only feasible and practical form of aircraft for trans-Atlantic service today. Also, the airship, from the point of view of cost per unit of load and distance, will retain its great economic advantage even when, as may happen, airplanes are eventually able safely to fly the Atlantic. Regular trans-Atlantic airship operations are now technically feasible." "Statement Regarding International Zeppelin Corporation," 4 December 1931, box 30, folder 4, 46, LTA Society, University of Akron archives, courtesy William Trimble.

7. Though the German organization felt that hydrogen could be used safely, it would have yielded to helium because of cooperation (or competition) with American rigids.

8. F. W. von Meister, interview by author, audio cassette, Peapack, N.J., 14 August 1976. Improbably, West Coast representatives of shipping lines proved supportive, seeing passenger airship travel as a means of stimulating American trade and commerce in the Pacific. See William Trimble, *Engineering the Air: A Biography of Jerome C. Hunsaker* (Washington, D.C.: Smithsonian Institution Press, in press), chap. 10. "I believe," Hunsaker observed, "all successful Atlantic flights may fairly be discounted as having been made by overloaded planes, without payload, by abnormally courageous pilots and in the most favorable summer weather that could be found." "Airships for Sea Routes," *New York Times,* 18 May 1930.

9. Concluding a 1930 visit to Goodyear-Zeppelin, a Royal Canadian Navy officer reported: "I was also informed that there was a scheme on foot to build two ships for commercial transportation on the Pacific, and also two for an Atlantic run. A lot of investigating had been taking place on trade routes to be followed particularly for the Atlantic run." Memorandum, Lt. Cdr. A. R. Pressey, RCN, to deputy minister, Public Archives of Canada, RG 24, vol. 2111, file H.Q. 921-28-32, vol. 2, courtesy Barry Jan Countryman. Already, though, the state of the economy was threatening *ZRS-5*. "The question as to whether a second ship would be taken in hand on the completion of this one seemed to be not definite. I was informed that the general trade depression at present might affect the decision." Ibid.

10. *Herald Tribune,* "Gigantic *ZRS-4* Awaits Cover at Akron Plant," 7 February 1931. The piece went on to note that approximately forty thousand square yards of fabric, or more than twice the area draping *Los Angeles,* would be required to cover *ZRS-4. ZR-3* was then flying with about 60 percent of her original hull fabric.

11. "This was one of the features of an airship which distinguished it from an aeroplane. You could climb all over every part of it in flight and carry out any repairs or maintenance that might be necessary." Nevil Shute, *Slide Rule: The Autobiography of an Engineer* (London: William Heinemann, 1954), 102. Shute was chief calculator for Vickers during the *R-100* construction project. "As the Germans often said, jokingly, one trouble with American-built airships is that they tried to make battleships out of airships. They tried to build them too strong." Cdr. F. W. Reichelderfer, interview by author, audio cassette, Washington, D.C., 18 June 1977.

12. It should not be assumed that interior "cars" were welcomed. Goodyear's proved uncomfortable in terms of temperature. Inside *Los Angeles*'s pods, comfort could be adjusted via the radiator shudders. Within the hull (unless one clambered onto the outrigger, which was done), there was little scenery for the engineers to enjoy. Lt. Cdr. William A. Baker, USN (Ret.), interview by author, audio cassette, Lakewood, N.J., 5 December 1976; Lt. David F. Patzig, USN (Ret.) interview by author, audio cassette, Spring Lake, N.J., 10 September 1977. Aviation machinist's mates, both men served aboard *Los Angeles, Akron,* and *Macon.*

13. Or so this feature was advertised. "In my opinion, the tilting propellers on AKRON-MACON were not worth what we hoped to get from them." Letter, Fulton to Keller, 18 June 1959. Slipstream interferences were greater in airships than in other types of craft. Owing to wandering of the propeller slipstreams as they flowed aft because of gusts and the undulations of *Akron*, widely differing propulsive efficiencies were obtained when differing engine combinations were operating. The four-in-line arrangement realized about 64 percent propeller efficiency. *Eighteenth Annual Report of the National Advisory Committee for Aeronautics,* 1932, 60; box 3, Mills Collection, NASM, Smithsonian.

14. Memorandum, Lt. (j.g.) Delong Mills to CO, NAS Lakehurst, 14 June 1927, NASM Archives, file A3U-721000-01, Smithsonian.

15. USS *Los Angeles*, log, 0400 to 0800 watch, 3 July 1929. This also was payday: "1005 The Disbursing Officer paid the ship's crew semi-monthly money." Ibid.

16. "The experiment was in the nature of a materials test as well as a test of the forces set up by such an operation," Wiley remarked. "As such it was successful. We found out a number of things that will be of value to us in carrying the experiment further." "Plane in Air Hooks to the *Los Angeles,*" *New York Times,* 4 July 1929.

17. Tests concluded, Wiley moored and refueled, only to be off again at 1839, for New England, with forty-four on board. *ZR-3* penetrated as far as Concord, New Hampshire.

18. "Results of tests showed that the plane reduced the air speed one knot, increasing the inclination of the ship 2 degrees." USS *Los Angeles,* log, 1600 to 2000 watch, 20 August 1929.

19. *New York Times,* 23 August 1929 (3:4).

20. "Steps from Airship on to Flying Plane," *New York Times,* 29 August 1929; Bolster. In a article published seven months later, Rosendahl remarked, "Since we demonstrated hooking on an airplane in flight at that meet, we have repeated the stunt successfully nearly every

time the ship goes out, and it is getting to be a routine matter." "Flying with an Airship Captain," *Popular Science Monthly*, March 1930, 41.

21. All shipments, personnel travel, fuel for *Los Angeles*, and related expenses were borne by the Air Race Committee. It provided the landing party as well.

22. Letters, Cdr. Fred T. Berry to Mrs. Berry, 16 July 1930, courtesy Tom Berry; Lt. F. W. Reichelderfer to Dr. G. W. Lewis, director of Aeronautical Research, NACA, 10 December 1930, courtesy Navy Lakehurst Historical Society.

23. On the 17th, a fire (from a short, presumably) started in the UO-1 hook-on plane, which was berthed close under the fin of *ZR-3*. "Prompt action of the LOS ANGELES riggers and engineers saved the situation and put the fire out in jig time." *Helium Cell*, 18 January 1930

24. Capt. Ralph S. Barnaby, USN (Ret.), "The Glider Flight from the *Los Angeles*," *U.S. Air Services* 40, no. 3 (March, 1955), 18; interview by author, audio cassette, Philadelphia, 7 April 1979.

25. On the second morning out, from 0730 to 0740, *Los Angeles* hove to over the field, altitude 325 feet, and took in a hundred pounds of drinking water in five-gallon tins. Why? The on-board supply had frozen.

26. Reports had reached the press that *ZR-3* would visit Havana on 15 February. Though permission was granted, eight hours over the tristate region was the month's only flying. The return found snow; during docking, much water was valved and eight men disembarked due to "accumulation of snow, particularly upon the fins." USS *Los Angeles*, log, 1200 to 1600 watch, 15 February 1930.

27. Audio cassette prepared for author, May 1995. Submariners have an intuitive grasp of airship handling. The fundamentals are similar: both are submerged in a fluid and derive their buoyancy from the weight of the fluid displaced. Accordingly, methods of control are almost identical—using ballast to vary buoyancy, for instance.

28. "Vince Clarke is most cordial these days," Berry observed a few months thereafter. "For sometime he did not know that I was assigned to the *L.A.* in a capacity that makes me Comrates when Harry is away. I was very much amused at the expression on his face when I told him." Letter, Cdr. Fred T. Berry to Mrs. Berry, 4 August 1930, courtesy Tom Berry.

29. At a June conference involving Fulton, the Operations Division of BuAer, and Capt. Abraham Claude, *Patoka*'s CO, complaints were made that *L.A.* had not ad-

equately exploited the tender. In reaction, a half-dozen moorings were flown that July–August. "We have a zero hour set for midnight," Berry recorded, "and if the ship gets out I am going along and will probably be gone for two or three days of operations around Newport. We didn't get out as expected Tuesday night." Letter, Cdr. Fred T. Berry to Mrs. Berry, 29 July 1930, courtesy Tom Berry. Total mast time with *Patoka* at year's close: 425 hours, twenty-four minutes.

30. Richard K. Smith, *The Airships* Akron *and* Macon, *Flying Aircraft Carriers of the United States Navy* (Annapolis, Md.: U.S. Naval Institute, 1965), 25. Short-wave apparatus had been installed on both the cruiser and *ZR-3*, whence traffic was relayed to listeners.

31. Rear Adm. Charles A. Nicholson, USN (Ret.), letter to author, 30 November 1976. "By such means," the navy commented, "it is possible to transfer supplies, including fuel, personnel, medical attention, etc. between an airship in flight and the ground or sea. This is certain to enhance the great value of airship travel." Press release, 22 May 1930, NASM Archives, file A3U-721800-01, Smithsonian.

32. Ingalls joined ship's company for a 16–17 July sortie. He was put up in the station BOQ—in Freddy Berry's room. "Yesterday was another hectic day with the Admiral, the Asst Secretary of the Navy and their respective good men Friday on the station for a *L.A.* flight. My room was used by the Secty, it being the most presentable in B.O.Q. That is what comes when one insists upon living in a reasonable degree of comfort. However he was duly appreciative so that is that." Letter, Cdr. Fred T. Berry to Mrs. Berry, 18 July 1930, courtesy Tom Berry.

33. Watson.

34. Letter, Cdr. Fred T. Berry to Mrs. Berry, 6 August 1930, courtesy Tom Berry; Joseph McBride, *Frank Capra: The Catastrophe of Success* (New York: Simon and Schuster, 1992), 223–25. Columbia Moving Pictures first appears in ship's log on 19 August; a five-hour sortie on the 26th concludes these entries. Premiered in April 1931, *Dirigible* was a major commercial success.

35. Hardly the first such performance, in September 1929 *ZR-3* had flown "formation maneuvers" with *J-3* and *J-4* over the harbor, then up the Hudson River. One of Wiley's four guest-passengers: Willy von Meister.

36. The primary mission of the Scouting Fleet was to search for, locate, and "develop" (i.e., information on) the forces of the enemy. In terms of its scouting capabilities, the ZR was most comparable to a heavy cruiser—yet exceeded its speed, range, and visibility.

37. Changing *Los Angeles*'s civil status had been explored in 1926. The disadvantages were many, such as the reopening of diplomatic issues raised against her procurement, and possible allocation to the army. The department, it was decided, could do all it wanted with *ZR-3*—except carry out scouting assignments and shoot guns. Memorandum, Fulton to Moffett, 27 August 1926, box 4008, Fulton Collection, NASM, Smithsonian.

38. "Fitchburg Boy Enjoys Flights on Los Angeles," *Fitchburg (Mass.) Sentinel*, 12 September 1930, Iannaccone Papers.

39. Wiley had initiated a panel-replacement program, exploiting intervals of marginal weather. Between commissioning and the close of 1930, *Los Angeles* logged 242 sorties. Of these, sixty-six were added during 1930 (716 hours, eleven minutes). Total flight time on the airframe as 1931 opened: 2,863 hours, three minutes.

40. Letter to Mrs. Berry, 28 January 1931, courtesy Lynn Berry. With orders to temporary duty as liaison officer during the maneuvers, Berry was en route to the battleship *Arkansas*, flagship of the Scouting Fleet. He reported aboard on the 27th, at Guantanamo Bay. Having conferred with Vice Adm. Arthur L. Willard, Commander, Scouting Fleet, Berry was directed to remain aboard *Patoka*, to assist in her mission, then fall in with *Arkansas* in Panama waters on or about 9 February.

41. First Assistant in Plans Division at BuAer, Rosendahl noted on the cover sheet, "I concur in this recommendation." Added Burgess, "The mast has no 'works,' so why worry?" CO *Los Angeles* to BuAer, RG 72, BuAer GenCor, box 5572, National Archives.

42. Watson. Clarke credited this innovation with eliminating the crew's dread of long flights. "Most of the *Los Angeles* crew will try to duck going to the *Akron* unless she puts in a smoking room," Clarke reported. The recommendation was heeded. Memorandum, Clarke to Moffett, 22 February 1931, Fulton Collection, courtesy Keller.

43. Unidentified clipping, "Fleets Converge as *Los Angeles* Nears Panama," 7 February 1931, Tyler Collection, courtesy Navy Lakehurst Historical Society. "There is considerable interest around these parts in her probable effectiveness," Berry wrote from *Patoka*, "and all hands seem to be looking forward to her arrival." Letter, Berry to Mrs. Berry, 4 February 1931, courtesy Lynn Berry.

44. C. R. Walker, "Sky Life," *Collier's*, 13 June 1931, 32, 64.

45. Ibid.

46. "Los Angeles Gets Panama Welcome," *New York Sun*, 7 February 1931.

47. "*Los Angeles* Is Moored to Ship at Panama," *New York Herald Tribune*, 7 February 1931.

48. Watson.

49. Calnan (aboard *Patoka*) to Fulton, 16 February 1931, Fulton Collection, Smithsonian.

50. "Officials View Entire Nation from Airship," unidentified, undated clipping, Scott E. Peck Collection, San Diego Aerospace Museum.

51. Calnan; letter, Clarke to Moffett, 22 February 1931, box 4008, Fulton Collection, Smithsonian. In the tropics, Zeppelin engineers fared similarly. "Flights to the south through the tropics," one remembered, "were actually very strenuous for the mechanics due to the high humidity, temperatures and rapid changes of climate. The hot air from the engine coolers and the heat given off by the engine raised temperatures in the engine cars up to 45 [degrees] C." Eugen Bentele, *The Story of a Zeppelin Mechanic*, trans. Simon Dixon (Friedrichshafen, Ger.: 1992), 61.

52. USS *Los Angeles*, log, 0000 to 0400 watch, 15 February 1931.

53. A scout's value is a function of communications. In 1930, *L.A.* boasted an upgrade, developed at the RCA laboratories. Chief Radioman Robert Copeland and Radio Electrician Thomas Thrasher had received orders for temporary duty at the plant. The apparatus consisted of a combination intermediate and high-frequency transmitter. A modified ship direction-finder (DF) receiver also was installed, in conjunction with a specially designed lightweight rotating loop within the hull, above the keel.

54. All possible fuel was wanted, as well as food. For her searches on the 15–16th, Clarke lifted off with 27,090 pounds of gasoline. "We stripped the ship down to pencils and sheets of paper, and to skeleton crew, and filled up with [helium] gas—97% at 83 degrees temp." Letter, Calnan to Fulton, 16 February 1931, Fulton Collection, courtesy Keller. Excess baggage and even the upper berths had come out; officers would sleep in relays in the lower bunks. Walker, *Sky Life*, 65.

55. The platform had a superb field of view for reconnaissance because the control car hung *beneath*, a virtue that few airplanes offered.

56. Walker, *Sky Life*, 65.

57. Ibid.

58. The information provided by the "destroyed" *ZR-3*, it is fair to note, was obtained at less cost than if a surface ship had been "lost" gaining it. Taut but fair, unwilling to dismiss any potential weapon without thor-

ough scrutiny, King disliked the contemptuous way in which the rigid airship often was dismissed. "He felt that any weapon that had been the object of so much experimentation both in the United States and Germany deserved at least some serious consideration." Ernest J. King and Walter Muir Whitehill, *Fleet Admiral King: A Naval Record* (New York: W. W. Norton, 1952), 221.

59. Watson.

60. Clarke; "Value of Dirigibles Proved in War Game," *New York Times*, 22 February 1932; letter, Berry to Mrs. Berry, 22 February 1931, courtesy Lynn Berry; Rosendahl, *Up Ship!*, 118; Smith, *Akron and Macon*, 5.

61. Rosendahl, *Up Ship!*, 117. Since his value was thought to be greater in Washington, at this moment Rosie had duty in BuAer. He had left in mid-1930; Lakehurst was not again his duty station until October 1931, when he delivered *Akron*.

62. Watson.

63. Letter to author, 11 January 1995.

64. "Certainly," Reichelderfer observed, "Rosendahl and Doc Wiley—all of them—were very much aware of the importance of press relations." The fact was reflected in, for example, Rosendahl's appearances, interviews, articles, and two books. In 1931, *Up Ship!* (history plus polemic) had just been released.

65. Notes prepared by Rosendahl, Fulton Collection, courtesy Keller. Two years hence, with *Macon* lost, the *Sun* concluded its inquisition by running a selection of its anti-airship editorials in the *Washington Post*. Because of certain information revealed in its criticisms, Moffett, among others, came to believe that the paper had a pipeline to certain factions within the department. Smith, *Akron and Macon*, 211.

66. Capt. M. M. Bradley, USN (Ret.), interview by author, audio cassette, Springfield, Va., 22 October 1977.

67. Had the absence introduced problems of morale? "Most of the crews in those times," Bradley rejoined, "were old sailors and had been to sea part of their lives, at least, and were accustomed to being away from home. There were no physiological problems in that regard." Ibid.

68. Watson.

69. Chief, BuAer, to BIS, 17 July 1931, RG 80, Office SecNav GenCor, box 4210, National Archives.

70. Commanding officers of naval air stations were required to be qualified naval aviators. With two ZRs in prospect, the need to train additional senior line officers was plain. "On the other hand, just imagine what you have in any organization—no matter how high-minded and congenial they are—when you bring in senior authorities over the people who have taken the activity, the program, from its pioneer days, when nobody believed in it." Cdr. F. W. Reichelderfer, USN (Ret.), interview by author, audio cassette, Washington, D.C., 3 July 1977.

71. Letter to author, 15 February 1977. Rosendahl found him wanting. "Dresel didn't believe in airships, didn't get as much as he could out of them. He flew as little as possible." Interview by author (notes), Toms River, N.J., 13 July 1975. Dresel's fitness reports ("Officers Record of Fitness") are exemplary throughout, e.g., "unusually able officer," "an excellent officer," "professionally of high ability. Inspires juniors in loyalty and industry." NASM Archives, files CD-646500-01 to 03, Smithsonian.

72. "I hereby request permission to make practice hook landings aboard USS LOS ANGELES at such time as is convenient to the commands concerned." Lt. (j.g.) D. W. Harrigan to CO, NAS Lakehurst, 12 August 1930. RG 72, BuAer GenCor, box 5582, National Archives.

73. Smith, *Akron and Macon*, 27.

74. "*Los Angeles* Damaged by Fire in Her Hangar," *New York Times*, 25 June 1931; *New York Herald Tribune*, 25 June 1931.

75. "Siamese Rulers Get Prize Thrill of U.S. Visit in Dirigible Cruise," *New York Herald Tribune*, 28 July 1931. Following lunch, the piece notes, charmingly, "the salt air and the gentle flight motion of the dirigible induced drowsiness, and the Siamese sovereigns, by that time taking the flight as a matter of course, retired for a nap in the commander's cabin."

76. Smith, *Akron and Macon*, 29.

77. Back on deck, Moffett remarked: "We expect to learn lessons far beyond those taught by the operations and experiences of the *Los Angeles* and the *Graf Zeppelin* because of the designed superiority of the *Akron*." *Akron Beacon Journal*, 24 September 1931.

78. Among the engineers attending the trials: Theodore von Karman, noted aerodynamicist. "Everything I had felt about the Zeppelins was crystallized here," he wrote of his flight. "The ship was comfortable, noiseless, and smooth. You had an abundance of space. You weren't restricted in movement as in an airplane. I recall sitting back in the vertical fin and enjoying through the window an unobstructed panoramic view of the Great Lakes and the beautiful surrounding countryside. It was exhilarating." Theodore von Karman. *The Wind and Beyond* (Boston: Little Brown, 1967), 162.

79. This double-circle system was needed because

the hangar's long axis was not parallel to the prevailing wind. In May 1932, Lakehurst requested a project order for a hauling-up circle at the east end, to be used when conditions were unfavorable for a west-end undocking. Estimated cost: twenty-five thousand dollars. The project was never executed. That same year, the stern beam was termed "invaluable" for shunting *Akron* in and out of berth. "A special wind-tunnel investigation into the forces acting on an airship when being handled near the ground has been planned for early conduct by the committee at Langley Field." *Eighteenth Annual Report of the National Advisory Committee for Aeronautics,* 1932, 60.

80. Weather remained a critical, unpredictable element. To assist ground operations, signaling gear was installed at Aerology. This consisted of a system of neon tubes, to form separate pairs of numbers plus an illuminated wind-direction arrow, all of which were visible from the field. The scheme provided much-needed wind and temperature data to handlers and ships' officers.

81. *Daily Mirror,* 23 October 1932.

82. "Because of the amount of money invested in one unit," *Airway Age* editorialized, "and because there have been several severe losses in the field, controversy as to the practicability and future value of lighter-than-air craft has, perhaps, been more bitter than controversy over heaver-than-air craft. Certainly there are sufficient possibilities [in airships] to justify further government expense and experiment." Vol. 12, no. 12 (20 June 1931), 4.

83. Devoted to a misunderstood cause, the ZR fraternity tended to be tightly knit, in the face of relentless doubters. So such losses as that of *Shenandoah* were all the more bitter. Further, Lakehurst's physical isolation nourished a sense of otherness, helping to stir internecine rivalries. Riven by turf battles, the prevailing atmosphere was one of gritty determination against the odds.

84. "Dirigible 'Akron' Is Believed Forerunner of Merchant Fleet," *U.S. Daily,* 3 November 1931.

85. USS *Los Angeles,* log, 0800 to 1200 watch, 2 November 1931.

86. "City Views Giant *Akron,*" *Daily Mirror,* 3 November 1931.

Chapter 6. Grounded

1. Audio cassette prepared for author, San Marino, Calif., 21 April 1976.

2. Rear Adm. Harold B. Miller, USN (Ret.) to Richard K. Smith, undated letter (carbon), courtesy Admiral Miller.

3. USS *Los Angeles,* log, 1200 to 1600 watch, 26 January 1932; Lt. Herbert R. Rowe, USN (Ret.), interview by author, audio cassette, Toms River, N.J., 17 April and 4 December 1977; Cdr. R. H. Ward, USN (Ret.), interview by author, Lakehurst, N.J., 11 June 1978. An inert gas, helium cannot combine with oxygen—hence the danger of helium asphyxiation. Crewmen were aware of the danger; however, complete immersion was improbable anywhere along the keel.

4. "The more I think of the [exchange] proposition," Moffett wrote Litchfield (unrealistically), "the more advantages I can see from it to organizations such as yours and International Zeppelin, which I understand to be interested in getting commercial airships established in this country." Letter, 10 February 1932, RG 72, BuAer GenCor, box 5587, National Archives.

5. "Los Angeles Sale Considered by U.S.," *Evening Star,* 29 January 1932; "Navy's Plan to Sell *Los Angeles* Barred," *Washington Post,* 30 January 1932; "U.S. May Fly *Los Angeles* in Commerce," *Washington Herald,* 2 February 1932; "Offers to Purchase Dirigible Numerous," *Evening Star,* 4 February 1932.

6. "Harry [Shoemaker], Alger [Dresel] and myself made our appearance this afternoon," he had written, "heard our orders read and received the congratulations of the board without any further formalities. We were rather surprised because we had expected a grilling as their farewell contribution." Letter, Cdr. Fred T. Berry to Mrs. Berry, 7 July 1930, courtesy Tom Berry.

7. Letter to author, 26 May 1995.

8. A flight to Maine authorized in July 1928 had been canceled because of thunderstorms. Portland area late-nighters had but "a brief view" as *ZR-3* passed over at 0353, moving like a ghost ship across the night sky. "The big craft circled the City and many who heard her and saw her twice during the circle were of the opinion there were two air craft over the City." *Portland Evening Express,* 19 April 1932.

9. Memorandum, Chief, BuAer to CNO, 3 February 1932, RG 72, BuAer GenCor, box 5586, National Archives. Night conditions were addressed through a double system of blinking lights.

10. Memorandum, Chief, BuAer to CO, *Los Angeles* via CO, NAS Lakehurst, 18 February 1932, ibid. "These markings," Moffett explained, "have been arrived at through attempting to secure maximum visibility without spoiling the general appearance of the airship, and without adding undue weight." Ibid.

11. This sum, broken down: pro rata share of Lake-

hurst's general overhead, two hundred thousand dollars; helium, thirty-five thousand dollars; gas cells, twenty-five thousand; miscellaneous experimental developments, ten thousand.

12. "Would Take Airship Out of Navy Service," *Evening Star,* 20 April 1932. These hearings approved nearly $1.5 million for completion of *Macon.*

13. Byrnes to SecNav, 28 April 1932, RG 72, BuAer GenCor, box 5573, National Archives. As early as 1930, gossip had it that *Macon* might be scrubbed. "There are newspaper reports from day to day to the effect that the ZRS-5 is to be cancelled and that a west coast base may never be built." Letter, Cdr. Fred T. Berry to Mrs. Berry, 6 August 1930, courtesy Tom Berry.

14. Memorandum, Chief, BuAer to SecNav via CNO, 9 May 1932, RG 72, BuAer GenCor, box 5573, National Archives.

15. Interview by author, audio cassette, 18 June 1977, Washington, D.C.

16. *Eighteenth Annual Report of the National Advisory Committee for Aeronautics,* 1932, 2.

17. Memorandum, Chief, BuAer to BIS, 20 May 1932, RG 72, BuAer GenCor, box 5579, National Archives.

18. Memorandum, President, BIS to CNO, 24 May 1932, RG 72, BuAer GenCor, box 5579, National Archives. Specimens of the original German lattices and channels had been submitted throughout 1932. "No [channel] specimen was found which had a tensile strength below design values." Though the lattice material indicated "a definite but slow" corrosive attack, "no specimen was found in which the strength had diminished sufficiently to lessen the strength of the girders." *Eighteenth Annual Report of the National Advisory Committee for Aeronautics,* 1932, 30.

19. ComRATES and CO, NAS Lakehurst, to Chief, BuNav, 17 June 1932, RG 24, BuNav GenCor (1925–1940), box 1407, National Archives. "To keep the ship in a satisfactory material state of readiness the most competent qualified and reliable airship men are required in this nucleus crew. It is recommended that flight orders be allowed these 15 men who are assigned for the upkeep of the LOS ANGELES." Ibid.

20. Letter to author, 6 May 1996.

21. USS *Los Angeles,* log, 0800 to 1200 watch, 30 June 1932. The entry was signed by Cdr. Frank C. McCord, who in September 1931 had replaced Scotty Peck as executive officer. The entry was approved via signature by Berry.

22. Letter to author, 26 May 1995. Framed by aluminum from one of her fuel tanks, *Los Angeles*'s "Final Colors" were presented to Fulton on the occasion of his retirement, December 1940. It is of interest to note that ZR-3 is, probably, the first and only Navy *aircraft* to have observed all the naval traditions of decommissioning.

23. Interview by author, audio cassette, Washington, D.C., 18 June 1977.

24. Dr. Karl Arnstein, "Research and Development Problems Arising in Airship Design," *Publication No. 1,* Daniel Guggenheim Airship Institute, 1933, 35.

25. Had *Akron* not been lost, there might have been a second decommissioning. In light of pressures to economize, the Navy Department had plans to ground ZRS-4 as soon as the sister was placed in commission.

26. Moffett's penchant for publicity remained undiminished. On takeoff morning for Guantanamo Bay (1335, 7 January), he arrived on station with a large group of visitors to see the airship—to McCord's annoyance. But McCord did a bit of media work himself, giving a radio interview the evening of the 6th. Letter, Cdr. Frank C. McCord to Mrs. McCord, 7 January 1933, courtesy Michael Fedosh.

27. Some instruments and other equipage were taken over by *Macon.* By spring 1934, the Maybachs and gas cells in Lakehurst storerooms were deteriorating, due to age and to the lack of upkeep funds and personnel.

28. Address before Women's Chapter of the National Aeronautic Association, Akron, Ohio, 11 March 1933; "Moffett Sought Congress Vote for 2 More Airships," *New York Evening Journal,* 5 April 1933.

29. "The Skipper sent word that because of the low-lying fog, he would not require the planes for the flight that night. We were scheduled to demonstrate some night hook-on landings for Admiral Moffett" who, along with Fred Berry and two other guests, had joined *Akron.* Rear Adm. D. Ward Harrigan, USN (Ret.), letter to author, 29 February 1972.

30. Fulton was to have been on board as well. "My father," a son reports, "left our Chevy Chase house that morning with his little overnight bag packed, and drove to his office in the old Navy Building on Constitution Avenue. He and his boss, Admiral Moffett, would proceed to Lakehurst that afternoon to join the scheduled flight. About noon, [both] were driven to nearby Anacostia NAS. As they were getting into their flight suits, a phone call to my father was relayed from the Navy Department. It was from Karl Arnstein at Goodyear, where *Macon* was nearing completion. There was a structural

problem with *Macon*'s lower fin, which would require my father's immediate attention. Accordingly, with Moffett's approval, my father changed plans and was later driven to Union Station where he took the late afternoon train to Akron." Langdon H. Fulton, letter to author, 7 December 1996.

31. Smith, 77–80.

32. As Berry's executive officer, Jessie L. Kenworthy assumed the responsibilities of air-station skipper, serving in that capacity until mid-1934. "Lost in USS *Akron* on last flight JLK," the lieutenant commander recorded in Berry's aviator's flight log book, then signed the entry. Total flight time: 1,508 hours, forty-five minutes. The enlisted survivors were Richard E. Deal, BM2c, and Moody E. Erwin, AMS2c.

33. Rear Adm. Calvin M. Bolster USN (CC), (Ret.), audio cassette prepared for author, San Marino, Calif., 21 April 1976.

34. "Sad Women Scan Skies as Ghost of *Akron* Rides," *New York Daily News*, 5 April 1933.

35. "Lakehurst Hearts Empty Like Void in Akron's Hangar," *New York Evening Journal*, 6 April 1933.

36. "*Akron* Praised by Rosendahl," *New York Evening Journal*. The accident had demonstrated, tragically, the importance of knowledge of the forces and moments acting on airships due to gusts and vortices.

37. "Eckener Anxious for *Akron* Crew," *New York Evening Journal*, 4 April 1933.

38. Chief, BuAer, to SecNav via CNO, 17 June 1933, box 4008, Fulton Collection, Smithsonian. Cost for an "enlarged *Macon* type" was put at $3,500,000; the trainer: three million dollars.

39. "A Costly Sacrifice" (editorial) and "*Akron* Probe Speeded by Roosevelt," *New York Evening Journal*, 5 April 1933. The *New York Times* was equally thoughtful. "If the rigid airship was of use before this disaster," it said, "it can not be argued now that the disaster has destroyed it." The NACA said more than it knew when, that November, it reported: "Speed is still the most important single factor in increasing the relative importance of aircraft for national defense and in extending their use for commercial purposes." *Nineteenth Annual Report of the National Advisory Committee for Aeronautics*, 1933, 1.

40. Seibel, *Helium: Child of the Sun* (Lawrence: University Press of Kansas, 1968), 85. In a further economy, *Patoka* was decommissioned that August. After six years in reserve, the vessel was recommissioned in 1939 as a seaplane tender.

41. Secretary Swanson to Roosevelt, 21 July 1933, box 4008, Fulton Collection, Smithsonian.

42. Other members were Capts. W. P. Robert (CC) and N. H. Wright; Cdr. C. B. Platt; and Lt. Cdr. J. J. Clark. Ordered to temporary duty in connection with the inspection: Lieutenants C. V. S. Knox, Karl Schmidt, and Cal Bolster—all Construction Corps. As in 1930, the board sought and received assistance from outside sources—in this instance the Bureau of Standards, BuAer, and Alcoa.

43. "Huge *Los Angeles* Is Barred from Air," *New York Times*, 11 July 1933. This left but one active ZR. A member of the joint committee found this "regrettable in view of the success Germany is having with airships and plans of combined German-Dutch interests to inaugurate regular airship service to various parts of the world." Ibid.

44. The sale of *Los Angeles* to private interests is an intriguing might-have-been. As early as March 1932 discussions had been held regarding possible sale. The following March, an inquiry was received relative to purchasing her for a movie, during the shooting of which ZR-3 would be destroyed. The sale was permissible, subject to certain restrictions. The Navy Department stipulated public advertisement and that the aircraft would go to the highest bidder. The navy reserved the right to reject all bids if the offers were not commensurate the airship's value. In this instance, the purchaser was not prepared to offer more than twenty-five thousand dollars; following discussions between SecNav, CNO, and the Chief of the Bureau, the matter was dropped.

45. BIS to CNO via BuAer, 12 July 1933, RG 80, Office SecNav, GenCor, box 4210, National Archives.

46. Memorandum for Chief, 19 September 1933, BuAer, RG 72, BuAer GenCor, box 5580, National Archives.

47. Memoranda, Chief BuAer to SecNav via CNO, 10 August and 19 September 1933, RG 80, Office SecNav, box 4208, National Archives. The latter was drafted by Fulton. In a handwritten note, King had requested a rewrite of Fulton's own memo, addressing the board's findings paragraph by paragraph. "Also, include mention of conditions (economic) which were the cause of LOS ANGELES being decommissioned—and *not* her material condition." Ibid.

48. Memorandum Endorsement for Assistant CNO, 23 September 1933, ibid.

49. Memorandum, CNO to SecNav, 3 October 1933, RG 80, Office SecNav, box 4208, National Archives; Memorandum, SecNav to CNO; All Bureau

Chiefs; Commandant, Fourth Naval District; CO, NAS Lakehurst, 13 October 1933, ibid.

50. *Akron Beacon Journal*, 21 April 1933.

51. "Macon Pleases Navy Men on First Flight," *New York Evening Journal,* 22 April 1933. Investigations on drag were rewarded by an increase in speed of *Macon* over that of *Akron*. A large part of the NACA's research programs consisted of specific requests of both the army and navy air organizations. At this time, airship work of the aerodynamics committee included speed trials with the army's *TC-11* and *TC-13* nonrigids. *Nineteenth Annual Report of the National Advisory Committee for Aeronautics,* 1933, 15.

52. Memorandum, SecNav to CNO; All Bureau Chiefs; Commandant Fourth Naval District; CO, NAS Lakehurst, 23 February 1934, RG 72, BuAer GenCor, box 5574, National Archives.

53. Requests for suggestions had gone to the NACA; the commanding officers of Lakehurst, Moffett Field (formerly Sunnyvale), and *ZRS-5*; to Goodyear-Zeppelin; BuAer's C. P. Burgess; Luftschiffbau-Zeppelin; the Daniel Guggenheim Airship Institute; and the Metalclad Airship Corp. All responded. Due to the press of work in connection with the East Coast flight of *Macon*, the COs of Moffett Field and *ZRS-5* deferred their responses till May and March, respectively.

54. Memorandum, Chief, BuAer to CO, NAS Lakehurst, 20 June 1934. RG 72, BuAer GenCor, box 5578, National Archives.

55. Memorandum, CO, NAS Lakehurst, to Chief, BuAer, 23 July 1934, ibid.

56. Quarterly hull board reports on material condition had been instituted early on. The practice continued after the decommissioning, although with less frequency. Self-interested to understand service problems with its product, the aluminum industry was a party to these studies.

57. Chief, BuAer (first endorsement) to CNO, 28 July 1934, RG 80, Office SecNav, box 4208, National Archives.

58. Memorandum, CO, NAS Lakehurst to Lt. Raymond F. Tyler, 31 August 1934, Tyler Collection, courtesy Navy Lakehurst Historical Society. "You are also assigned additional duties as instructor in the Officer's Ground School and nonrigid pilot under the Officer-in-Charge of Training and the Operations Officer, respectively." Ibid. Knox was a graduate of Lakehurst Class IX (1934–35), Bradley and MacIntyre of Class V (1928–29).

59. Lt. Cdr. William A. MacDonald, USN (Ret.), audio cassette prepared for author, Akron, Ohio, 10 March 1976.

60. Experimental projects using *Akron* had been under way or planned, so the bureau had wanted her close by. *Macon*, an improvement over her sister, was deemed more suitable for fleet operations. Conditions for ground handling proved favorable on the West Coast. Also, the charm of California delighted most of *Macon*'s complement.

61. "*Los Angeles* Flying from Mooring Mast," unidentified clipping, 19 December 1934, Mills Collection, box 28 (scrapbook), Smithsonian; *Philadelphia Record,* 22 October 1933. The latter went on to label the program as a "Disastrous Dirigible Experiment Marked by Great Heroism, Colossal Blunders."

62. "Air Base Hopeful as *Macon* Leaves," (Philadelphia) *Evening Public Ledger,* 12 October 1933.

63. Memorandum, CNO to All Bureaus and Offices; Commandant, Fourth Naval District; CO, NAS Lakehurst, 12 October 1933, author files.

64. Clark L. Bunnell Collection. From 1930 to 1961, Bunnell was a civilian employee in the Assembly and Repair (later Overhaul and Repair) Department.

65. CO, NAS Lakehurst to Chief, BuAer, 20 December 1934, RG 72, BuAer GenCor, box 5586, National Archives. By mid-1936, save for small sections near the wing cars all of the original outer cover had been replaced with American fabric.

66. "*Los Angeles* Flying from Mooring Mast," unidentified clipping, 19 December 1934, Mills Collection, box 28 (scrapbook), Smithsonian. NACA was cooperating by making available instruments and personnel for an investigation of pressure distribution on the hull and fins. *Twentieth Annual Report of the National Advisory Committee for Aeronautics,* 1934, 5.

67. Report of Investigation to CO, NAS Lakehurst, 21 March 1935, box 3, Mills Collection, Smithsonian; memorandum for Chief of Bureau, 16 March 1934, RG 72, BuAer GenCor, box 5574, National Archives.

68. Letter, 24 August 1935, courtesy Navy Lakehurst Historical Society.

69. As to Wiley's inspired use of the F9C-2s. "My father," son Gordon said, "was sent to command the *Macon* by Adm. King to 'prove' the airship as a scout and to do it quickly." Letter to author, 26 March 1977. In October 1934, King arranged for Rosendahl and Anton Heinen to observe *Macon*'s operations and perhaps suggest improvements. "By this time," Richard K. Smith writes, "Wiley, his officers, crew, and the *Macon* had prac-

tically rewritten the book. Rosendahl and Heinen could not help but be favorably impressed." Smith, *Akron and Macon*, 137. Rosendahl's 8 October sortie was his last aboard a ZR.

70. Lt. George W. Campbell, USN, "Five o'Clock, Off California," *The Saturday Evening Post*, 15 May 1937, 20.

71. *Akron* carried no life vests and but one rubber raft. Bolster had made "firm resolve, that as long as I was First Lieutenant of the *Macon*, we would carry life preservers for all hands and that inflatable life rafts would be on board in sufficient numbers to handle anyone on board. We also made up some emergency repair material for rapid repair to broken girders, consisting of wooden boards, properly cut and drilled to permit them to be rapidly bolted around a girder if need should arise." Bolster. On 12 February the "need" arose. But everything had happened too fast and in a location difficult to access.

72. Coulter.

73. Quoted in "Dirigibles in Disrepute After Macon's Loss," *The Literary Digest*, 23 February 1935, 9.

74. The short name of the Special Committee on Airships of the National Academy of Sciences, chaired by Dr. William F. Durand of Stanford University, a NACA committee member from 1918. Mandate? To study not only *Macon*'s loss, but to consider the broader engineering aspects of design and construction and to make recommendations for future policy. The committee proved a unwitting instrument for Swanson to sidetrack airships. Insisting on a new program, King expressed his view that there was no reason to await a further report. Swanson, though, continued to beg off, with the excuse that he could do nothing until he had received all of the committee's reports. Richard K. Smith, *The Airships* Akron *and* Macon, *Flying Aircraft Carriers of the United States Navy* (Annapolis, Md.: U.S. Naval Institute, 1965), 157.

75. "Pressure-Distribution Measurements at Large Angles of Pitch on Fins of Different Span-Cord Ratio on a 1/40-Scale Model of the U.S. Airship "Akron," 1937, *NACA Report No. 604*.

76. "Let *Los Angeles* Replace *Macon*, Rosendahl Asks," *New York Herald Tribune*, 16 February 1935.

77. Theodore von Karman, with Lee Edson. *The Wind and Beyond: Theodore von Karman, Pioneer in Aviation and Pathfinder in Space* (Boston: Little, Brown, 1967), 167; Special Committee on Airships to SecNav, *Report No. 1*, 16 January 1936, Stanford University Press.

78. Capt. M. M. Bradley, USN (Ret.), interview by author, audio cassette, Springfield, Va., 24 October 1977.

79. "Notes on Research into the Structure of Wind Gusts and Violent Air Currents for Use in LTA Design and Operation," 21 April 1928, 3, courtesy Navy Lakehurst Historical Society. In the quest for more accurate data, efforts were renewed to determine a "standard" squall or gust, comparable to the "standard wave" used by naval architects.

80. Cdr. F. W. Reichelderfer, USN (Ret.), interview by author, audio cassette, Washington, D.C., 18 June 1977.

81. C. P. Burgess, "Comparative Strength and Factors of Safety of U.S.S. LOS ANGELES as Designed and in Service in 1930," *Design Memorandum No. 94*, February 1930, 1.

82. In January 1937, Lakehurst had a dozen Maybachs in storage, seven received from Sunnyvale. The five spares for *ZR-3* were decommissioned for disposal that year.

83. Donald E. Woodward and William F. Kerka, "The 'Cut-Down' *Los Angeles*," *Aerostation* 6, no. 1 (Spring 1979), 10–11, 22.

84. "Comments by Several Officers on Proposal to Convert *Los Angeles* to Small Size Training Ship," box 4, folder 25, Mills Collection, Smithsonian; "Comment on Proposed Modernization of U.S.S. *Los Angeles*," Tyler Collection, courtesy Navy Lakehurst Historical Society.

85. "The seriousness of the [commercial airship] proposals seemed 'up and down' in the years 1934–37," Fulton later complained. "Administration support for a commercial project also varied. FDR was favorable up to a point. Airlines were actively against. Navy was coy, but tried to help, so long as their appropriations were not affected." Letter, Fulton to Keller, 10 March 1962.

86. *Aerostation* 12, no. 1 (Spring, 1989), 6–9. The design included a hook-on trapeze, for the delivery or discharge of mail—"of particular value in a coastal service such as that proposed from the New York area to Rio de Janeiro." "The possibility of military conversion," Goodyear added, "has been kept in mind so that this airship can be very readily be converted into a naval scout or airplane carrier." Ibid.

87. "Four Blimps Fly in Formation over Manhattan Island," *New York Herald Tribune*, 18 October 1935.

88. Memorandum, F. W. Reichelderfer to Lt. Cdr. J. B. Anderson, 17 January 1936, courtesy Navy Lakehurst Historical Society.

89. Memorandum, Chief, BuAer to CO, NAS Lakehurst, 7 February 1935, RG 72, BuAer GenCor, box 5578, National Archives.

90. CO, NAS Lakehurst to BuAer, attachment to letter NA4/ZR/A4–3, 16 December 1935, Fulton Collection, NASM, Smithsonian.

91. Of Hugo Eckener, von Meister would write, "He was one of the very few international Germans who, in addition to his accomplishments in the field of trans-oceanic air transportation, also represented the 'other Germany' during the Nazi period." Letter to W. Averell Harriman, 5 January 1968, courtesy Langdon H. Fulton.

92. At her Friedrichshafen and Frankfurt termini (the latter under construction into 1936), the ground-handling gear used by *Hindenburg* was similar to that in New Jersey—a mast and tracks for a riding-out car, and trolleys.

93. Letter, Eckener to Roosevelt (trans. von Meister), box 4016, Fulton Collection, NASM, Smithsonian. "Count Zeppelin left me the vision of a world-wide system of airship routes, uniting nations into closer contacts and better understanding. It has been my life-long ambition to bring this vision into realization." Ibid.

94. "The German's [*sic*] use of Lakehurst was readily agreed to—all the way up to FDR. The view was frequently expressed that the success of the German operations would point to what our long-range policy should be." Letter, Fulton to Keller, 10 March 1962. "Had it not been for the trust and respect felt for Dr. Eckener by Franklin Roosevelt," Gordon Vaeth believes, "it is doubtful whether the *Hindenburg* would have been permitted to use the Navy facilities at Lakehurst as a port of entry." Quoted in F. W. von Meister to W. Averell Harriman.

95. Memorandum (copy), Lt. Cdr. Scott E. Peck to Adm. E. J. King, 20 February 1936, author files.

96. One (Durand) committee report had just been released. This paid "high tribute" to the practicality of lighter-than-air craft and recommended continuation of the program that the Navy had been prosecuting with *Akron* and *Macon*.

97. Letter to author, 21 May 1936.

98. "Report of observations on Flights Nos. 41 and 42 on German Airship HINDENBURG," Lt. R. F. Tyler to CNO (Office of Naval Intelligence) via CO, Lakehurst, 23 October 1936, Tyler Collection, courtesy Navy Lakehurst Historical Society.

99. Interview by author, Washington, D.C., 18 June 1977. Lehmann had been a guest in Reich's quarters. "He used to spend hours at the piano—we had a piano out on the sun porch—and it was his release, his relief." Ibid.

100. In a "Thrills Aplenty for Milady" column in the *New York American,* Lady Drummond Hay remarked that the shower bath "was simply a great luxury. There was lots of hot water and a grand flow of it. The staterooms are not as roomy as on a boat, of course, but fully as comfortable." A German countess disagreed: "The shower was a thin trickle and very cold. The airship is not at all a cozy place." A French journalist complained, "The greatest gift America can give me is a good hot bath! During the day we were princesses, but at night we were like paupers because the quarters were quite cramped." Others said simply, "I just loved it," or, "The trip was very enjoyable." Mrs. Charles B. Parker, of Cleveland, summarized the virtues of this, her sixth airship flight. "Traveling this way is a wonderful beauty secret. It is so absolutely calm and effortless. There's no nervous strain. Any women knows what that does for her appearance." *New York American*, 10 May 1936.

101. Memorandum, Lt. Cdr. A. F. Heinen, USNR, to CNO, "Observation Report on Flights No. 34 and 35 of German Airship 'Hindenburg,'" 16 September 1936, Fulton Collection, NASM, Smithsonian. Eckener had brought up the matter of placing the "*129*" at a slight angle, to allow a slightly greater length. (As yet, the ship was incomplete at the stern.) He had been informed of the schedule of operations for *ZR-3* and "that it might be necessary to have LOS ANGELES in the hangar at the same time with the *LZ-129*. The station will check against the practicability of housing the *LZ-129* at an angle at the same time LOS ANGELES is in the hangar." Memorandum, CO, NAS Lakehurst, to Chief, BuAer, 7 November 1934, HOAC.

102. Capt. George F. Watson, USN (Ret.), audio cassette prepared for author, Litchfield Park, Ariz., 1–7 November 1981.

103. Letter to author, 21 May 1996.

104. Letter, William H. Sitphen to Chief, BuNav, 13 May 1936, RG 80, Office SecNav, box 4208, National Archives.

105. *New York Sunday News,* 10 May 1936.

106. Nonetheless, the Germans had begun experimentation with water-recovery apparatus.

107. Rear Adm. Richard S. Andrews, USN (Ret.), audio cassette prepared for author, Hillsborough, Calif., 20 December 1976. The Navy at this time had but forty-nine naval aviators (airship) on its active list, and attrition of experienced enlisted personnel had been high. In 1935, Moffett Field was turned over to the army; *Patoka* was already in reserve.

108. Chief, BuAer, to CO, NAS Lakehurst, 31 July

1936; CO, NAS Lakehurst, to Chief, BuAer, 12 August 1936, RG 72, BuAer GenCor, box 5578, National Archives.

109. Memoranda, Experimental Board to CO, NAS Lakehurst, 9 January 1937 and Chief, BuAer, to CO, NAS Lakehurst, 22 January 1937, ibid.

110. Andrews. To some airmen, perhaps, the grounded aircraft was a reminder of failure, a reproach to their expertise and the value of their naval careers.

111. Among the invited: Goodyear's Paul Litchfield and Tom Knowles; Eddie Rickenbacker, General Manager, Eastern Air Lines; and Juan T. Trippe, President, Pan American Airways. Trippe regarded commercial airships as "serious competition" to flying boats (the platform Pan Am was then committed to) and zeppelin service as "threatening" to his vision of transoceanic transport. He conspired accordingly. See Marylin Bender and Selig Altschul, *The Chosen Instrument: Pan Am, Juan Trippe; The Rise and Fall of an American Entrepreneur* (New York: Simon and Schuster, 1982), 197 and 290.

Chapter 7. Denouement

1. By November 1936, lighter-than-air represented, in money and personnel, about 1 percent of the total naval-aviation effort.

2. CO, Lakehurst, to Chief, BuAer, 20 January 1937, RG 72, BuAer GenCor, box 5580, National Archives.

3. Andrews.

4. Letters to author, 6 May (Owen) and 21 May 1996.

5. "My job was to assist the officer in charge of the operation and personally see that landing parties were in position, that they tended and walked their lines properly and had their lines properly tensioned at all times. The officer in charge was generally in a position ahead or where he could watch for treacherous wind shifts. We knew his orders and signals and acted accordingly. When the ship was properly moored and riding free and safe, the lower vertical fin was rested on a padded flat car which was free to run right or left on a track which was a circle with the mooring mast at its center." Crewmen could egress via a long ladder shipped onto the control car, passengers "by a very short ladder [stairway] extending down from the lowest part of the keel which moved with the ship and was a few inches free of the ground." Andrews.

6. F. W. von Meister, interview by author, audio cassette, Peapack, N.J., 14 August 1976; Lt. Herbert R. Rowe, interview by author, audio cassette, 11 January 1976, Toms River, N.J.; Reichelderfer; Andrews.; Owen Tyler and Louise Tyler Rixey.

7. Logbook, NAS, Lakehurst, N.J., 0000–0400 watch, 7 May 1937, box 146, HOAC.

8. Andrews.

9. "Next year we will come to America every two weeks with new helium-inflated Zeppelins," Capt. Max Pruss predicted from a hospital bed. Informed of a proposal by Rosendahl to recommission *Los Angeles*, his swollen face lit up. "She is a good ship," he said. "She should be used. We must carry on." *New York Times*, 10 August 1937.

10. Letter, Fulton to Keller, 10 March 1962. "FDR, at the showdown, went along with Ickes," Fulton added, "but I think reluctantly and only because of the worsening international situation." Ibid.

11. Save for noting an investigation of boundary-layer control on airship forms, the 1938 report of the NACA Subcommittee on Airships offers little. Results by Goodyear-Zeppelin on the fatigue strength of aluminum-alloy girders had been issued, along with translations of German papers dealing with airship problems. "The subcommittee has kept informed of the latest developments in connection with airship design, construction, and operation, particularly the activities of Germany, where interest in the airship remains active, in spite of the unfortunate disaster to the *Hindenburg*." *Twenty-fourth Annual Report of the National Advisory Committee for Aeronautics*, 1938, 19.

12. Memorandum, "1939 Estimates—Lighter-than-Air," 13 July 1937, box 4008, Fulton Collection, NASM, Smithsonian.

13. American Legion Annual Report (Aeronautics Commission), 2–3, 1937, box 4008, Fulton Collection, NASM, Smithsonian.

14. Undated telegram (February 1938), courtesy Langdon H. Fulton; letter, CNO to CO, NAS Lakehurst, 24 November 1937, box 4008, Fulton Collection, NASM, Smithsonian; Jacob Klein, "Up Ship, Gentlemen!," *Popular Aviation*, December 1936, 16.

15. 1939–40 position paper, box 4008, Fulton Collection, Smithsonian. A rigid of moderate displacement represented, in Fulton's view, "an airship of appropriate size on which to carry forward airplane carrying developments and to train personnel," a means "of holding and perfecting the gains made to date in this field." Ibid.

16. Letter, King to Fulton, 12 April 1939, courtesy Langdon H. Fulton.

17. USS *Arizona* (San Pedro, Calif., 28 August 1937): "When directed by comdg officer in August Ensign Marion H Eppes detached proceed Lakehurst N.J. Report comdg officer NAS temporary duty under instruction in lighter than air training." Departing Long Beach 6 September, the ensign reported on board on the 21st. Destined to command the Lakehurst station during 1959–61, Eppes was advanced one month's pay in connection with these orders—$125. Courtesy Captain Eppes.

18. Capt. M. H. Eppes, USN (Ret.), interview by author, audio cassette, Bethesda, Md., 27 November 1977; Capt. Howard N. Coulter, USN (Ret.), letter to author, 12 October 1981.

19. Letter (with attachments) to author, 27 January 1982, Eppes.

20. In his report to SecNav for fiscal year 1939, Towers recommended amending the law to allow qualification on the basis of nonrigid experience. Andrews received No. 6749, Antrim No. 6750. Designation was granted for Class XI—e.g., Eppes, (No. 6753)—on 20 September 1940.

21. G. V. Whittle to CO, NAS Lakehurst, 13 October 1937, courtesy Navy Lakehurst Historical Society.

22. Rosendahl directed that the tests be conducted at the earliest practicable date, so as to avoid ice and snow conditions with *ZR-3* moored out. "Completion of the tests at an early date will also minimize the freezing problem with the water ballast." Ibid.

23. Letter, President to Representative John D. Dingell, 25 April 1938, box 4008, Fulton Collection, NASM, Smithsonian.

24. Memorandum, Lt. Cdr. Howard N. Coulter to CO, NAS Lakehurst, 6 July 1938, box 3, Mills Collection, NASM, Smithsonian.

25. Assistant Secretary since 1937, Edison was appointed Acting Secretary when Swanson died in July 1939; that December, he was designated as successor. "It will carry 3.7 times more useful load and have approximately 3.0 times the range of the 325 ft. ships," Burgess wrote of the favored proposal, a three-million-cubic-foot, 650-foot-long ship, "and will cost only 47% more. Besides being far more efficient, it will also be much safer, having its gas space subdivided into 14 cells, instead of only 7 or 8. It is also suitable for carrying airplanes." C. P. Burgess, *Design Memoranda No. 312,* box 4010, Fulton Collection, NASM, Smithsonian.

26. Letter, Fulton to Col. R. G. Elbert, 10 August 1939, box 4008, Fulton Collection, Smithsonian. "As nearly as I can find out," Fulton added, tellingly, "Senator Byrnes fully realized what he was doing. He knew the President had reversed his position, but thought the President was wrong on size of airship. He stated, so I am told, he had his own sources of information in the Navy Department and he knew the Admirals in control did not want any such airship." Ibid.

27. Letter, 9 July 1938, box 4016, Fulton Collection, NASM, Smithsonian. Rosie had come to Friedrichshafen to attend the hundredth birthday celebration of Count Zeppelin.

28. Bentele, *Zeppelin Mechanic,* 57–59. Postwar technique came to include the pickup of seawater ballast prior to landing, producing a heavy (nonrigid) ship. "Pilots simply landed like an airplane with approach speeds that ensured good control. I always touched down well in front of the landing officer and taxied up to him." Lt. Cdr. R. W. Widdicombe, USN (Ret.), memorandum to Norman L. Beal, USNR, CND Office of Naval Warfare, 1 November 1985, courtesy Commander Widdicombe.

29. Letter, F. W. von Meister to Garland Fulton, 15 April 1939, Fulton Collection, NASM, Smithsonian.

30. Letter, Fulton to King, 4 October 1938, box 4008, Fulton Collection, NASM, Smithsonian.

31. Even at this hour, airmen held to the faith. "It is recommended," the report notes, "that the ship be completely equipped with an entirely new ballast system in event ship is placed on an operating status." Hull Board to CO, NAS Lakehurst, 30 September 1938, 4, RG 72, BuAer GenCor, box 5580, National Archives.

32. Quoted in Hull Board to CO, NAS Lakehurst, 30 September 1938, RG 72, BuAer GenCor, box 5580, National Archives.

33. The 1939 board consisted of Rear Adm. H. L. Brinser, president; Capt. H. S. Howard (CC); Fulton; and Lt. Cdr. R. E. Jennings. Howard was detached in January, when the preliminary report was submitted. The board was assisted by Lakehurst's Cdr. T. G. W. Settle, Lt. Cdr. George V. Whittle, and Lt. Charles W. Roland. Lt. Clinton S. Rounds, as well as others, also took part.

34. "Remarks on Airship Situation," 25 January 1939, box 4008, Fulton Collection, NASM, Smithsonian. As for the 131st zeppelin, it never advanced beyond the earliest stages.

35. Letter, Rosendahl to Fulton, 30 December 1938, box 4008, Fulton Collection, NASM, Smithsonian.

36. The following were forwarded to BuStandards: twenty-two pieces of girder channel, thirty-six lattice crosses, two pieces of brass tubing, four pieces of aluminum alloy tubing, and fourteen specimens of steel wire.

All had been cut from *ZR-3*. In addition, two five-foot sections of girder (one longitudinal, one transverse) were shipped to Goodyear-Zeppelin for fatigue tests.

37. As part of a major overhaul, the following were recommended: new gas cells and outer cover, a new fuel system and water-recovery apparatus, a new ballast system, a new control system, a new electrical system and an overhauled radio, and new cord and wire nettings over the gas cells. Further, the engines needed reconditioning, and the shear and bulkhead wires required general overhauling. The entire structure would receive a thorough inspection and cleaning. As well, the interior spaces wanted refurbishing, and numerous miscellaneous items of equipage needed checking for possible replacement.

38. Report of Material Inspection, BIS to CNO, 27 May 1939, RG 80, Office SecNav, GenCor, box 4210, National Archives.

39. Winston S. Churchill, *The Gathering Storm* (Boston: Houghton Mifflin, 1948), 156.

40. Chief BuAer to CO, NAS Lakehurst, 23 October 1939, RG 72, BuAer GenCor, box 5579, National Archives.

41. Why the need for deck space? BuAer still nursed hope for another ZR. Also, naval air station blimps were an active part of the prewar neutrality patrol.

42. The rudder and elevator systems were the major controls. Minor controls were the maneuvering valve control system, the engine-telegraph system, the water ballast and fuel dump controls, and the mooring gear controls, cable, and wire.

43. This went to Rosendahl; it remained among the admiral's effects until his death in 1977. The signed photograph, *ZR-3*'s original Telefunken radio, and the incomparable Rosendahl Collection were lost to New Jersey; today they are held by the University of Texas at Dallas.

44. CO, NAS Lakehurst, to Chief BuAer, 27 October 1939, RG 72, BuAer GenCor, box 5579, National Archives.

45. In 1940, Paul Garber, the primary figure who shaped the Aeronautical Collection of the National Museum, Smithsonian Institution, was an assistant curator in the Section of Aeronautics in the Division of Engineering, Department of Arts and Industries. That June, he suggested that the institution inform the secretary of the navy of its interest in an engine and, possibly, some instruments and structural specimens from *ZR-3*. "This was one of the most famous airships of all time and had a longer period of service, I believe, than any other." Memorandum, Garber to Taylor, 24 June 1940, NASM Archives, file A3U-722400–01, Smithsonian. Subsumed by war preparations, the matter came to nothing.

46. *New York Herald Tribune*, 3 December 1939.

47. CO, NAS Lakehurst, to Chief BuAer, 8 December 1939, RG 72, BuAer GenCor, box 5578, National Archives.

48. These had been installed in September–November 1934. Five additional VL-1s were in storage, along with seven VL-2s from the inventory of *Macon*. The hull board reports for the decommissioned years include a list of Maybachs to hand, along with a summary of their respective operational and overhaul histories, and their prevailing conditions. The 1939 inspection found nine VL-1 type suitable for use if overhauled. The design was obsolete, however; lighter-weight, air-cooled engines were recommended should a reconditioning be ordered.

49. Letter to author, 6 July 1981.

50. Memorandum, Chief, BuAer, to CO, NAS Lakehurst, 13 February 1940, RG 72, BuAer GenCor, box 5578, National Archives. The likelihood of such tests actually being executed seems problematic.

51. The scrap comprised three lots: twenty-eight thousand pounds of duralumin; three thousand pounds of wire and steel; and one thousand pounds of aluminum. Letter, Mills to Fulton, 9 May 1940, Mills Collection, box 8, folder 18, NASM, Smithsonian. *J-4* was struck as well. "Too bad about *J-4* but she has had a useful and honorable career and might as well die in shed. I suppose we might get a new envelope in a years time but I imagine it is better not to fool with so old an airship any longer." Fulton to Mills, 19 February 1940, ibid.

Epilogue

1. Smith, *Akron and Macon*, 169. In the end, the villain of this drama was, ironically, its foremost player. "This new airship [ZRN] was never funded and never built because C. E. Rosendahl was loathed by the bureaucracy within the Navy Department circles due to his czar-like temperament and total egotism." Capt. Douglas Cordiner, USN (Ret.), letter to author, 23 December 1977. Cordiner (Lakehurst Class XIV, No. 8141) arrived during the *ZR-3* dismantling ("loud noises coming from the big hangar"). Part of the wartime expansion, he became chief staff officer for Capt. George Mills.

2. *Akron Beacon Journal*, 20 October 1939.

3. Lt. Cdr. C. E. Rosendahl, "Bigger and Better Blimps," *World's Work*, July 1929, 66–71.

4. Rear Adm. Harold B. Miller, USN (Ret.), letter to author, 2 February 1972.

5. "Scouting for Raiders," *Flight* 36, no. 1610 (2 November 1939), 341. The pocket battleships *Deutschland* and *Admiral Graf Spee* had slipped in secret from German ports to assigned stations well before war was declared, after which merchantman after merchantman failed to gain port. Surface raiders, understandably, were causing extreme consternation in the Admiralty.

6. *New York Herald Tribune*, 3 December 1939.

7. Letter, Cdr. G. H. Mills, USN, CO, NAS Lakehurst to Director, Department of Debating and Public Discussion, University of Wisconsin, Madison, 2 December 1941, courtesy Navy Lakehurst Historical Society.

8. Letter to author, 21 May 1996.

9. Smith, *Akron and Macon*, 170. In Weyerbacher's view, Fulton was one of the most brilliant minds in BuAer, if not the Navy Department. If the ZR project had prevailed, he would have been a big name—and prominent anywhere under any other circumstances. He had simply ridden "the wrong horse." Cdr. Ralph D. Weyerbacher, USN (Ret.), conversation with Richard K. Smith, Booneville, Ind., 8 July 1960, courtesy Keller.

Selected Bibliography

Primary Unpublished Sources

Research for this book has relied in significant part on major archival sources. Particularly informative were the old navy records held by the National Archives and Records Administration (NARA) as Record Group 72, "Bureau of Aeronautics General Correspondence 1925–1942." Details of the airship's acceptance are in Record Group 38, "Office of CNO, BIS Reports of Acceptance Trials of Naval Aircraft 1919–1932," and of its military operations in Record Group 80, "Office of the Secretary of the Navy, General Correspondence 1926–1940." The airship's operating log is in Record Group 24. NARA research and correspondence conducted by Mr. Charles L. Keller, combined with that of the author, constituted the bedrock of this project.

The National Air and Space Museum, Smithsonian Institution, holds a wealth of archival information pertaining to *Los Angeles* and the ZR experiment. The Annual Reports of the National Advisory Committee for Aeronautics offer context as well and are invariably excellent for technical matters. In addition to these published items there are C. P. Burgess's *Design Memoranda*, 401 in number, copies of which are held by the NASM Archives Division. The NASM-held papers of Cdr. Garland Fulton, USN (Ret.), Dr. Jerome Clarke Hunsaker, and Commodore George H. Mills, USN (Ret.), were invaluable sources. Each collection holds day-to-day correspondence as well as technical and miscellaneous materials pertaining to *ZR-3* specifically, and generally to lighter-than-air aeronautics in the U.S. Navy.

The incomparable Charles E. Rosendahl Collection, lost to New Jersey, is held by the University of Texas at Dallas as part of its History of Aviation Collection. The library of the San Diego Aerospace Museum also has much useful material.

More than two decades of interviews and correspondence inform this work. From 1975 to 1998, one-on-one sessions were audio-recorded with officer, enlisted, and civilian personnel associated in one way or another with *ZR-3*, Naval Air Station Lakehurst, and lighter-than-air. In a few instances, queries drew cassette-recorded responses.

Published information from primary and selected secondary sources were essential to corroborate background detail and supporting information and also to help round out the narrative.

Letters and Statements

Rear Adm. Calvin M. Bolster, USN (CC) (Ret.)
Capt. Howard N. Coulter, USN (Ret.)
Capt. Garland Fulton, USN (CC) (Ret.) (to Charles L. Keller, courtesy Mr. Keller)
Mr. Langdon H. Fulton
Rear Adm. Harold B. Miller, USN (Ret.)
Cdr. Frank C. McCord (to Mrs. McCord, courtesy Michael Fidosh)
Mrs. Louise Tyler Rixey
Vice Adm. T. G. W. Settle, USN (Ret.) (and interview notes)
Mr. Owen F. Tyler
Capt. George F. Watson, USN (Ret.)
Capt. Gordon S. Wiley, USN (Ret.)

Family Papers

Cdr. Fred T. Berry, courtesy Ms. Lynn Berry and Mr. Thomas Berry
Cdr. Garland Fulton, USN (Ret.), courtesy Mr. Langdon H. Fulton
Audio-recorded interviews by author and/or notes:
[* = audio-cassette prepared for author]
Rear Adm. Richard S. Andrews, USN (Ret.)*
Capt. Ralph S. Barnaby, USN (Ret.)
Mr. Kurt Bauch
Rear Adm. Calvin M. Bolster, USN (CC) (Ret.)*
Capt. M. M. Bradley, USN (Ret.)
Capt. Howard N. Coulter, USN (Ret.)
Mr. Melvin J. Cranmer

ADC John A. Iannaccone, USN (Ret.)
Mr. William P. Kramer
Mr. John Lust (former Seaman 2c)
Lt. Cdr. William A. MacDonald, USN (Ret.)
Mr. Norman J. Mayer
Mr. Walker C. Morrow*
Cdr. Joseph P. Norfleet, USN (Ret.)
Lt. David F. Patzig, USN (Ret.)
Rear Adm. Scott E. Peck, USN (Ret.) (with Michael C. Miller)
Rear Adm. George E. Pierce, USN (Ret.)
Cdr. F. W. Reichelderfer, USN (Ret.)
Rear Adm. Charles E. Rosendahl, USN (Ret.) (interview notes)
Lt. Herbert R. Rowe, USN (Ret.)
Lt. Cdr. Leonard E. Schellberg, USN (Ret.)
Capt. Earl K. Van Swearingen, USN (Ret.)
Mr. F. W. von Meister
Cdr. R. H. Ward, USN (Ret.)
Capt. George F. Watson, USN (Ret.)*

Published Sources

Books

Allen Hugh. *The Story of the Airship.* 8th ed. Akron, Ohio: Goodyear Tire and Rubber Company, 1932.

———. *The Story of the Airship [Non-Rigid].* Akron, Ohio: Goodyear Tire and Rubber Company, 1943.

Allen, Frederick Lewis. *Only Yesterday: An Informal History of the Nineteen-Twenties.* New York and London: Harper and Brothers, 1931.

Althoff, William F. *Sky Ships: A History of the Airship in the United States Navy.* New York: Orion, 1990.

Anderson, John D., Jr. *A History of Aerodynamics and Its Impact on Flying Machines.* Cambridge: Cambridge University Press, 1997.

Bender, Marylin, and Selig Altschul. *The Chosen Instrument: Juan Trippe, Pan Am—The Rise and Fall of an American Entrepreneur.* New York: Simon and Schuster, 1982.

Bentele, Eugen. *The Story of a Zeppelin Mechanic: My Flights 1931–1938.* Translated by Simon Dixon. Friedrichshafen, Ger.: Wolfgang Meighorner-Schardt, 1992.

Burgess, Charles P. *Airship Design.* New York: Ronald Press, 1927.

Countryman, Barry. *Helium for Airships and Science: The Search in Canada, 1916–1936.* Toronto: privately printed, 1992.

Churchill, Winston S. *The Gathering Storm.* New York: Houghton Mifflin, 1948.

Crouch, Tom Day. *To Ride the Fractious Horse: American Aeronautical Community and the Problem of Heavier-than-Air Flight, 1875–1905.* Ph.D. dissertation, Graduate School of the Ohio State University, 1976.

Eckener, Hugo, *My Zeppelins.* Translated by Douglas H. Robinson. London: Putnam, 1958.

Fritzsche, Peter. *A Nation of Fliers: German Aviation and the Popular Imagination.* Cambridge, Mass.: Harvard University Press, 1992.

Hansen, James R. *Engineer in Charge: A History of the Langley Aeronautical Laboratory, 1917–1958.* NASA History Series (NASA SP-4305). Washington, D.C.: 1987.

King, Ernest J., and Walter Muir Whithall. *Fleet Admiral King: A Naval Record.* New York: W. W. Norton, 1952.

Lehmann, Capt. Ernst A., with Leonard Adelt. *Zeppelin: Story of Lighter-than-Air Craft.* Translated by Jay Dratler. London: Longmans, Green, 1937.

Lehmann, Capt. Ernst A., and Howard Mingos. *The Zeppelins.* New York: J. H. Sears, 1927.

Litchfield, P. W. *Industrial Voyage.* Garden City, N.Y.: Doubleday, 1954.

McBride, Joseph. *Frank Capra: Catastrophe of Success.* New York: Simon and Schuster, 1992.

Piercy, N. A. V. *Aerodynamics.* London: English Universities Press, 1937.

Reynolds, Clark G. *Admiral John H. Towers: Struggle for Naval Air Supremacy.* Annapolis, Md.: Naval Institute Press, 1991.

Robinson, Douglas H. *The Zeppelin in Combat,* rev. ed. Sun Valley, Calif.: John W. Caler, 1966.

———. *Giants in the Sky: A History of the Rigid Airship.* Seattle: University of Washington Press, 1973.

Robinson, Douglas H., and Charles L. Keller. *Up Ship! U.S. Navy Rigid Airships, 1919–1935.* Annapolis, Md.: Naval Institute Press, 1982.

Rosendahl, Charles E. *Up Ship!* New York: Dodd, Mead, 1931.

Seibel, Clifford W. *Helium: Child of the Sun.* Lawrence: University Press of Kansas, 1968.

Shute, Nevil. *Slide Rule: Autobiography of an Engineer.* London: William Heinemann, 1954.

Sinclair, J. A. *Airships in Peace and War.* London: Rich and Cowan, 1934.

Smith, Richard K. *The Airships* Akron *and* Macon, *Flying Aircraft Carriers of the United States Navy.* Annapolis, Md.: U.S. Naval Institute, 1965.

SELECTED BIBLIOGRAPHY

Trimble, William, *Engineering the Air: A Biography of Jerome C. Hunsaker.* Washington, D.C.: Smithsonian Institution Press, in press.

Turnbull, Archibold D., and Clifford L. Lord. *History of United States Naval Aviation.* New Haven, Conn.: Yale University Press, 1949.

Vaeth, J. Gordon. *Graf Zeppelin: Adventures of an Aerial Globetrotter.* New York: Harper and Brothers, 1958.

Van Fleet, Clarke, and William J. Armstrong, eds. *United States Naval Aviation 1910–1980.* Washington, D.C.: U.S. Government Printing Office, 1981.

Vissering, Harry. *Zeppelin: The Story of a Great Achievement.* Chicago: privately printed, 1922.

von Karman, Theodore, with Lee Edson. *The Wind and Beyond: Theodore von Karman, Pioneer in Aviation and Pathfinder in Space.* Boston: Little, Brown, 1967.

von Schiller, Hans. *Kapitän, Zeppelin: Wegbereiter des Weltluftverkehrs.* Bad Godesberg, W. Ger.: Kirschbaum Verlag, 1966.

Wittemann, A. *Die Amerikafahrt Des Z.R. III.* Wiesban, Ger.: Amsel-Verlag, 1925.

Technical Publications

Arnstein, Karl. "The Development of Large Commercial Rigid Airships." *Aeronautical Engineering* 50, no.6 (January–April 1928), 1–14.

———. "Research and Development Problems Arising in Airship Design." *Publication No. 1,* Daniel Guggenheim Airship Institute, 1933, 23–35.

Blakemore, Thomas L., J. F. Boyle, and Norman Meadowcroft. "Design, Construction, and Handling of Non-Rigid Airships." *Aeronautical Engineering* 1, no. 2 (April–June 1929), 29–42.

Bottoms, R. R. "The Production and Uses of Helium Gas." *Aeronautical Engineering* 1, no. 3 (July–September 1929), 107–16.

C. P. Burgess, "Factors of Safety in Airship ZR-1 by the Zeppelin Method of Calculation." *Design Memorandum No. 25,* 15 October 1923.

———. "Stresses Observed on Sixth Trial Flight of *ZR-1*, September 27, 1923." *Design Memorandum No. 26,* 20 October 1923.

———. "Forces on an Airship in Gusts." *Design Memorandum No. 32,* 31 March 1924.

———. "Comparison of Dynamic Lift and Pressure Distribution on the Airship LOS ANGELES by Observation and Experiment." *Design Memorandum No. 44,* December 1924.

———. "Strain Gage Tests on U.S.S. *Los Angeles* May 25 to June 4, 1925." *Design Memorandum No. 55,* 23 July 1925.

———. "Strength and Pressure Tests on U.S.S. LOS ANGELES, April and May, 1926." *Design Memorandum No. 63,* November 1927.

———. "Stresses in U.S.S. *Los Angeles* Held across a 20 Mile Wind." *Design Memorandum No. 77,* April 1927.

———. "Comparative Strength of Airships." *Design Memorandum No. 80,* June 1927.

———. "Forces in Airplane Landing Device for U.S.S. LOS ANGELES." *Design Memorandum No. 83,* October 1927.

———. "Forces on U.S.S. LOS ANGELES during Wind Shifts at Mooring Mast." *Design Memorandum No. 86,* January 1928.

———. "Static Bending Moments and Stresses in U.S.S. LOS ANGELES in Present Normal Flight Condition." *Design Memorandum No. 92,* December 1929.

———. "Speed and Deceleration Trials of U.S.S. LOS ANGELES, September, 1927." *Design Memorandum No. 82,* December 1927.

———. "Flight Tests on U.S.S. *Los Angeles*, Part II: Stress and Strength Determination." *Report No. 325, Fifteenth Annual Report of the National Advisory Committee for Aeronautics,* 1929, 485–98.

———. "The Application of the Principle of Least Work to the Primary Stress Calculations of Space Frameworks." *Aeronautical Engineering* 1, no. 3 (July–September 1929), 131–39.

——— "Static Bending Moments and Stresses in U.S.S. LOS ANGELES in Present Normal Flight Condition." *Design Memorandum No. 92,* December 1929.

———. "Comparative Strength and Factors of Safety of U.S.S. LOS ANGELES as Designed and in Service in 1930." *Design Memorandum No. 94,* February 1930.

———. "Water Recovery Apparatus for Airships." *Design Memorandum No. 96,* April 1930.

———. "Progress in Airship Design from U.S.S. SHENANDOAH to U.S.S. AKRON." *Design Memorandum No. 102,* October 1930.

———. "Forces on the U.S.S. LOS ANGELES at a Mooring Circle." *Design Memorandum No. 192,* March 1935.

———. "Strength of the U.S.S. LOS ANGELES." *Design Memorandum No. 310,* December 1938.

———. "Discussion of Four Airship Proposals Submitted by the Goodyear-Zeppelin Corp." *Design Memorandum No. 312,* March 1939.

Crowley, J. W., and S. J. DeFrance. "Pressure Distribution on the C-7 Airship." *Report No. 223, Eleventh Annual Report of the National Advisory Committee for Aeronautics,* 1925, 329–67.

DeFrance, Smith J. "Flight Tests on U.S.S. *Los Angeles.* Part I—Full Scale Pressure Distribution Investigation." *Report No. 324, Fifteenth Annual Report of the National Advisory Committee for Aeronautics,* 1929, 451–81.

DeFrance, S. J., and C. P. Burgess. "Speed and Deceleration Trials of U.S.S. *Los Angeles.*" *Report No. 318, Fifteenth Annual Report of the National Advisory Committee for Aeronautics,* 1929, 305–24.

Freeman, Hugh B. "Force Measurements on a 1/40-Scale Model of the U.S. Airship 'Akron.'" *Report No. 432, Eighteenth Annual Report of the National Advisory Committee for Aeronautics,* 1932, 591–606.

———. "Measurements of Flow in the Boundry Layer of a 1/40-Scale Model of the U.S. Airship 'Akron.'" *Report No. 430, Eighteenth Annual Report of the National Advisory Committee for Aeronautics,* 1932, 567–79.

Fritsche, Carl B. "A Comparative Examination of the Airplane and the Airship." *Aeronautical Engineering* 50, no. 26 (September–December 1928), 9–20.

———. "The Metalclad Airship." *Aeronautical Engineering* 1, no. 4 (October–December 1929), 245–66.

———. "Some Economic Aspects of the Rigid Airship." *Aeronautical Engineering* 3, no. 1 (January–March 1931), 25–40.

Fulton, Cdr. Garland USN (CC). "Current Airship Problems." *Publication No. 1,* Daniel Guggenheim Airship Institute, 1933, 36–39.

Hooper, A. G., and C. P. Burgess. "Mark III Condenser for Airships." *Design Memorandum No. 85,* January 1928.

Hunsaker, Cdr. J. C., USN (CC). *The Development of Naval Aircraft: Rigid Airships* and *Non-Rigid Airships for the Navy,* 1923, Charles E. Rosendahl Collection, box 150, History of Aviation Collection, University of Texas at Dallas.

———. "Airship Problems Arising in Commercial Aviation." *Publication No.1,* Daniel Guggenheim Airship Institute, 1933, 40–44.

McHugh, James G. "Pressure-Distribution Measurements at Large Angles of Pitch on Fins of Different Span-Chord Ratio on a 1/40-Scale Model of the U.S. Airship 'Akron.'" *Report No. 604, Twenty-third Annual Report of the National Advisory Committee for Aeronautics,* 1937, 585–604.

Operation Manual for U.S.S. Macon. Goodyear-Zeppelin Corp., undated ("Received Feb. 1 1934, Inspector of Naval Aircraft, Akron, Ohio"), courtesy Lt. Gordon H. Cousins, USN (Ret.).

Pagon, W. W. "Stresses in Rigid Airships: Effect of Indeterminateness on Their True Value." *Aeronautical Engineering* 4, no. 3 (July–September 1932), 123–39.

Reichelderfer, F. W. "Some Aerological Principles Applying to Airship Design and Operation." *Aeronautical Engineering* 1, no. 3 (July–September 1929), 171–78.

Rigid Airship Manual (1927). Washington, D.C.: U.S. Government Printing Office, D.C., 1928, courtesy Capt. Earl K. Van Swearingen, USN (Ret.).

Rosendahl, Lt. Cdr. Charles E., USN. "The Mooring and Ground Handling of a Rigid Airship." *Aeronautical Engineering* 5, no. 1 (January–March 1933), 45–52.

Silverstein, Abe, and B. G. Gulick. "Ground-Handling Forces on a 1/40-Scale Model of the U.S. Airship 'Akron.'" *Report No. 566, Twenty-second Annual Report of the National Advisory Committee for Aeronautics,* 1936, 405–18.

Special Committee on Airships [Durand Committee], "Report No. 1: General Review of Conditions Affecting Airship Design and Construction with Recommendations as to Future Policy." Stanford, Calif.: Stanford University Press, January 1936, 12.

———. Report No. 2. "Review and Analysis of Airship Design and Construction Past and Present." Stanford, Calif.: Stanford University Press, January 1937, 127.

Troller, Dr. Theodor. "The Vertical Wind Tunnel of the Daniel Guggenheim Airship Institute." *Publication No. 1,* Daniel Guggenheim Airship Institute, 1933, 11–22.

U.S. National Advisory Committee for Aeronautics, *Annual Reports:* Sixth (1920), Tenth (1924), Eleventh (1925), Thirteenth (1927), Fifteenth (1929), Sixteenth (1930), Eighteenth (1932), Nineteenth, 1933, Twentieth (1934), Twenty-second (1936), Twenty-third (1937) and Twenty-fourth (1938). Washington, D.C.: U.S. Government Printing Office.

von Karman, Dr. Theodor. "Some Aerodynamic Problems in Airships." *Publication No. 1,* Daniel Guggenheim Airship Institute, 1933, 45–52.

War Department. *Aerostatics.*" Technical Manual No. 1–325. Washington, D.C.: 1 October 1940.

Miscellaneous Documents

Baker, Lt. Cdr. (Ret.) William A. Personal notebook, Airship Training School, Naval Air Station Lakehurst, N.J., July–December 1931.

Eckener, Dr. Hugo. "Remarks of Dr. Hugo Eckener Before Annual Meeting of National Advisory Committee for Aeronautics. Washington, D.C., 16 October 1924" (extract from minutes). Fulton Collection, National Air and Space Museum, Smithsonian Institution.

Hunsaker, Jerome C. "Some Lessons of History." Second Wings Club "Sight" Lecture, New York City, 26 May 1965, courtesy Langdon H. Fulton.

Lust, John A. Personal notebook, Airship Training School, Naval Air Station Lakehurst, N.J., July–December 1930.

Navy Relief Society. "Airships." 2nd ed. Naval Air Station Lakehurst, New Jersey, spring 1929, unpaginated.

"Photographs ZR-3" (binder of dated/labeled photographs). Office of Inspector of U.S. Naval Aircraft, Friedrichshafen a/B, Germany, courtesy Lt. Gordon H. Cousins, USN (Ret.).

Reichelderfer, F. W. "Aviation Weather Service, Particularly for Ocean Route." 1934 Aviation Commission Hearing (report to), courtesy Navy Lakehurst Historical Society.

Roland, Capt. Charles William, USN (Ret.). "Handling Rigid Airships on the Ground: With Emphasis on the Use of Mechanical Equipment." Privately printed and distributed, 1978, courtesy Captain Roland.

Rowe, Lt. (Ret.) Herbert R. Personal notebook, Airship Training School, Naval Air Station Lakehurst, N.J., July–December 1930.

Shock, James R. "A Research of Rigid Airship Ground Handling." Privately printed and distributed, July 1998, 39, courtesy Mr. Shock.

———. "U.S. Navy Airship Pilots, 1917 through April 15, 1942: A Research of Naval Aviators (Airship) by Qualification Numbers and Dates including Data on Balloon Pilots and Select Enlisted Men." Privately published and distributed, August 1998, 151, courtesy Mr. Shock.

Steele, Capt. George W. "The Airship in War and Peace, Beginning and Early History." Garland Fulton Collection, National Air and Space Museum, Smithsonian Institution.

Stewart, Andrew. *About Helium.* U.S. Bureau of Mines Information Circular 6745, September 1933.

U.S. Naval Air Station, Lakehurst, New Jersey, log, Record Group 24, National Archives and Records Administration.

USS *Los Angeles* (ZR-3), log, Record Group 24, National Archives and Records Administration.

Wicks, Zeno. "A Short History of Naval Aviation, Particularly of Lighter-than-Air, as Experienced by Z. W. Wicks." Undated carbon copy, courtesy Langdon H. Fulton.

Periodicals

"Aboard the Airship '*Los Angeles*' to Porto [*sic*] Rico and Back." *The Literary Digest,* 4 July 1925, 58–62.

Airway Age 12, no. 12 (20 June 1930).

Aero Digest 5, no. 5 (November 1924); 5, no. 6 (December 1924); 6, no. 1 (January 1925); 6, no. 4 (April 1925); 8, no. 1 (January 1926).

Aviation, 29 September 1923, 24 December 1923, 22 September 1924.

Althoff, William F. "The Decommissioned USS LOS ANGELES Part 11: Material Condition and Dismantling." *American Aviation Historical Society Journal* 25, no. 1 (Spring 1980), 29–41.

———. "Balloon Training at NAS Lakehurst." *American Aviation Historical Society Journal* 28, no. 1 (Spring 1983), 2–18.

Barnaby, Capt. Ralph S., USN. "The Glider Flight From the LOS ANGELES." *U.S. Air Services* 40, no. 3 (March 1955), 17–18.

"By Airship to Bermuda." *Scientific American,* May 1925, 344–45.

Campbell, Lt. George W. "Five O'Clock, off California." *The Saturday Evening Post,* 15 May 1937, 20–21, 122–27.

Coletta, Paolo E. "Admiral Marc A. Mitscher, USN: A Silhouette." *American Aviation Historical Society Journal* 38, no. 1 (Spring 1993), 66–74.

Flight 8, no. 23 (9 June 1921); 36, no. 1610 (2 November 1939).

Fulton, Capt. Garland, USN (CC). "The General Climate for Technological Developments in Naval Aeronautics on the Eve of World War 1." In Doyce B. Nunis, Jr., ed., "Recollections of the Early History of Naval Aviation: A Session in Oral History." *Technology and Culture* 4, no. 2 (Spring 1963), 149–76.

Hamlen, W. L. "The First Lighter-than-Air Class at Akron." *Naval Aviation News,* November 1967, 26–29.

Hubbard, Walter Whitely. "The Transoceanic Dirigible: Graf Zeppelin Starts Transatlantic Passenger Service." *American Aviator: Airplanes and Airports* 1, no. 4 (November 1928), 9.

Hughes, Patrick. "Francis W. Richelderfer, Part 1: Aerologists and AirDevils." *Weatherwise* 34, no. 2 (April 1981), 52–59.

———. "Francis W. Reichelderfer, Part 11: Architect of Modern Meteorological Services." *Weatherwise* 34, no. 4 (August 1981), 148–57.

Iannaccone, John A. *Buoyant Flight: Bulletin of the Lighter-than-Air Society* 28, no. 2 (January–February 1981), 9.

Klein, Jacob. "Up Ship, Gentlemen!" *Popular Aviation*, December 1936, 16.

"The Last Zeppelin—Ours." *The Literary Digest,* 25 October 1924, 12–14

Mackey, Capt. Donald M., USN. "An Overnight Free Balloon Flight." *Buoyant Flight: Bulletin of the Lighter-than-Air Society* 57, no. 3 (March–April 1977), 2–5.

Pollock, Capt. Edwin T., USN. "Studying the Eclipse from the *Los Angeles.*" *McClure's Magazine* 1, no. 2 (June 1925), 9–22.

Rosendahl, Lt. Cdr. Charles E., USN. "Dirigibles." *Science and Invention*, May 1928.

———. "Flying with an Airship Captain." *Popular Science Monthly*, March 1930.

Scherz, Walter. "In Three Days to America." *Living Age*, 15 August 1925, 347–50.

Tinker, Clifford A. "Riding the Clouds in an Air Pullman." *Popular Mechanics*, July 1927, 104.

von Schiller, Hans. "Log of the *ZR-3.*" *Aero Digest* 5, no. 5 (November 1924), 266–68, 300.

Walker, C. R. "Sky Life." *Collier's*, 13 June 1931, 32, 64–65.

Winans, Dave. "USS *PATOKA*: An Airship Tender." *Model Ships and Boats*, May/June 1977, 34–35, 55–56.

Woodward, Donald E., and William F. Kerka. "The Cut-Down *Los Angeles.*" *Aerostation* 6, no. 1 (Spring 1979), 10–11, 22.

Newspapers

Akron Beacon Journal
Asbury Park (N.J.) *Evening Press*
Daily Mirror
Evening (Philadelphia) Public Ledger
Fitchberg (Mass.) Sentinel
Fort Worth Record-Telegram
Newark (N.J.) *Evening News*
New York American
New York Evening Journal
New York Herald Tribune
New York Sun
New York Sunday News
New York Telegram
New York Times
Philadelphia Record
Portland (Me.) *Evening Express*
Rochester (N.Y.) *Democrat & Chronicle*
Rochester Union & Advertiser
Sunday Call
The Evening Bulletin—Philadelphia
The Evening Star
The (Hamilton, Bermuda) *Royal Gazette*
Washington Herald
Washington Post

Index

A-1, xiii
ACR (aviation chief rigger), *85*
Adams, Charles Francis, 65, 131, 151, 158–59
aerodynamic force testing, *93*
Aerographer's School, 66
aeronautical design, 89
aerostatics, 70
Airdock, 124
air exploration tests, 127, *127*
airplanes, 95
Airship Board, Joint Army and Navy, xvi, xx–xxi, 3, 7
airships: criticisms of, 144–45, 166; development of, 1–38, 178; employment of, 49, 52–53, 103, 162; evaluation of, 166; hazards to, 94–95; *Hindenburg* loss and, 195–98; length of missions, 85; *Macon* loss and, 178; obsolescence of, 119; organization of, 85–86; passenger spaces, *112, 202*; possible contributions to World War II, 219–20; reaction against, 165. *See also* commercial airships
Airship School, 69–71, *216*; graduates of, 225–27
Akron (ZRS-4), 55, 62, 152, *152*, 153–54, *154*; commissioning of, 151, *151*; construction of, 124, 126; crew of, 146, 155; damage to, 158; loss of, 75, 158, *163*, 163–64, *164*, 164–65, 180; model of, *178, 199–200*; specs, 126–27, *153*; tests and trial run, 151, 157, 162
alcohol, denatured, 56, 122
Aluminum Company of America, xvi
Anderson, Bruce, 49, 95
Andrews, Richard S., 189–91, 194, 196
antifreeze, 56
antisubmarine warfare, xvi–xvii, xviii, xx
Antrim, Richard N., 189, 194–95

Army, U.S., xxi; airship fleet, xiv; airship training program, 80–81; and *Los Angeles,* 20–21
Arnold, Joseph C., 52, *92*
Arnstein, Karl, 9, 46, 124, 151, 167; at National Air Races, 130
Ashley, Sylvia, 186
Ashton, Horace D., 60
aviation chief rigger (ACR), *85*

Baker, Bill, 212–14
Baldwin, Thomas, xv
ballonet, *xx*
balloons. *See* free balloons; kite balloons
Barnaby, Ralph S., 131–33
Barrage, G. H., 47
Bauch, Charles E., 5, 8, 52, *92*
Bauch, Kurt, 11–12, 27, 30, 34
B class, xvii–xviii, *xviii*; training with, 79
Berkey, R. S., 138
Berry, Fred T.: character of, 158; in command, 157–58, 160–61, 237–38; on fire, 147; and grounding, 159–60; on Lakehurst, 65; loss of, 163; on movie, 136; on National Air Races trip, 130; and Panama trip, 137–38, 144; on testing, 124, 131; and training, 71, 73, 75, 81
berthing, 86
Bingham, Hiram, 129
blimps, xx, 217; term, xvii. *See also* airships
Board of Inspection and Survey, 47, 204
Bodensee (LZ-120), 1–3, *2*
Bolster, Calvin: and *Akron,* 164–65; and Berry, 65; on Dresel, 146; on explosions, 78–79; on free balloons, 76; on golden age, 123; on ground handling, 50, 95, 101; on hydrogen release, 47; and *Macon,* 177; on *Saratoga* rendezvous, 102–3; and testing, 114, 127–28, 130–31, 155; and training, 198; and upending, 96, 98
Bolster beam, *153*
Boyd, T. S., 4–5
Bradenstein-Zeppelin, Hella von, 31, 34
Bradley, Maurice M. "Mike," 76, 81, 117, 133, 145, 162, 168, 179
Brandemeuhl, Don, 65–66
break time, *85*
Britain: and airship development, 3, 5; airship fleet, xvi–xvii
Britten, Fred A., 147, 157
Buckley, William A., 42, 84, 212, 215
budget issues, 50, 67–68, 102, 120–21, 126, 145, 158, 162, 171
Bureau of Aeronautics (BuAer), 5, 8, 50, 95, 100–101, 129, 166–67
Bureau of Construction and Repair, xv, xvii, 3
Bureau of Mines, xvi, 5, 110–11
Bureau of Navigation (BuNav), 50
Burgess, C. P., 46, 62, 89, 151, 179, 217
Bushnell, Wilford, 73
Business Advisory Coundil, 189
Butler, Thomas S., 108
Byrd, Richard E., xv, 136
Byrnes, James F., 159, 202

calcium chloride, 56
Calnan, George, 140, 150, 156, 162
Campbell, George W., 177
Capra, Frank, 136
Carter, James B., 52, *92*
C class, xx, *xx*; training with, 79
Champion, C. C., *103*
Charteris, Leslie, 186
Chrysler, Walter, 113
Churchill, Winston, xvi, 208

circumnavigation of globe, by *Graf Zeppelin,* 119, 125, 130
civilian staffing, at Lakehurst, 171–72
Clarke, Vincent A., Jr., *133;* in command, 133–36, 138–46, 234–36; on trips, 89, 92, 95, 107
Clements, Rell, 65
Cochrane, lt. cdr., 177
Cockrane, Edwin F., 171
Cody, William, *120*
Colsman, Alfred, 6
commercial airships, 180, 187
designs for, *112*
development efforts with, 124–26; *Akron* and, 151; *Hindenburg* loss and, 196
Graf Zeppelin's record, *187*
promenade deck, *112*
Committee on Naval Affairs, 157
Cook, Arthur B., 191
Coolidge, Calvin, 45, 51, 65
Coolidge, Grace, 50–51, *51–52,* 209
Cope, Alfred L., 198
Coulter, Howard N., 86, 122, 178, 198
Council of Ambassadors, 3, 5–6
Cranmer, Melvin, 98
crawler mast, *100,* 100–101, *149*
Curtiss, Glen, xiii
Curtiss F9C-2, 155

Daniel Guggenheim Airship Institute, 162
Davis, Robert J., 135
Davis, Roy T., 140
Day, George C., 166
D class, 79
Deem, Joseph M., 39, 52
Dennett, Rodney R., 103
Deutsche Luftschiffart A. G. (DELAG), xiii, 1
Deutsche Zeppelin Reederei, *185,* 189, 191
dirigibles: criticisms of, xv; definition of, 70; development of, xiii. *See also* airships
DN-1 project, xvii
Douglas, Donald W., xv
drag chain, 63
Dresel, Alger H., *141;* character of, 146; in command, 146–48, 150–51, 153–57, 236–37; and *Macon,* 162, 167, 175; on maneuvers, 141, 143; in training, 73
duralumin, xvi

Durand Committee, 178–79
Dürr, Ludwig, 9

Eckener, Hugo, 3, 7–8, 12, *24, 27,* 30, 108, 112–13, *113–14;* advocacy for airships, 44–45, 49, 113–14, 125–26, 191; and *Akron,* 151, 165; docking at Lakehurst, 39–40, 42–43; and *Hindenburg,* 182, *183,* 186–87, 196–97; and LZ-130, 202; at National Air Races, 130; and ocean crossing, 33–38; and trial run, 24, 26–27, 31–32
Edison, Charles, 202, 217
Edwards, E. I., 100
Ellyson, T. G., xiii, 107
Empire State Building, 138
Eppes, M. Henry, 198–99
expeditionary mast, 108

Fairbanks, Douglas, Sr., 186
fleet maneuvers, 136–45, *141–43,* 175–78
Flemming, Hans Curt, 26, 33–34, 55
flight organization chart, *104*
flight pay, 76
flying boats, xx
Ford, Henry, 94
Fordney, C. L., *172*
Foulois, Benjamin D., 6, 8
France, and airship development, xiii
Free and Kite Balloon Hangar, *70*
free balloons, 66, *71–72;* landing, 78–79; specs, *78;* training with, 70, 76–77, *77*
Friedrichshafen, construction of *Los Angeles* at, 1–38, *13–16, 17–22*
Fritsche's Lake, 69
Fulton, Garland "Froggy," xiv–xv, 3, 8–9, *22,* 46, 62, 102, *113;* on accident, 174; and airship development, 166, 197–98, 204; and *Akron,* 151, 157; on commercial development, 124; and construction, 12–13, 21–22; and disassembly, 203, 208; and *Hindenburg,* 190–91; on maintenance, 59; on operations, 64; retirement of, 219; and Rosendahl, 90, 122; and trial run, 24, 30
Fulton, Jack, 24
funding. *See* budget issues

G-1, *81*
Geiger, Harold, 8, 30
General Board, xiii, xv, xvi, 52, 198

Germany: and airship development, xiii–xiv, xv, 7–8, 31, 33, 37, 45, 49, 186, 193; airship fleet of, xvii; in Nazi era, 197, 203, 208
Ghormley, Robert L., 58
gliders, *125,* 131–32, 134–35, *135*
Goering, Hermann, 189, 218
Goodyear Tire and Rubber Company, xviii, 20–21, 33, 69; flying field, xix, *xix*
Goodyear-Zeppelin Corporation, 45–46, 68, 124, 152, 180, 202
Gorton, A. W. "Jake," 127–30
Graf Zeppelin (LZ-127), 45, 92, *114, 119,* 123, 134–35, *135,* 191; circumnavigation of globe by, 118–19, 125, 130–31; commercial record for, *187;* disassembly of, 218, *218;* ocean crossing by, 110, 112–13; refilling of, *115*
Graves, Ralph, 136
Great Depression, 80, 117, 155
Griffen, lt.cdr., 168
ground handling techniques, 95–96, 100–101, *153, 170*
Ground School, 71–73; requirements of, *75*
Guggenheim gust experiments, *200,* 200–201, *201*

Hagaman, Allen, 196
Hamlen, Warner L., 69–70
Hampton Roads NAS, 5
Hancock, Lewis J., Jr., 50
handling techniques, 95–96, 100–101, *153, 170*
Hanza (LZ-13), *xiv*
Harding, Warren G., 6, 45
Harrigan, Daniel Ward, 146, 148, 150–51
Heinen, Anton, 19, *190*
helium, 5, 64, 197, 203; effects of, 156; extraction of, xvi; production of, 48–49, *49,* 111; reclamation of, 210; storage of, *48;* warming, 140
Hensley, William N., Jr., 4
Herndon, William H., *176*
Herrick, Myron T., 6
Hicks, George, 134
high mast, 92, *93,* 96, 119–20
Hill, Malvin M., 156
Hindenburg (LZ-129), 182–93, *183, 185*–86, *188,* 190–92; destruction of,

194–96, 194–97; performance summary for, *189*
Holt, Jack, 136
homing pigeons, 66, *79*
Honey, Robertson, Mrs., 58
hook-on experiments, 114, 127, *127–28,* 128–31, 136, 146–51, *148–49*; design for, 127, 155
Hoover, Herbert, 116–17, 134
Hughes, Howard, 157
Hull Board, 193, 203
Hunsaker, Jerome C., *xviii*; and aeronautical engineering course, xiv; and airship development, xv, 3, 5, 9; and commercial development, 125–26; and construction, 12; on DN-1 project, xvii; on Maxfield, xix; at National Air Races, 130
hydrogen, 40, 46, 76, 185, 187, 196, 204; for *Graf Zeppelin,* 115; production of, 76

Iannaccone, John, 117, 136–37
ice removal, 181–82
Ickes, Harold, 197
Ingalls, David S., 118, 136, 140–41, 144, 146, 151
Inspector of naval aircraft (INA), 8
intelligence, airships and, xv–xvi
International Zeppelin Transit Corporation, 126, 163–64

J-3, 96, 109, 113, 134, 164
J-4, *79,* 109, 113, *137,* 198
Jackson, Edward S., 90, *113*
Jahncke, Ernest Lee, 152
Jandick, Paul A., 162
J class, 79
Johnson, Walter, 98

K-1, *81, 181,* 198
K-4, *220*
Karman, Theodore von, ix, 162, 179
Karpinski, Frank T., *84, 143*
Kennedy, Frank M., 8, 16, *22,* 24, 30, 33, 42
Kenworthy, Jesse L., 169, 177, 208–10
Kepner, William E., 71, 80
Kilgallen, Dorothy, 186
King, Ernest J., xv, 143, 165–68, 179, 182, 198
kite balloons, xiii, xv, *70, 181*; training with, 75–76

Klein, Jacob H., Jr., 30–31, 33, 42, *42, 52,* 59, 198; in command, 229; and commissioning, 50–51; and demonstration flights, *55,* 55–56; on employment of airships, 49, 52; reputation of, 90
Knorr, Ludwig, 34
knot, definition of, 27
Knowles, Thomas, 125
Knox, C. V. S. "Connie," 168, 198
Kraus, Sidney M., 13, 16, *22,* 24, *27,* 30, 33, 42, 49

L-1, *81,* 199
L-72, 1, 3
Lakehurst, 4, 39–68, *71,* 94, 171–72, *172, 204, 219*; after disassembly of *Los Angeles,* 219; characteristics of, 64–68; location of, 53–54; and loss of *Akron,* 164–65
Land, Emory S., 58, 67
landings, 47, *47, 58, 59,* 94, 107–8; balloons, 78–79; casualties in, 135–36; duties for, *132*; training in, 85
Lange, Karl, *190*
Lansdowne, Zachary, 3, 5, 49, 64–65
latex, 59
Leahy, William D., 198
Lehmann, Ernst: advocacy for airships, 191; and construction, 8, 12, 21, 26, 33, 35; and *Hindenburg,* 183, 194, 196; on landing, 40
Leviathan, 134
Lexington, 102
lighter-than-air (LTA) craft: *Hindenburg* loss and, 195–98; *Macon* loss and, 178; training with, 69–90. *See also* airships
Lindbergh, Charles E., *120,* 131
Lipke, Donald L., 108
Litchfield, Paul W., 20, 45, 125, 130
Lloyd George, David, 5–6
London Protocol, 3
Los Angeles (ZR-3)
 accidents with, 147, 156–57, 174, *175*
 chronology of, xi–xii
 commissioning of, *50,* 50–51, *51*
 construction of, 1–38, *13–16, 17–22*
 cost of maintenance and operation of, 157
 crew of, 91–92, 156, 239–40; organization of, *104*

 decommissioning of, *149,* 160–62
 disassembly of, 208–16, *209–15*
 end of career of, 193–216
 evaluation of, 120–21, 146, 159–60, 166, 199–200, 203–7, *205–7*
 in fleet maneuvers, 136–45, *141–43*
 flight log, 229–38
 flight preparation, *129*
 grounding of, 155–92
 keel corridor, *10*
 keel-laying, 9
 at Lakehurst, 39–68; arrival, 39–44, *40–46*
 maintenance of, 56, 59–60, 116, *118, 125,* 135, 156–57; cost of, *157*; progress report, *117*
 markings, *156,* 158
 meals aboard, 86–88, 147
 navigational equipment, 241–42
 ocean crossing, 32–38, *33*; log of, *40*
 reconditioning of, 168–69, 172–74, *173*
 riding-out scheme, 174, *174*
 Rosendahl's command of, 90–123
 specs, 10–11, *153,* 221–23; distribution of useful load, *117*; engines, 12–15, 31; fuel system, 82–83, *83*; galley, *87*; gas cells, 14, *83,* 92, *137,* 203; keel corridor, 82, *82*; keelway, *86*; passenger coups, *14, 87*; radio gear, 35; water-recovery apparatus, 53; water tanks, *82*; wiring, *11*
 and testing, 114–15, 124–54, *125, 149*
 and training, 69–90, 156
 trial runs, 22–32, *23–25, 28–30, 32*
 upending of, 95–99, *97*
Luftschiffbau-Zeppelin, xv; and construction of USS *Los Angeles,* 1, 3, 6–32; and guarantee period, 47–48
LZ-10 *(Schwaben),* xiv
LZ-13 *(Hanza),* xiv
LZ-102, xvi
LZ-120 *(Bodensee),* 1–3, *2*
LZ-127. *See Graf Zeppelin*
LZ-129. *See Hindenburg*
LZ-130, 187–89, *202,* 202–4, 208, 218
LZ-131, 189

MacCracken, Alan, 125–26
MacDonald, William, 169
MacIntyre, Alexander, 168

Macon (ZRS-5), 55, 62, 155, 169–71, *170*, *177*; construction of, *164*, 164–65, *165*; crew of, 160; and fleet maneuvers, 175–78; loss of, 177–78, 180; maiden voyage of, 167–68; size of, 157; specs, 126–27, *163*, 167
Maitland, Edward, 6
Marines, at Lakehurst, 66, *171*
Marx, Ludwig, 33–34
Massachusetts Institute of Technology, xiv, 69
masts: crawler, *100*, 100–101, *149*; expeditionary, 108; high, 92, *93*, 96, 119–20; mobile, 100–101, *101*, *149*; short, 98; stub, *99*, 99–100. *See also* mooring
Maxfield, Lewis H., xix, 3, 5, *6*, 69–70
May, Ben, 151, 161, 195
Maybach, Carl, 12–13
Maybach Motorenbau, 12, 42
Mayer, Roland G., 52, 90, 99, *113*, 131
McClintock, Bruce, 108
McCord, Frank C., 89, 94, 136, 162
Meister, E. W. von "Willy," 42, 44, 98, 113–14, 203; advocacy for airships, 191; and *Akron*, 151; and *Hindenburg*, 195
Merchant Airship Act, 126
Metalclad Airship Corporation, 202
Meyers, admiral, 59
Miller, Harold B. "Min," 78, 155, 217
Miller, Martin O. "Mo," 98, 162
Mills, Delong, 127
Mills, George H., 180, *190–91*, 215, 219, *220*
Mitchell, William "Billy," 4
mobile mast, 100–101, *101*, *149*
Moffett, George, *152*
Moffett, William A., 5, 7, 19–21, 31, 49, 58, 62, *103*, *114*, *152*; advocacy for airships, 162–63; and *Akron*, 124, 144, 152; and budget, 67–68; and command, 90; and commissioning, 51; on deterioration, 120; on employment of airships, 53, 103; and evaluation, 146; on Fulton, 8; and grounding, 155, 159–60; loss of, 163; and Rosendahl, 122; on Steele, 13; and testing, 129, 131, 136
Moffett, William A., Jr., *152*
moment, definition of, 83
Montgolfier, Joseph Michael, 69

mooring, 47, *47*, *57*, 62–63, 174; gear for, 116; in Texas, 111; upending incident, 95–99, *97*. *See also* masts
mooring out, 174–75, 181–82, *184*, *192*
mooring-out circle, *173*, *181*, 219
mooring tower, 53
Morris, Edward R., *176*

National Advisory Committee for Aeronautics (NACA), 5, 44, 67, 89, 119
National Air Races, 129–30
Naval Aeronautic Stations (NAS): Hampton Roads, 5; Pensacola, xv, xvi, 71, 75; Rockaway, xx; Sunnyvale, 169, *169*, 171
Naval Helium Production Plant, 110–11
navigator, duties of, 84–85
Navy, U.S.: Aircraft Division, xv, xvii; and airship development, xiii–xxi, 3, 67–68, 126, 144–45, 197, 201; airship fleet, xiv, xxi; and commercial development, 124–25; fleet maneuvers, 136–45, *141–43*, 175–78; fleet of, *123*; and grounding, 166–68, 178–79; infighting in, 122. *See also specific bureau*
Nicholson, Charles A., 129, 135
Nordstern, 3
Norfleet, Joseph P., 51, 69

O'Brien, John J., 131, 136, 146
officer of the deck (OOD), 84
Oliver, A. M., 156
Olson, A. G., 49

Pacific, expansion into, 62, 126
Panama. *See* fleet maneuvers
Parachute Material School, 66, *66*
passenger spaces, *14*, *87*, *112*, 202
Patoka, 47, 53, *53*, 55, 61–62, 103, 106, *106*, 115, *150*; in fleet maneuvers, 137–39, 141–42, *142*
Patrick, Mason M., 4, 8
Patzig, David E., 48, 55, 96, 98, 211–12
Peak, Richard S., 87–88, 108, 138–39, 147
Peck, Scott E. "Scotty," xvii, 65, 133, 180; on infighting, 122; and *Macon*, 177; on Panama trip, 103–4, 106, 108; and reserach, 182
Pennoyer, Ralph G., 8, 16
Pensacola NAS, xvi, 71, 75
Peterson, A. R., 56
Pierce, George E., 198

Pierce, Maurice R., 39, 49, 51, 90, *92*, 93, 118, 131
pigeons, homing, 66, *79*
pilot, duties of, 85
Pollock, Edwin T., 55, *55–56*
Prajadhipok, king of Siam, 147, *147*
Pratt, William V., 143–44
presidential review, 134, *134*
Prüfling glider, *125*, 131–33, *135*
Pruss, Max, 33, 193–94, *197*, 218
Pye, William S., 191

R-34, 1, *2*, 53
R-38, 3, *4*
R-100, *109*
radar, 208
radio-compass station, calibration of, 62
Rambaibarni, queen of Siam, 147, *147*
Reber, E. E., 119–20
Reichelderfer, Francis W., 39, 57–58, 66, 77, 113, 115, 130; on advocacy, 159; on decommissioning, 161–62; on funding, 102; and *Hindenburg*, 184, 186, 195; on research, 84, 89, 91, 101, 131, 179–80
Richardson, Ann, 65
Richardson, Jack C., 52, 131
riding out. *See* mooring out
rigid airships. *See* airships
Riley, E. J., 95
Rixey, Louise Tyler, 65, 184, 186, 195, 219
Robinson, Assistant Secretary of the Navy, 58
Rockaway NAS, xx
Rodgers, Bertram J., *84*, 103, 116
Rodgers, John, xiii
Rogers, Will, 122
Roland, Charles, 100
Roosevelt, Franklin Delano, xix, 4, 158, 164, 178, 182, 201–2
Roosevelt, J. R., 58
Rosendahl, Charles E., 50–51, 58, 60–61, *92*, *113*, 145, 217; and *Akron*, 151, 165; character of, 91–92, 101–2, 122; in command, 90–123, *91*, 230–34; on deterioration, 120; and fleet maneuvers, 144; and grounding, 168, 173–74, 189–90, 204–5; on helium, 64; and *Hindenburg*, 186; and hook-on experiments, 127; on LZ-130, 202; and mooring, 62; and research, 180, 182; on upending, 98

INDEX

Rounds, Clinton S., 154
Rowe, Monty, 156, 195
Ruefus, R. E., 94
Russell, William, 58

Sachs, director, 186
Sammt, Albert, 34
Saratoga, 102, *102*
Schellberg, Leonard, 40
Scherz, Walter, 34
Schiller, Hans von, 26, 34, 36–38, 218
Schmidt, observer, 24
Schofield, admiral, 144
Schwaben (LZ-10), xiv
Settle, Faye, 98, 215
Settle, T. G. W. "Tex," 39, 52, 73, 76–78, *78*, 96, *135*, *172*; and testing, 124, 134–35
shear tests, 205–6, *205–6*
Shenandoah (ZR-1), 3–4, *47*, 48, *61*, *66*, 79–80; loss of, 62, 64, 66–67, *67*; repairs to, *49*; specs, *153*; trial run, 16–19
Sheppard, Edgar W., 50
Shoemaker, Harry E., 131, 136, *151*, 160–61
short mast, 98
Sitphen, William H., 186
smoking, issues with, 88, 138
snow removal, 181–82
Solar, Charles S. "Chick," *176*
Special Committee on Airships, Science Advisory Board, 178–79
spoons, term, 53
Standley, William H., 167, 191
Steele, George W., Jr., 13, 16, 24, *24*, *27*, 27, 31, 33, 36, 42, *54*; in command, 229–30; and commissioning, 50–51; and demonstration flights, 55, 58–64; on maintenance, 56; and test flights, 88–90
Stevens, Leslie C., 129
Stimson, Henry L., 136
stress testing, 89
stub mast, *99*, 99–100

Sunnyvale NAS, 169, *169*, 171
surface cruisers, specs, *163*
Swanson, Claude A., 164, 166–68, 178, 203–4

taxi car, *99*, 99–100
Taylor, David W., xvi, xvii, *xvii*
TC-13, 95
TC-14, 198, *199*
testing: aerodynamic force tests, *93*; air exploration tests, 127, *127*; of *Akron*, 162; Berry and, 124, 131; Bolster and, 114, 127–28, 130–31, 155; Moffett and, 129, 131, 136; Settle and, 124, 134–35; shear tests, 205–6, *205–6*; stress tests, 89; water recovery apparatus tests, 135, *148*; Wiley and, 127–28, 132–33. *See also Los Angeles:* evaluation of
Thornton, John M., 52, 94
Thurman, Emmett C. "Casey," 51
Tobin, Frederick J. "Bull," 51, *54*, 58, *71*, 98
Towers, John H., xvii, 30, 208, 217
training, 66, 69–90, 117–18, 122, *156*, 198; with airships, *74*; curriculum for, *73*; with free balloons, *72*; graduates of, 225–27; hazards of, 78
trapeze, *127*, 127–31, *128*, 146–51, *148–49*, 176; design for, 155
Tyler, Owen, 161, 194–96
Tyler, Raymond F., 51, 73, 77–78, *92*, 168, 180, 184; ticket for, 185

Upham, captain, 7
Upson, Ralph, xviii, 68, 80

Vallarino, J. J., 140
Vinson, Carl, 165, 178, 203
Vinson Naval Expansion Act, 202

Walker, C. R., 138, 142
Walker, Joe, 136
Walker, Myron J., 89, 94
Ward, Reginald H. "Ducky," 96, 156, 212

Warner, Edward P., 113–14
water recovery apparatus tests, 135, *148*
Watson, George F., 39, 103, 128, 136, 183; on Berry, 158; on Clarke, 133, 135; and fleet maneuvers, 144–46; on grounding, 161
weather forecasting, 33, 56–58, 77, 96, 163
Weed, lt. cdr., 136
weighing off, 24; procedure, *140*
Weyerbacher, Ralph, 16, 93, 151
Whittle, George V., 124, 198
Wicks, Zeno, 5, 49, 67, 89, 92, 118, 130
Wilbur, Curtis D., 45, 67
Wiley, David, *109*
Wiley, Herbert V., 52, 92, *92*, 103, *109*, 113; and *Akron*, 151, 164; in command, 118–20, 130–31, 232–34; and *Macon*, 175–76; in temporary command, 108–12; and testing, 127–28, 132–33
Willard, Arthur Lee, 140, 142
Wilson, Woodrow, xvii
wind studies, 179, *200*, 200–201, *201*
Wingfoot Lake, 69
Witteman, Anton, 31, 34, 36, *218*
Wood, R. F., 5
World War I, xv–xxi
World War II, 217–19; airships and, 219–20
Wright brothers, xiii, 131

YE16, 68
Young, Howard L., 148, 150
Young, William, 20

Zeppelin, Ferdinand von, xiii
zeppelins. *See* airships
ZMC-2, 80, *80*, 129, 134, *137*, 198
ZR-1. *See Shenandoah*
ZR-2, 3, 4, 6
ZR-3. *See Los Angeles*
ZRN, 198, 201–2
ZRS-4. *See Akron*
ZRS-5. *See Macon*

About the Author

William F. Althoff has written extensively for aviation publications. His first book, *Sky Ships: A History of the Airship in the United States Navy,* received critical and popular acclaim. During 1999-2000, Althoff held the Ramsey Fellowship in Naval Aviation History at the National Air & Space Museum. He lives in Whitehouse Station, New Jersey.